THE GEORGE FISHER BAKER
NON-RESIDENT LECTURESHIP IN CHEMISTRY
AT CORNELL UNIVERSITY

ORGANIC CRYSTALS AND MOLECULES

By J. Monteath Robertson

ORGANIC CRYSTALS AND MOLECULES

Theory of X-Ray Structure Analysis with

Applications to Organic Chemistry

J. MONTEATH ROBERTSON

Gardiner Professor of Chemistry in the University of Glasgow

❖❖❖

CORNELL UNIVERSITY PRESS

Ithaca, New York, 1953

548.5
R 651

COPYRIGHT 1953 BY CORNELL UNIVERSITY

First published 1953

PRINTED IN THE UNITED STATES OF AMERICA

BY GEORGE BANTA PUBLISHING CO., MENASHA, WISCONSIN

IN MEMORY OF
NEVIL VINCENT SIDGWICK,
WHO WAS PRESENT AT THESE
LECTURES DURING HIS LAST VISIT
TO THE UNITED STATES

Preface

THIS BOOK is based on the course of lectures which I gave at Cornell University in the autumn of 1951 during my tenure of the George Fisher Baker Non-resident Lectureship in Chemistry. I wish to express my warmest thanks to Professor F. A. Long and his colleagues in the Department of Chemistry for providing me with this opportunity, and for the excellent facilities placed at my disposal in the Baker Laboratory during the period of my lectureship.

I am also indebted to the staff of the Chemistry Department and to many other members of the University for the friendship and hospitality that made my residence at Cornell a most memorable and happy period. I am especially grateful to the young men of Telluride House, who acted as my hosts for the greater part of this time and who looked after me so well.

In the preparation of the book itself I am greatly indebted to Professor Kathleen Lonsdale, Professor Ray Pepinsky, and Dr. J. D. Loudon, who have read certain chapters of the manuscript and made many valuable suggestions. To Professor J. L. Hoard, who has read the whole of the manuscript, I am particularly grateful for much help and encouragement. Dr. J. C. Speakman, Dr. V. Vand, and Dr. T. H. Goodwin have kindly helped to read the proofs. In the preparation of the diagrams, in addition to the acknowledgments made in the text and references, I have to thank Mr. Dave Miller, of the Cornell University Photographic Science Laboratory, and Mr. H. M. M. Shearer and Mr. J. B. Findlay, of the Chemistry Department of Glasgow University, for much valuable assistance.

PREFACE

It is also a pleasure to make acknowledgement to the following bodies for permission to reproduce certain figures, or parts of figures: the Chemical Society for figures 37, 77, 78, 87–91, 103, 104, 107–113; *Acta Crystallographica* for figures 55, 56, 65, 66, 81, 83–85, 115–117, 130; the Royal Society for figures 79, 80, 82, 92, 93; *Nature* for figures 51, 94, 105, 106; the *Physical Review* for figures 57, 58; the *Journal of the American Chemical Society* for figures 74, 75; the National Academy of Sciences, U.S.A., for figure 127; *Naturwissenschaften* for figure 26; Messrs. G. Bell and Sons, Ltd., London, for figures 30, 70; Princeton University Press for figures 118–120; the Clarendon Press, Oxford, for figure 73; and Interscience Publishers, Inc., New York, for figure 132.

J. MONTEATH ROBERTSON

Chemistry Department
The University
Glasgow, Scotland
October, 1952

viii

Contents

CONTENTS

CONTENTS

CONTENTS

Organic Crystals and Molecules

I

Introduction

1. SCOPE OF MODERN CRYSTALLOGRAPHY

Von Laue's discovery of the diffraction of X-rays by crystals in 1912 had consequences in two main directions. The first concerned the nature of the radiation itself, and with the development of X-ray spectroscopy this led to immediate and fundamental advances in the theory of atomic structure. In the second place, a powerful new tool became available for investigating the structure of matter on an atomic scale and elucidating the details of the arrangements present in various molecules and crystals. This second aspect, with which we are concerned in the present book, is generally referred to as crystallography, although the scope of the subject is now so great that it bears little resemblance to the studies formerly embraced by this term. It is now found that a great range of substances, from metals to the most complex biological structures, display a sufficient degree of order in their atomic arrangement to give diffraction effects, and so lie within this field of study. Even liquids and gases, which cannot in any sense be classed as crystals, are capable of investigation by somewhat analogous methods.

This new and enlarged crystallography has already had most far-reaching effects on many quite diverse branches of science. In physics itself, and especially in metallurgy, the relation between the properties of matter and its ultimate structure can now be studied in detail, and is found to depend very largely on the size, arrangement, degree of perfection, and method of growth of the constituent crystals. In geology and mineralogy, the traditional field for the crystallographer, the deeper insight gained regarding the atomic arrangement in minerals has succeeded in explaining many of the baffling complexities in chemical composition.   In biology, the most complex field of all, the possibilities for the future application of the new science are perhaps greater than in any other area.   Actual crystals are comparatively rare in living matter, but fibre structures, of one type or another, are fundamental constituents capable of detailed examination by the X-ray method.   Great progress has already been made, but explanations in terms of exact atomic arrangements are still sketchy.   In the field of protein structure, where complete three-dimensional crystals can be prepared, striking advances are now being made in spite of the forbidding complexities of the problem.   Here and in many related fields it is certain that X-ray methods will become of ever-increasing importance when the intricacy of the structures and the instability of the molecule combine to defy the traditional methods of chemistry.

2. IMPACT OF THE NEW DISCOVERIES ON CHEMISTRY

In this book we are chiefly concerned with the more simple problems of X-ray analysis, as applied to the structures of some of the better-defined molecules of organic chemistry.   It is therefore appropriate that we should examine the relationship of crystallography to chemistry, and in particular to organic chemistry, in rather more detail.

Some of the earliest and most precise demonstrations of a particulate theory of matter are to be found in the crystal studies of Robert Hooke and Christiaan Huygens.   Their theories of crystal structure, based on the idea of a lattice, demanded the indefinitely continued repetition of small, identical repeating units.   These, of course, need not be atoms, but the realisation that matter must be split up into small identical units of some kind represents one of the essential ideas of atomic theory. About a century later Haüy's *molécule intégrante* and *molécules élémentaires* still preceded Dalton's atomic theory by a number of years.   From that time onwards the relationship of classical crystallography to chemistry remained one of great practical and theoretical importance, although this was perhaps not always fully appreciated.

With the advent of X-ray methods of analysis in 1912, however, the new and enlarged science of crystallography immediately became an integral and indispensable part of chemistry. By this time chemistry was already an advanced and very exact science. The consequences of Dalton's atomic theory had been fully explored and applied. It was realised that the properties of a compound depended not only on the number and kind of atoms present in the molecule but especially on their spatial arrangement. Great progress had been made with the well-defined and yet easily volatile compounds of organic chemistry, and the relative positions of the atoms in several hundreds of thousands of different kinds of molecules were known with precision.

In inorganic chemistry, however, the situation was different, and the subject had almost reached an impasse. The exact composition of a vast number of compounds was known, but owing to the non-volatile and intractable nature of the substances the exact number of atoms in the molecule was often open to doubt, and little or nothing could be guessed about their spatial arrangement. Further progress and understanding of the subject clearly demanded guiding principles of the kind provided by Kekulé and Pasteur for the carbon compounds, but this could not be achieved without further exact structural knowledge.

It is natural, then, that the first great triumphs of the new science of X-ray analysis were in the field of inorganic chemistry. Determination of atomic positions in crystals is the central problem of X-ray analysis, and in a short time the structures of many simple inorganic substances were determined, with revolutionary results. The crystal rather than the molecule often became the essential unit of structure, because it was discovered that in most cases the "molecule" of the chemical formula did not in fact possess any individual existence at all. For the first time it became possible to measure accurately the size of the various ions and assess the importance of this factor in structure. Later, with the further development of the science of structure analysis in the school of W. L. Bragg, the bewildering complexity of the silicate structures became understood for the first time and they were shown to conform to a few relatively simple principles of a generality equal to those discovered earlier for the compounds of carbon.

In the field of organic chemistry, however, the progress made by the new science was at first slow. The structures of diamond and graphite had been elucidated, and so the fundamental inter-atomic distances to be expected in organic molecules were known. But at first the most that could be said was that the dimensions of the unit cells in organic crystals as measured by X-rays were such as to accommodate molecules of the shape and size expected from the chemical formulas, and from the dimensions of the rings of atoms found in the diamond and graphite crystals. It was shown that, unlike the majority of the inorganic structures, these were molecular crystals in which the building unit generally con-sisted of a complete chemical molecule.

It is only within about the last twenty years that the methods of X-ray analysis have become sufficiently refined to examine the structures of organic crystals in real detail, and to determine the positions of all the atoms. A few structures had been determined earlier, and these, like the later results, have served in the first place to confirm most fully the detailed structural formulas of the organic chemist. The perfectly precise yet non-mathematical logic of organic chemistry which has enabled the relative spatial positions of all the atoms in about a million different compounds to be determined with certainty is undoubtedly one of the greatest triumphs of modern science. It may well be asked, then, what the science of X-ray analysis has got to add to this knowl-edge and what is to be gained from the detailed examination of these complex structures by the new method. The second part of this book is mainly concerned with the answer to this question, but the objects of the work can here be stated briefly under three rather different headings.

In the first place, the method of X-ray analysis translates the structural formulas of the chemist into exact metrical terms and provides quantitative measurements of bond lengths, valency angles, and the general electron distribution in the molecule. All these measurements lie beyond the scope of ordinary chemical methods, but they are of the utmost importance in providing the necessary connecting link with quantum mechanical theory. Although most of the measurements are not yet sufficiently pre-cise and the problems are exceedingly complex, there is no doubt that the work will ultimately succeed in transforming the subject from an empirical to a sound theoretical basis.

In the second place, as well as giving accurate bond length measurements, the X-ray method provides equally precise information about the way in which organic molecules are linked together in the solid state. This may not seem of much interest to chemistry, and in fact when the attractions are solely of the weak, residual types, ordinary packing considerations are often found to determine the crystal structure. But special intermolecular attractions like the hydrogen bond are of vital importance in many structures and often determine the behaviour of the substance. This is especially true in the more complex biological materials, and advances in this field may largely depend upon a careful study of such factors. The necessary quantitative structural information can usually only be derived from X-ray studies.

Finally, with the development of certain new methods of X-ray analysis, it is now possible to attempt to elucidate the structures of molecules which are so complex that they have not so far yielded to the ordinary methods of chemistry. Rapid progress has recently been made in this direction, largely by the use of the heavy atom and isomorphous substitution methods. In the future, with the development of new and improved mathematical and computational techniques, it is to be expected that progress will become more rapid and that the new methods will outstrip the standard analytical and degradative procedures of organic chemistry. The energies of the chemist will then be freed to concentrate on his second major problem, that of synthesis, which at present lies beyond the scope of crystallography. Meanwhile, an unlimited field lies ahead for the structure analyst as he advances towards the challenge of the more and more complicated molecules associated with living matter.

3. ARRANGEMENT OF THE SUBJECT MATTER

The time has long passed since the science of X-ray analysis could be dealt with adequately in a single text, and numerous books dealing with specialized aspects of the subject are now available. It has, however, been thought advisable to devote the first part of this book to a survey of general principles. In this, attention has been focused on the more geometrical aspects of single crystal structure analysis, where the ideas and calculations involved are relatively simple. A great deal has of necessity been omitted, and in particular there is no discussion of what may be termed the physical theory of X-ray diffraction in crystals, nor has any attempt been made to deal with the various types of lattice imperfections which may occur in crystals.

Many modern accounts of X-ray analysis begin by assuming the lattice structure of crystals and using this as the basis of the explanation of the diffraction effect. Generalisations such as the law of constancy of angles and the law of rational indices, which are really the experimental foundations on which space lattice theory as applied to crystals must be based, can then be "explained" and briefly dismissed. While there may be something to commend this type of "rational approach" to the subject in certain circumstances, a great deal of imaginative interest is necessarily lost and there is also some danger of presenting a completely distorted view of scientific method. This approach also does little justice to the labours of those early workers who succeeded in bringing the geometrical side of the subject to a state of perfection some time before the discovery of X-rays by Röntgen in 1895. We therefore prefer to devote the first two chapters to giving a brief and largely historical account of the geometry and symmetry of crystals, introducing nothing that was not known before X-rays were discovered.

It was not until 1912 that it became possible to explore the intimate structure of crystals and verify experimentally that the conclusions of space group theory actually did apply in all their detail to real crystals. All this can be done from a study confined mainly to the *positions* of the diffracted X-ray beams, and this matter is considered in the next chapter. Much wider possibilities open out when the *intensities* of the different reflections are considered. The possibility then arises of determining the exact atomic arrangement and even the details of the electron distribution within the ultimate building unit of the crystal. There are, however, certain fundamental difficulties which prevent a complete solution of this problem in the general case. These questions, and methods of overcoming the difficulties, are dealt with in the remaining chapters of Part I.

In Part II some results obtained from the X-ray analysis of certain organic molecular structures are described. This branch of the subject has expanded rapidly in recent years, and no attempt has been made to give anything like a complete picture or systematic account of all the structures that have now been

8

determined. Such a task is being undertaken, but it would have been quite unsuitable material to present in the series of lectures on which this book is based.

Instead, I have attempted to illustrate the main trends of research in this field by examples, drawn, I am afraid, very largely from the work with which I am most familiar, and no doubt with the neglect of many other equally suitable examples. After some of the early and important structure determinations have been dealt with, a fairly full account of the condensed ring hydrocarbons is given. For the most part these structures are relatively simple in the crystallographic sense, and reasonably accurate bond length measurements can be made. Chemically they are also sufficiently simple to enable theoretical predictions of the bond lengths to be made with some confidence, and a comparison between theory and experiment in this field is a matter of some general interest.

Illustrations of various types of molecular arrangement in crystals are then given, with some discussion of the very important problem of hydrogen bonding.

Most of the structures described in these chapters are chemically well known, and their X-ray analysis can usually proceed quite conveniently from trial models based on the chemical formulas. The method of successive approximations to the true structure involved in these analyses, however, makes the final conclusions independent of the initial chemical assumptions. In the later chapters some structures are considered which were not fully established by chemical methods before the X-ray work began. Analytical methods largely independent of chemical knowledge are employed for their solution, and the application of these methods to compounds of unknown or partly unknown chemical formula is illustrated by a number of recent examples. These are chosen to illustrate the analytical methods employed, and also to take advantage of the chemical and biochemical interest of the compounds themselves.

This leads naturally to some description in the last chapter of the much more complex structures of direct biological interest which are now attracting the attention of an increasing body of workers in many laboratories. A number of physical methods, including both X-ray analysis and electron microscopy, are being brought to bear on these problems, and many new and exciting results are being obtained. This rapidly growing and rapidly changing field is difficult to review adequately, but some of the better established and more important conclusions are summarized.

Part One

THE NATURE OF CRYSTALS;
GEOMETRICAL ASPECTS

II

The Crystal
as a Geometrical Figure

1. DEFINITION OF A CRYSTAL; DIRECTIONAL PROPERTIES

As naturally occurring objects, crystals have excited interest and curiosity from the earliest times and evoked speculation regarding their ultimate structure. Most things in nature are rounded or have curved outlines; sharp edges and corners are generally avoided. In contrast, crystals may exhibit faces which are geometrical planes to a high degree of precision, and the edges are sharp and straight. Well-developed symmetry is also a feature common to many crystals.

It would be difficult, however, to base any definition of a crystal on these external features. We might grind away the faces and edges, but the matter remaining would still be crystalline and even capable, under suitable conditions, of growing fresh faces. The important feature of a crystal which distinguishes it from a liquid or from amorphous matter is that certain of its properties are different in different directions. If we cut rods from a crystal in different directions and test properties such as cohesion and elasticity, or optical, thermal, electric, and magnetic properties, we find that they may differ in the different directions. In spite of this fact, the internal texture of a crystal is obviously homogeneous, and so it is clear that these directional properties, as well as the symmetry displayed by the crystal, must be related to some feature of the ultimate structure of the matter from which the crystal is composed. Theories regarding this structure are described in the next chapter.

For the moment it is sufficient to define a crystal as solid

homogeneous matter in which some of the usual physical properties are different in different directions. Modern experiment
shows that the crystalline state is certainly the normal state for
solid matter, although the individual crystals may often be small
and imperfectly formed. However, when crystalline matter is
allowed to grow slowly in an unconfined space, it is usually
bounded by plane surfaces forming a polyhedron. These surfaces, or faces, may be differently developed in different specimens, owing to the external conditions of growth, the aspect which
a particular specimen assumes by virtue of the relative sizes of
its faces being called its *habit*. Sometimes the habit, particularly
in the case of certain minerals, may be quite deceptively different
for specimens formed under even slightly different conditions.

To afford a basis for the exact study of the essentials of crystal
symmetry, the crystal may be idealised into a geometrical figure
if we represent faces which have the same physical properties by
planes of identical shape and size equidistant from a fixed point
O, which may be called the *centre* of the crystal.   The planes and
edges of the polyhedron so formed will be parallel to the planes
and edges of the actual crystal. Or again, the geometrical
properties of the crystal may be represented by drawing from *O*
a set of lines representing the normals to the crystal faces.   The
stereographic projection, in which the points of intersection of
these normals with the surface of a sphere about *O* are projected
from a point *A* also on the surface of this sphere to a plane
parallel to the tangent plane at *A*, affords a further useful construction in crystallography.

2. CONSTANCY OF ANGLES AND RATIONAL INDICES

Early in the history of crystallography it was discovered that
for a given crystalline structure, in spite of variations in habit,
the angles between corresponding faces are constant. This was
first demonstrated by Nicolaus Steno[1] in 1669, who proved the
law of constancy of angle in the case of quartz, and was extended
to other substances by Guglielmini[2] in 1688, who also announced
the constancy of cleavage directions in crystals. About a century

---

[1] Nicolaus Steno, *De solido intra solidum naturaliter contento disserta-
tionis prodromus*, Florentiae, 1669. (English translation, London, 1671.)

[2] D. Guglielmini, *Riflessioni filosofiche dedotte dalle figure de sali*, Padova,
1706; *De salibus dissertatio epistolaris*, Venet, 1705.

later, Romé de l'Isle,[3] with better instruments, extended this work to a large variety of natural mineral crystals and so began the exact study of crystallography.

Abbé René Just Haüy[4] followed up this work and discovered the fundamental law of rational intercept ratios, or the law of rational indices as we now call it, which is described below and on which all the arguments in this chapter are based. He was also able to show how the exceedingly numerous crystal varieties could be constructed from a few fundamental types of symmetry, and thus provided a sound experimental guide for later developments in the geometry of crystals.

Much of Haüy's work dealt with theories of crystal structures and lattices, with which we are not at present concerned, but it is interesting to note in passing that he regarded the crystal unit, or *molécule intégrante*, as composed of *molécules élémentaires* or chemical atoms of definite and constant form, and that these views preceded Dalton's atomic theory by a number of years. Haüy's theories were largely based on a study of cleavage in crystals, and although his speculations as to the shapes of the ultimate particles were soon forgotten, the essential geometrical ideas survive.

The fact that all crystals can be described, as Haüy first showed, in terms of a small number of symmetry types is at first very puzzling and seems to imply a certain artificiality. This, of course, is not so, and we shall see that the existence of this limited number of types is a mathematical necessity arising from the essential nature of crystal structure. It can be shown to follow from the generalisation known as the law of rational indices, which in turn finds its explanation in the lattice structure of crystals. Before these matters are dealt with it is necessary to adopt some system of describing crystal faces.

As reference axes let us choose the intersections of any three non-parallel faces, represented by *OA*, *OB*, *OC* (Fig. 1). Let any other face (called the *parametral* or *standard* face) meet these lines at *A*, *B*, and *C*, making intercepts *a*, *b*, and *c*. We then have a system of reference axes of defined relative lengths to

[3] Romé de l'Isle, *Essai de cristallographie*, Paris, 1772; *Cristallographie, ou description des formes propres à tous les corps du règne minéral*, Paris, 1783.

[4] René Just Haüy, *Traité de minéralogie*, Paris, 1801; *Essai d'une théorie sur la structure des crystaux*, Paris, 1784.

which every face of the crystal may be referred. The ratios $a:b:c$ are called the axial ratios. It is important to notice that *any* set of non-parallel faces on the crystal may be chosen for this purpose and the same general laws will apply. It is, however, a matter of convenience in practice to choose the intersections of certain simple planes of the crystal, and it will sometimes be possible to adopt an orthogonal system of axes.

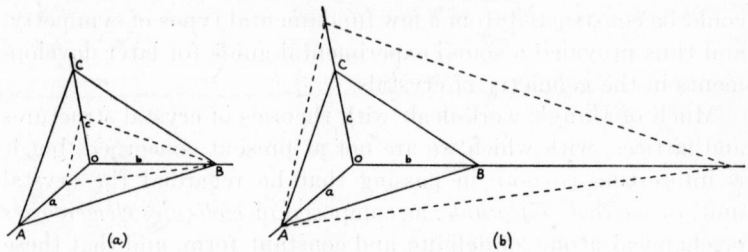

Fig. 1. Intercepts and indices.

If now any other face on the crystal makes intercepts on these axes of $\dfrac{a}{h}$, $\dfrac{b}{k}$, $\dfrac{c}{l}$, it is said to have the Miller[5] indices (hkl).

The indices are thus the reciprocals of the intercepts of any face, or a plane drawn parallel to it, on the calibrated set of axes. The indices of our standard face are (111), and of that shown by the dotted lines (312). In Fig. 1b a plane is drawn (dotted) parallel to this latter plane, making intercepts a, $3b$, $1\frac{1}{2}c$, and having therefore the indices $(1, 1/3, 2/3)$. Clearing of fractions, we obtain as before (312) as the indices for the dotted plane. If a plane is parallel to any axis, the intercept is at infinity and the corresponding Miller index is zero. The reciprocal notation involved in this system is of the greatest service in crystallography and is now generally adopted.

The fundamental law of crystallography may now be stated. It is an observed fact that the ratios of the indices of any face of any crystal are rational, and in general are the ratios of small whole numbers. The emphasis here is on *small* integers, and in this form the law holds generally. Were this not so, experimental limits of accuracy in measurement would, of course, make the

[5] W. H. Miller, *A Treatise on Crystallography*, London, 1839.

statement meaningless. In practice, indices as high as 6 are extremely rare for naturally occurring crystal faces.

The physical explanation of the law of rational indices is dealt with in the next chapter. For the moment we may accept it as an experimental generalisation and make it the basis of our examination of possible types of crystal symmetry.

3. SYMMETRY

When we say that a solid body or a finite geometrical figure has symmetry, we mean that if some movement or operation is applied to the figure it will bring every point of it into some position which was previously occupied by some other point of the figure. Such symmetry operations may be of several kinds. Thus, a figure may have various axes of symmetry, 2-fold, 3-fold, $\cdots n$-fold, such that each rotation of $2\pi/n$ about the axis brings the whole figure into self-coincidence, that is, into a position which cannot be distinguished from the original position. These particular symmetry elements, known as *rotation axes*, are conveniently denoted by symbols representing their multiplicity, e.g., 1, 2, 3, $\cdots n$, the symbol 1 denoting the identity operation, which is, of course, common to all figures. The simple cube, for example, has various 2-fold, 3-fold, and 4-fold rotation axes, while the sphere has an infinite number of n-fold axes where n is infinite.

A different kind of symmetry element is represented by the *plane of symmetry* (symbol m). If this is present, one-half of the figure is the reflection of the other half in the plane. Again, a figure may have a centre of symmetry, or *centre of inversion* (symbol $\bar{1}$), such that when a line is drawn from any point on the figure through this centre, and produced an equal distance on the other side of the centre, it meets an identical point. This really represents reflection through a point instead of reflection in a plane.

These last two symmetry elements, m and $\bar{1}$, represent operations which are essentially different in kind from the pure rotation axis. This can be seen by considering the effect of the various operations on a completely unsymmetrical object, such as the left hand. The effect of any pure rotation is simply to produce another left hand in a different position, i.e., a *congruent* figure in which the relative positions of all the points are un-

changed.   But if the left hand is acted on by a plane of symmetry, or by an inversion, a right hand is produced, i.e., an
enantiomorphous figure in which the relative positions of corresponding points are different.   In both cases, however, the operations are symmetry operations, and the effect when applied to
an unsymmetrical figure is to leave the distance between any
two given points unchanged, so that exact coincidence can be
achieved.   These two different kinds of operations are usually
referred to as operations of the "first sort" (producing congruent
figures) and operations of the "second sort" (producing enantiomorphous figures).

All the symmetry which any figure can have may be described
by means of these operations, or by combinations of them. The
general operation of the second sort is taken to be an axis of
rotatory inversion, i.e., a combination of rotation about an axis
and inversion through a point on the axis, for this includes
reflection in a plane as a special case.   Axes of rotatory inversion
are indicated by placing a bar over the number representing the
rotation axis, e.g., $\bar{1}, \bar{2}, \bar{3}, \cdots \bar{n}$.

The meaning of these latter symbols must be interpreted with
some care.   The combination of the identity operation and centre
of inversion, $\bar{1}$, is clearly just the centre of inversion alone.   With
$\bar{2}$, rotation of 180° and inversion back through the centre of
symmetry is equivalent to the pure reflection plane m.   By
continuing these operations, we find that, in general, when $\bar{n}$ is
odd the result is equivalent to a pure rotation axis n combined
with a centre of symmetry.   When $\bar{n}$ is even, however, the multiplicity of the pure rotation axis involved is reduced to $n/2$;
and when $\bar{n}$ is even but $\bar{n}/2$ odd, we have in addition a plane
of symmetry perpendicular to the axis, as in $\bar{2}$ and $\bar{6}$.   These
relations are illustrated by stereographic projections for the
cases $\bar{1}, \bar{2}, \bar{3}, \bar{4}$, and $\bar{6}$ in Fig. 5 (p. 23).

It is obvious that the collection of symmetry operations present
in any given body must form a self-consistent set.   The result of applying any one of these operations, or of applying
any number of them together, is to bring the body into self-
coincidence, i.e., to leave it unmoved.   This collection of operations forms a *group*, in the mathematical sense.   Thus, the product of any two, or the square of any one, of these operations is
equivalent to some member of the series, and also the series

18

always contains the operation A^{-1} (which is the reverse operation of A) if it contains the operation A, the product $A \cdot A^{-1}$ being the identity operation 1. A simple example is illustrated in Fig. 2 for the group $2m\bar{1}$, i.e., for a 2-fold axis, a plane of symmetry perpendicular to it, and a centre of inversion at the point where the axis cuts the plane. It is clear that the following relations then hold:

$$2 \cdot m = \bar{1}, \qquad 2 \cdot \bar{1} = m, \qquad m \cdot \bar{1} = 2,$$

the product sign indicating that the operations denoted by the symbols are applied in succession. Also

$$m^2 = 1, \qquad \bar{1}^2 = 1, \qquad 2^2 = 1.$$

It may be noted that all the operations of such a group leave one point unmoved. In a finite geometrical figure all the axes of symmetry and any planes of symmetry must pass through such a fixed point, which is called the centre of the group, a group of this kind being called a *point group*. If there were two parallel axes of symmetry, or two parallel

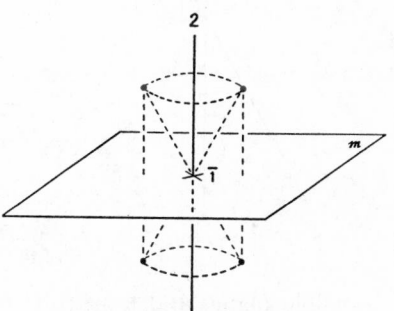

Fig. 2. The group $2m\bar{1}$.

planes of symmetry at a fixed distance apart, the operation of one plane on the other, or of one axis on the other, would produce a third, and so on to infinity, the figure thus becoming indefinitely extended.

4. LIMITATION OF SYMMETRY POSSIBLE IN A CRYSTAL

A solid body or a geometrical figure can be constructed to display any desired degree of symmetry. Rotation axes may range in multiplicity from 1 to ∞, and these may be combined with planes of symmetry or with an inversion. In a crystal, however, which we may for the present regard as idealised into a geometrical figure *conforming to the law of rational indices,* there is an important limitation. It may have planes of symmetry and a centre of inversion, but only 1-fold, 2-fold, 3-fold, 4-fold, and 6-fold rotation axes or axes of rotatory inversion are possible.

This very fundamental feature of crystal structure may be demonstrated in the following way. In the first place, in any crystal with an n-fold rotation axis of symmetry there are always possible faces, i.e., faces which obey the law of rational indices, both parallel and perpendicular to the rotation axis. The proof of this statement is somewhat involved in the general case, and even more so when $n=3$, but it is readily seen to be true when n is even. Let OA (Fig. 3a) be this rotation axis, and let OB be a line drawn through O parallel to any arbitrarily chosen edge of the crystal. Rotation of 180° about OA produces another possible edge OB', and the plane containing these two edges is a

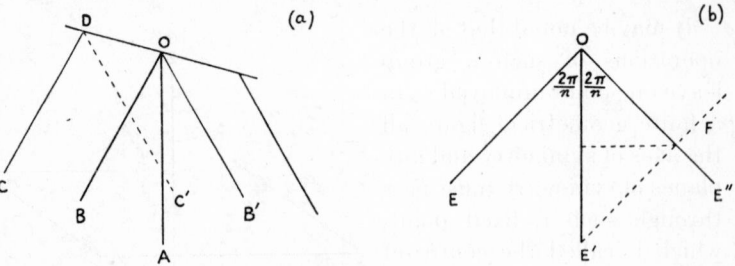

Fig. 3. Symmetry limitations.

possible plane, and is parallel to the rotation axis. Also, draw the plane $OBCD$, parallel to any arbitrarily chosen crystal plane, and the rotation of 180° gives a plane of which $OB'C'D$ is a part, and this is a possible plane. The line of intersection of these two planes gives a possible edge OD, which is perpendicular to the rotation axis, and two such edges will involve a plane perpendicular to the rotation axis.

Next, let us draw a projection of a crystal with an n-fold rotation axis through O, perpendicular to the plane of the paper (Fig. 3b). OE is drawn parallel to any plane which is parallel to the rotation axis. OE' and OE'' are the successive positions to which OE is brought by the operation of the rotation axis. We wish to find the Miller indices of the plane of which OE is the trace. As reference axes we may chose the symmetry axis through O, OE', and OE'', because these lines can be the intersections of possible crystal planes; and we may take the axes OE' and OE'' to be of unit length (by symmetry they are equal). Through E' draw a plane $E'F$ parallel to OE. Then the intercepts of this plane on the reference axes are

∞ on the axis through O, perpendicular to the paper,

1   on OE',

$$\frac{1}{2 \cos \dfrac{2\pi}{n}} \text{ on } OE''.$$

The indices of the plane of which OE is the trace are therefore $\left(0,\ 1,\ 2 \cos \dfrac{2\pi}{n}\right)$, and these must be integers.   Hence

$$2 \cos \frac{2\pi}{n} = \pm 2,\ \pm 1,\ 0,$$

$$\cos \frac{2\pi}{n} = \pm 1,\ \pm\tfrac{1}{2},\ 0,$$

$$n = 1, 2, 3, 4, \text{ or } 6.$$

5. THE 32 POINT GROUPS APPLICABLE TO CRYSTALLOGRAPHY

We have now seen that the number of distinct symmetry operations applicable to crystals, considered as idealised, finite geometrical figures, is strictly limited.   They are, in fact, only 1, 2, 3, 4, 6 and $\bar{1}$, $\bar{2}$ (or m), $\bar{3}$, $\bar{4}$, $\bar{6}$.   It is obvious that only a comparatively small number of self-consistent sets containing these operations will be possible, and they may be derived by combining the operations in all possible ways in three dimensions.

This derivation requires a long and careful analysis, which is beyond the scope of this book.   It was first performed by Hessel[6] in 1830, who showed that 32 distinct symmetry classes are possible.   This early work, deduced from Haüy's law of rational indices, was remarkable, especially as at that time only a few of these classes were known to be represented by natural crystals. Gadolin,[7] in 1867, arrived at the same result independently.   The

[6] J. F. C. Hessel, *Krystallometrie oder Krystallonomie und Krystallographie*, Leipzig, 1831; article "Krystall" in Gehler's *Physikal Wörterbuch*, 1830, pp. 1023–1340; also L. Sohncke, *Z. Kryst. Mineral.*, 1890, **18**, 486.

[7] Axel Gadolin, *Acta Soc. Sci. Fennicae*, 1867. **9**, 1–71.

proofs, however, are perhaps not quite general, owing to the difficulty in the case of a trigonal axis mentioned above, and require some confirmation from structure theory, dealt with in the next chapter.

Without attempting any rigorous derivation or proof of the 32 possible point groups, they may now be briefly enumerated. They are illustrated in each case by stereographic projections of a system of equivalent points, ● denoting a point on the upper hemisphere and ○ a corresponding point derived by some symmetry operation on the lower hemisphere. The points in which the sphere is cut by rotation axes are indicated in the projections by the symbols for these axes, and the lines in which it is met by symmetry planes by full lines in the projections. The diagrams are in principle similar to those employed by Hilton.[8] Each point group is denoted by the now generally accepted Hermann-Mauguin symbol,[9] which enumerates the essential symmetry elements present. This is followed by the older Schoenflies' symbol,[10] which is still useful for many purposes of classification.

There will first be one point group corresponding to each of the possible pure rotation axes, viz., 1, 2, 3, 4, and 6, as illustrated in Fig. 4. Similarly, there is a point group corresponding to each axis of rotatory inversion $\bar{1}$, $\bar{2}$, $\bar{3}$, $\bar{4}$, and $\bar{6}$ (Fig. 5). Next there are the groups containing one n-fold axis with 2-fold axes perpendicular to it. When $n=2$, this involves two 2-fold axes perpendicular to it (or a system of three mutually perpendicular 2-fold axes, sometimes referred to as the quadratic group, Q). When $n=3$, there are three 2-fold axes perpendicular to n, and so on, giving the four groups illustrated in Fig. 6. Those remaining groups which are built solely of rotation axes contain a set of four 3-fold axes along the diagonals of a cube with (1) 2-fold axes perpendicular to its faces (group T) and (2) 4-fold axes perpendicular to its faces, this latter case involving six additional

[8] H. Hilton, *Mathematical Crystallography*, Oxford, 1903.

[9] *Internationale Tabellen zur Bestimmung von Kristallstrukturen*, Berlin, 1935; *International Tables for X-Ray Crystallography*, Birmingham, Eng.: Kynoch Press, 1952, published by the International Union of Crystallography.

[10] Arthur Schoenflies, *Krystallsysteme und Krystallstruktur*, Leipzig, 1891.

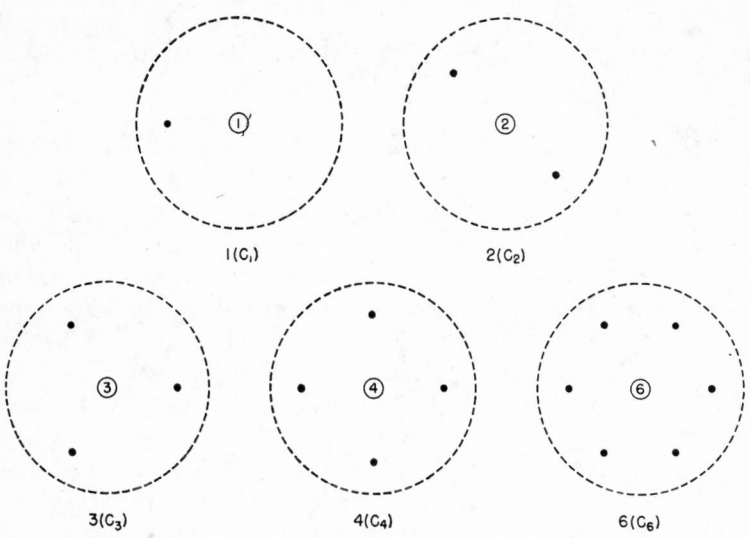

Fig. 4. Point groups, 1–6.

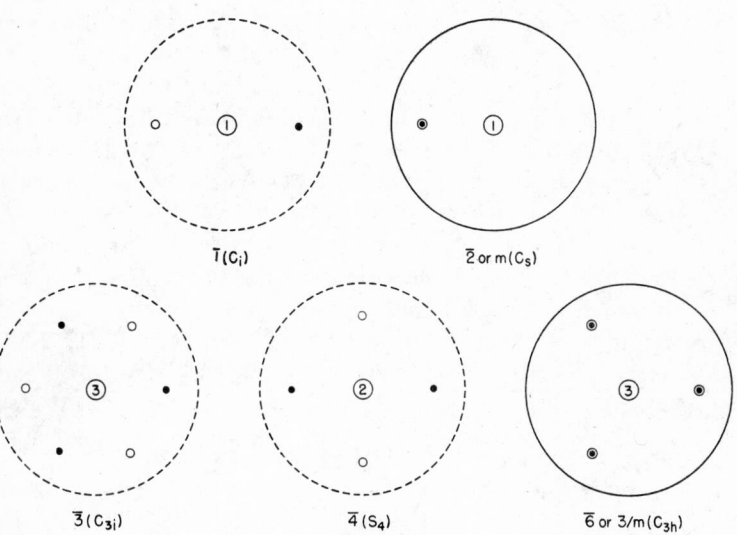

Fig. 5. Point groups, $\bar{1}$–$\bar{6}$.

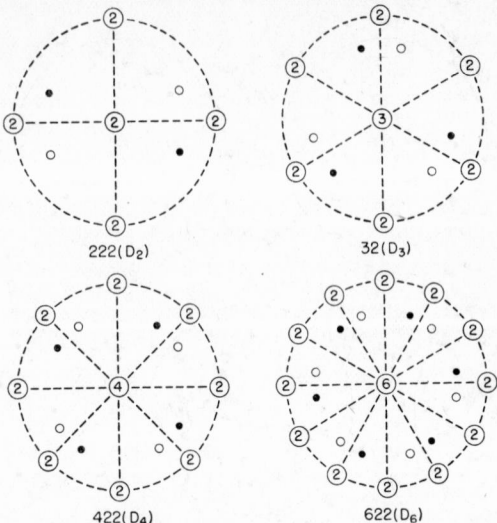

Fig. 6. Point groups, D_2–D_6.

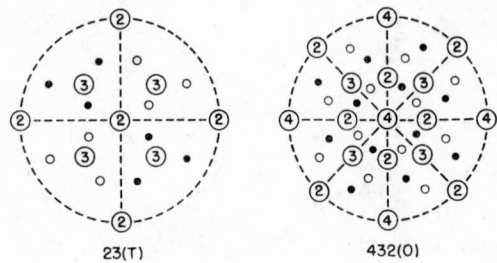

Fig. 7. Point groups, T and O.

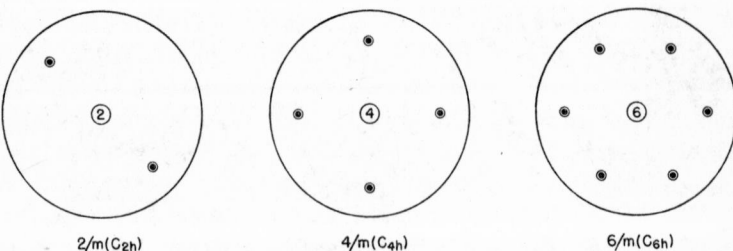

Fig. 8. Point groups, C_{2h}–C_{6h}.

2-fold axes bisecting the angles between the 4-fold axes and also the angles between neighbouring 3-fold axes (group O). These two cubic groups are shown in Fig. 7.

The introduction of a reflection plane perpendicular to the n-fold axis brings it into self-coincidence and gives rise to the groups of Fig. 8 and one, $3/m = \bar{6}$ (C_{3h}), already mentioned above (Fig. 5). The introduction of a reflection plane *parallel* to the n-fold axis gives rise to a set of such planes equal in number to the multiplicity of the axis. These groups are illustrated in

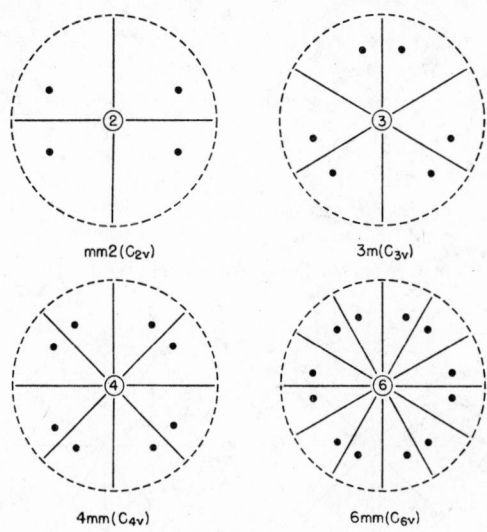

mm2(C_{2v}) 3m(C_{3v})

4mm(C_{4v}) 6mm(C_{6v})

Fig. 9. Point groups, C_{2v}–C_{6v}.

Fig. 9, and are analogous to those of Fig. 6. A reflection plane can also be introduced perpendicular to the n-fold axis of the groups $D_2 - D_6$ (Fig. 6), giving rise to the four further groups of Fig. 10. In the case of D_2 and D_3 (Fig. 6), reflection planes can be introduced bisecting the angles between the horizontal 2-fold axes, providing the two further groups shown in Fig. 11. Finally, the complicated system of axes of the cubic groups T and O (Fig. 7) can be brought into self-coincidence by the introduction of certain sets of reflection planes. This gives rise to the three further cubic groups shown in Fig. 12.

For practical purposes it is convenient to classify the 32 point groups according to the most simple choice of reference axes.

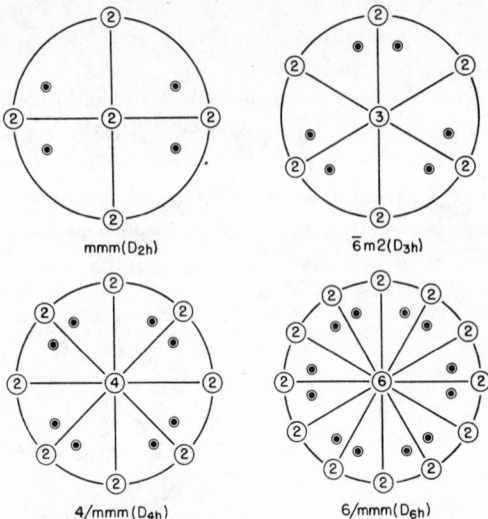

Fig. 10. Point groups, D_{2h}–D_{6h}.

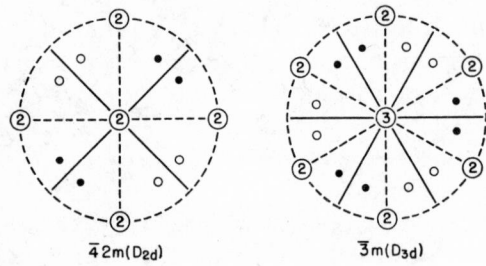

Fig. 11. Point groups, D_{2d} and D_{3d}.

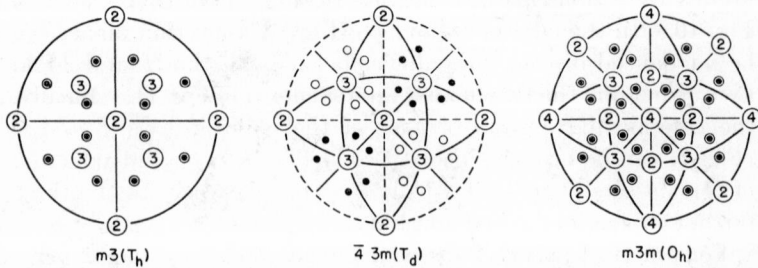

Fig. 12. Point groups, T_h, T_d, and O_h.

TABLE I. THE 32 CRYSTAL CLASSES

(The eleven classes of distinct Laue symmetry are separated by vertical lines)

System	Axes and angles			Crystal classes (point groups)							
Triclinic	a	b	c	1	$\bar{1}$						
	α	β	γ	C_1	C_i						
Monoclinic	a	b	c	2	m or $\bar{2}$	$2/m$					
	90°	β	90°	C_2	C_s	C_{2h}					
Orthorhombic	a	b	c				222	$mm2$	mmm		
	90°	90°	90°				D_2	C_{2v}	D_{2h}		
Tetragonal	a	a	c	4	$\bar{4}$	$4/m$	422	$4mm$	$\bar{4}2m$	$4/mmm$	
	90°	90°	90°	C_4	S_4	C_{4h}	D_4	C_{4v}	D_{2d}	D_{4h}	
Trigonal (rhombohedral)	aaa or $\alpha\alpha\alpha$	a 90°	a 90°	c 120°	3 C_3	$\bar{3}$ C_{3i}		32 D_3	$3m$ C_{3v}	$\bar{3}m$ D_{3d}	
Hexagonal	a	a	c	6	$\bar{6}$	$6/m$	622	$6mm$	$\bar{6}m2$	$6/mmm$	
	90°	90°	120°	C_6	C_{3h}	C_{6h}	D_6	C_{6v}	D_{3h}	D_{6h}	
Cubic	a	a	a	23	$m3$		432	$\bar{4}3m$	$m3m$		
	90°	90°	90°	T	T_h		O	T_d	O_h		

This gives rise to the seven well-known crystallographic systems, and the arrangement generally adopted is shown in Table I. Reference axes are also indicated, and these range from the triclinic system, where the three axes may be of any length, inclined at any angle, to the cubic system, with three equal axes mutually perpendicular.

While the 3-fold axis characterises the trigonal or rhombohedral system, it should be noted that this system cannot be based exclusively on the rhombohedral lattice (see Chapter III). When the space groups (Table III) are developed from these point groups, it is found that the simplest lattice in many cases is hexagonal rather than rhombohedral, so that the alternative hexagonal description must be made available for this system. Because of this duality the name trigonal is to be preferred to rhombohedral.

III

The Crystal

as a Lattice Structure

1. EARLY STRUCTURE THEORIES

All the preceding arguments have been based on a study of the external shapes of crystals, using the experimental generalisation known as the law of rational indices to limit the number of possible symmetry elements. These purely morphological considerations have not required any reference to physical theories regarding the structure of crystals, although it has been pointed out that the proofs are not quite general and require confirmation. Nevertheless, the subject had progressed as far as this some time before the structure theory was at all fully developed.

The fundamental physical idea that crystals must be based on some kind of lattice structure did, however, arise much earlier in the history of science. By lattice is meant some regular, geometrical repetition in space of identical uints (Fig. 15). Such an idea immediately explains many of the peculiar features of crystals, their smooth plane faces, constant angles, and above all, as was clearly recognised by Haüy, the law of rational indices. Long before Haüy's work this basic idea of crystalline structure had been discussed. One of the earliest references, perhaps, is to be found in *Micrographia*, by Robert Hooke, published in 1665. Fig. 13 is taken from that work and illustrates Hooke's attempts to imitate the various shapes of alum crystals by the close packing of spherical objects such as bullets. Referring to the regularity of figure exhibited by diamonds, minerals, precious stones, and salts, Hooke goes on to say:

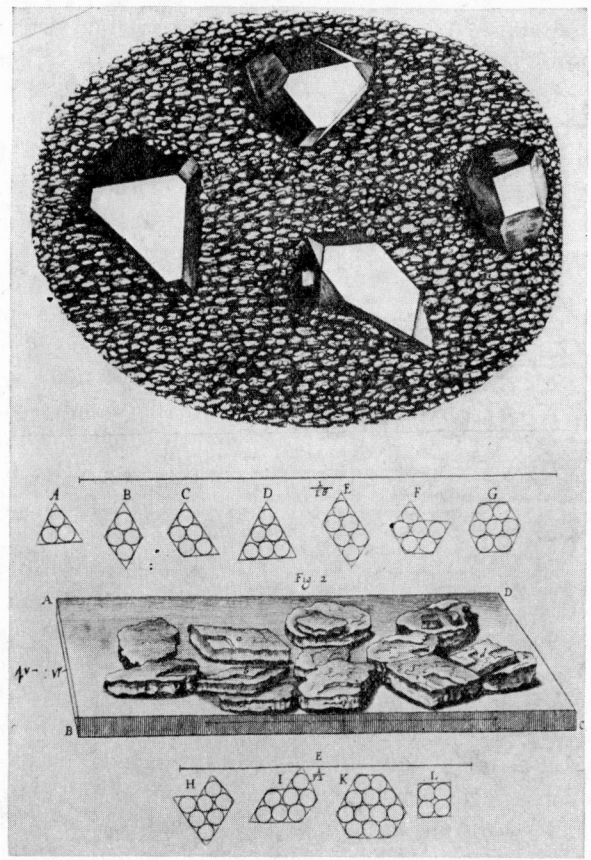

Fig. 13. Hooke's explanation of the regular figures formed by crystals (from *Micrographia*, 1665).

I think, had I time and opportunity, I could make probable that all these regular Figures that are so conspicuously various and curious . . . arise only from three or four several positions or postures of Globular particles, and those the most plain and obvious, . . . the coagulating particles must necessarily compose a body of such a determinate regular Figure, and no other.[1]

An even more striking illustration of a lattice structure, which might well occur in any modern textbook on crystallography, is shown in Fig. 14. This is taken from Huygens' famous treatise

[1] Robert Hooke, *Micrographia*, London, 1665, p. 85.

on light, published in 1690, but probably written some twelve years before that.   Like Hooke, Huygens refers to various well-known crystals, including rock crystal (hexagonal bars), diamonds, various salts, sugar, snow flakes, and ice, and says, "It seems to me that in general the regularity which occurs in these productions comes from the arrangement of the small invisible equal particles of which they are composed."[2]   He then goes on to describe in detail a model for the Iceland (spar) crystal, which is illustrated in the figure. This he pictured as built of small rounded corpuscles, not spherical but flattened spheroids, lightly stuck together, and in this way he explains the cleavage and angles of the crystal.

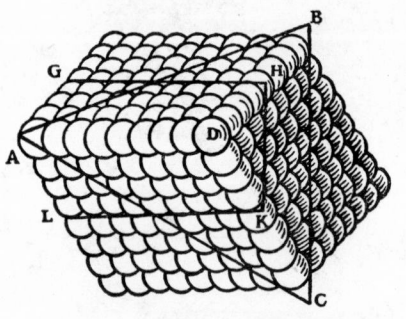

Fig. 14. Huygens' model of Iceland spar crystal (1690).

About a century later these ideas were given much greater precision by the work of Haüy, to which we have already referred.   Haüy's conception of crystal structure was based on a study of the cleavage figure, and he pictured the crystal as being built of small parallelopipedal units stacked together in a regular fashion.   If some of these units were regularly omitted during growth of the crystal, the occurrence of secondary forms could be explained.   On this picture, the law of rational intercept ratios, the discovery of which was Haüy's greatest contribution, immediately becomes apparent.

It soon became clear that the actual shapes of Haüy's molecules were really geometrical abstractions, and no doubt this was the view of Haüy himself; their dimensions give the intervals which separate the repeating units of the crystal along the three axial directions of the unit parallelopiped.   Interest then soon became focused on the purely geometrical investigation of lattice structures, without regard to the shape or form of the ultimate particles, or the chemical or physical properties of crystals.   The

[2] Christiaan Huygens, *Traité de la lumière*, Leiden, 1690.   Eng. trans. by S. P. Thompson, London: Macmillan and Co., 1912.

first results, obtained by M. L. Frankenheim,[3] were not entirely satisfactory, but rigid geometrical proofs were obtained soon afterwards by Auguste Bravais,[4] who demonstrated that only 14 distinct types of space lattice are possible.

2. LATTICES AND NETS

To construct the systems of points regularly distributed in space, which are the subject of Bravais's investigation, we may proceed in three steps. A rectilinear system of equidistant points, extending indefinitely in both directions, is called a *row*

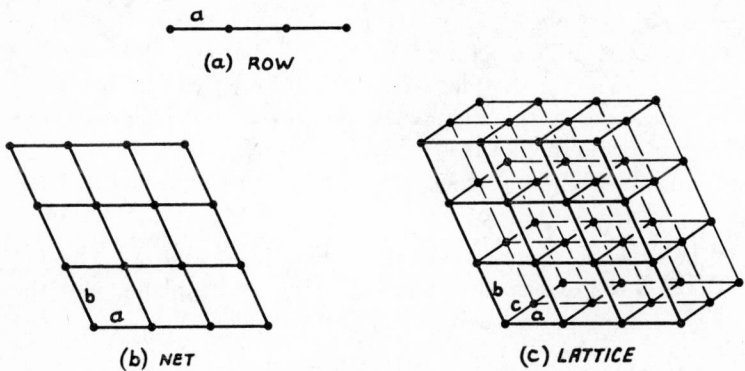

(a) ROW

(b) NET   (c) LATTICE

Fig. 15. Lattices and nets.

(Fig. 15a), and is fully described by one parameter, a, the fundamental interval separating two neighbouring points. A regular series of such parallel rows, lying in one plane, can be described by two fundamental intervals, a and b, and an arbitrary angle γ. This is called a *net* (Fig. 15b). The system of points regularly distributed in three dimensions consists of an indefinitely extended regular series of parallel nets, and can be described by three fundamental intervals, a, b, and c, together with three arbitrary angles, α, β, and γ. Such a system is called a *lattice* (Fig. 15c). The smallest parallelopiped which is identically

[3] M. L. Frankenheim, *Die Lehre von der Cohäsion*, Breslau, 1835; also *Nova Acta Acad. Caes. Leopoldino-Carolinae Nat. Cur.*, 1842, **19**(2), 471–660.

[4] Auguste Bravais, "Mémoire sur les systèmes formés par des points distribués regulièrement sur un plan ou dans l'espace, *J. école polytech.* (Paris), 1850, **19**, 1–128. Eng. trans. by A. Shaler, Cryst. Soc. America, Memoir no. 1, 1949.

repeated in this system (such as is outlined in the diagram) is called the generating parallelopiped or *unit cell*.

All lattices constructed in this way possess the important property that "none of the individual points can be distinguished from any other by any uniqueness of relative position." The surroundings of every lattice point are exactly the same.

With regard to symmetry it is clear that every net has at least a plane of symmetry in the plane of the net, a centre of symmetry at each point, and a 2-fold axis perpendicular to the plane of the net through each point. Similar centres of symmetry and 2-fold axes at half-way positions between the net points are also involved. The net may also have various 2-fold axes of symmetry lying in the plane of the net, but it cannot have any other axes not perpendicular to the plane, for such rotations could not bring the net into self-coincidence.

In all, five different types of net can be distinguished by their symmetry (Fig. 16).

(a) The lowest symmetry type is that given above. The net contains no axis of symmetry lying in its plane, and the generating parallelogram has unequal sides.

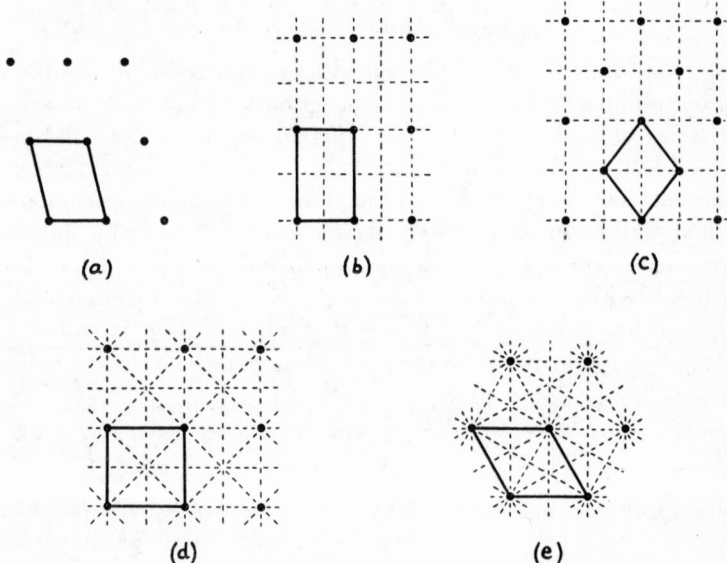

Fig. 16. Nets. The five symmetry types.

(b) The net has two systems of symmetry axes (dotted lines) lying in its plane, mutually perpendicular. The generating parallelogram is a rectangle.

(c) The net has two similar systems of symmetry axes, but the generating parallelogram is a rhombus. This can also be regarded as a centred rectangular net, just as (b) can be regarded as a centred rhombic net.

(d) The net has four systems of symmetry axes lying in its plane. The generating parallelogram is a square.

(e) The net has six systems of symmetry axes lying in its plane. The generating parallelogram is a rhombus with 60° and 120° angles.

3. LIMITATION OF SYMMETRY IN A LATTICE

With regard to lattices the situation is rather more complex, but Bravais demonstrated that only 14 distinct types are possible. It is first necessary to establish that the symmetry axes in any lattice can only be of the 1-, 2-, 3-, 4-, or 6-fold types. This has already been shown to follow from the law of rational indices, but it is, of course, a direct consequence of the lattice structure of crystals, and Bravais's proof is interesting. We take an n-fold axis of symmetry (drawn perpendicular to the plane of the paper) through some point O,

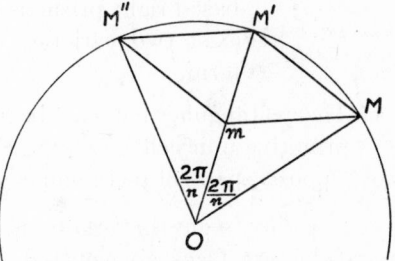

Fig. 17. Bravais's construction.

which need not be assumed a lattice point (Fig. 17). Take M to be the *nearest* lattice point, and let the operation of the axis produce M', M''. Complete the generating parallelogram $MM'M''m$, which in this case is a rhombus, because $MM' = M'M''$. m is a lattice point, and it can readily be shown that

$$Om = OM' \left(1 - 4 \sin^2 \frac{\pi}{n} \right).$$

Now the distance Om cannot be less than OM' (because M was taken as the nearest lattice point) except for the special case when m coincides with O, which then becomes a lattice point. Hence, the only possible values of n are 1, 2, 3, 4, and 6.

4. THE 14 BRAVAIS LATTICES

The next step calls for a thorough investigation of all possible combinations of these symmetry axes in lattices. This was carried out by Bravais, who thereby showed that 14 and only 14 different types of lattice, belonging to 7 systems, can be distinguished. We shall not attempt to establish the various theorems required to prove this, but give only the final results.

First there is the triclinic lattice with no axis or plane of symmetry. The unit cell is defined by three arbitrarily chosen axes, a, b, c, inclined to each other at angles, α, β, γ, which are in general unequal (Fig. 18).

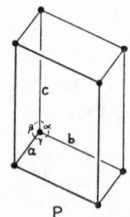

Fig. 18. Triclinic lattice.

Next we can have one 2-fold axis of symmetry, which involves a plane of symmetry normal to it. One axis of the unit cell, the symmetry axis b, is therefore unique, and is perpendicular to the other two axes, a and c, which can be chosen arbitrarily (Fig. 19). The net planes parallel to the symmetry axis are rectangles, giving a parallelogram-based right prism as unit cell. In this lattice there exist two varieties, or "modes," to use Bravais's term.

(a) The rectangular net can be uncentred, giving a simple, or primitive unit cell.

(b) Opposite pairs of rectangular nets can be centred.

It is readily seen that there is no other possibility. If the parallelogram faces are centred, the lattice is of the type (a) and can be referred to a unit cell of half the size, because the choice of the axes a and c is arbitrary. Again, body centring is equivalent to (b) if the a and c axes are chosen differently. The remaining case with *all* the rectangular nets centred does not

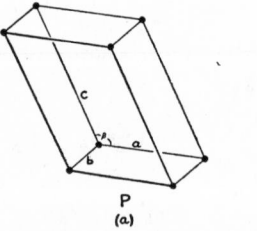

P
(a)

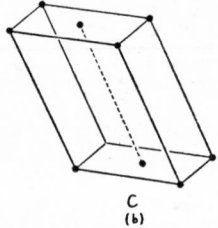

C
(b)

Fig. 19. Monoclinic lattices.

constitute a lattice because then the surroundings of each point are not all equivalent.

The next higher symmetry class of lattice involves three mutually perpendicular 2-fold axes and corresponding symmetry planes. The reference axes of the unit cell, taken as these symmetry axes, are mutually perpendicular, but of arbitrary lengths, a, b, and c (Fig. 20). In this class there are four distinct modes of arrangement.

(a) The primitive unit cell, a right prism with rectangular base.

(b) Two opposite faces of the unit cell may be centred. In this case a right prism with rhombic base may be chosen for a primitive unit cell.

(c) All the faces of the unit cell may be centred. If the right prism with rhombic base is chosen for the unit cell, it becomes body centred in this case.

(d) The rectangular cell, body centred.

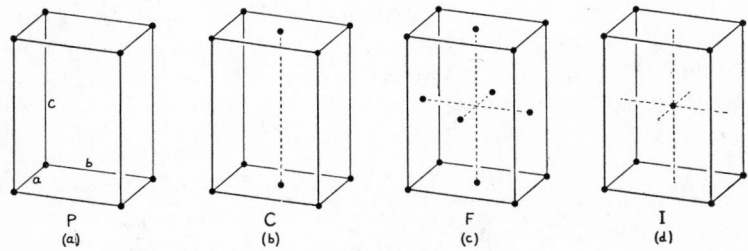

Fig. 20. Orthorhombic lattices.

The introduction of a single 4-fold axis involves a square net perpendicular to it, with two 2-fold axes parallel to the sides of the square and two 2-fold axes parallel to its diagonals, as well as symmetry planes normal to all the axes. Only two distinct modes exist in this class.

(a) The right prism with square base.

(b) The same, body centred.

It is easy to see that centring the square base nets gives a unit cell equivalent to (a), with the two equal axes rotated 45° (Fig. 21).

The 3-fold axis involves triequiangular nets perpendicular to it, but the axis only passes through a lattice point in every third

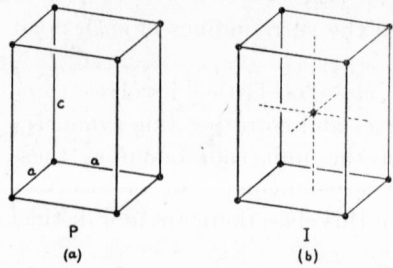

Fig. 21. Tetragonal lattices.

net. (A centre of inversion lies on this axis, which makes it $\bar{3}$.) There are also three 2-fold axes and three symmetry planes. There is only one mode, where the unit cell is a rhombohedron, with sides a and angle α (Fig. 22).

The 6-fold axis also involves triequiangular nets perpendicular to it, but it passes through a lattice point in each successive net. Unlike the 3-fold axis, it involves a plane of symmetry perpendicular to itself. Six 2-fold axes are also involved, with six other symmetry planes. The unit cell is a right prism, based on a rhombus with 60° and 120° angles (Fig. 23), or the lattice may be referred to a rectangular cell, centred on the c-face.

Finally, there is the combination of three 4-fold, four 3-fold, and six 2-fold axes giving cubic symmetry. These axes join the centres of opposite faces, opposite apices, and the mid points of opposite edges. Nine planes of symmetry are involved, normal to the 4-fold and 2-fold axes.

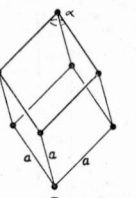
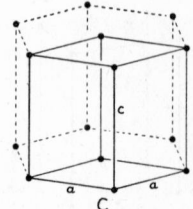

Fig. 22 (Left). Rhombohedral lattice.
Fig. 23 (Right). Hexagonal lattice.

The unit cell is a cube, and there are three distinct modes of arrangement (Fig. 24).

(a) The simple cube.
(b) The cube with all faces centred. In this case a rhombohedron, angle 60°, represents the true primitive cell.
(c) The body-centred cube. In this case a rhombohedron with angle 109° 28′ represents the true primitive cell.

We find therefore that, in all, seven classes of lattice exist (Figs. 18–24), and in symmetry these correspond to the crystal class of highest, or holohedral, symmetry in each of the seven crystal systems already mentioned in the last chapter.

In his treatment of lattices Bravais employs the mathematical concept of a coincidence movement in establishing his various

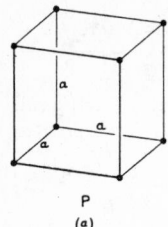 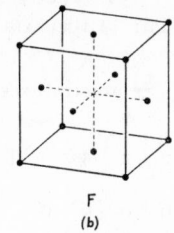 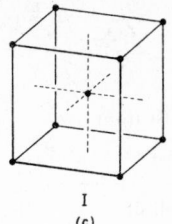

Fig. 24. Cubic lattices.

theorems. A more general treatment by group theoretical methods is now possible, the lattice being regarded as a translation group. A translation of the whole lattice along the repeat distance, or some multiple of this, in any lattice direction, brings it into self-coincidence. The smallest translation of this kind in any direction, involving only one repeat distance, is called a *primitive translation*. We have seen that in some of the Bravais lattices it is convenient to employ a "unit cell" which is not the true generating parallelopiped of the lattice, but contains one or more additional lattice points within it. If the unit cell chosen does correspond to the true generating parallelopiped with no internal lattice points, the lattice is called *primitive* (symbol P). If a multiply-primitive cell is chosen with an opposite pair of faces centred, the symbols A, B, or C are employed; if all faces are centred the symbol is F, while I denotes a body-centred cell. In the trigonal (rhombohedral) system the special symbol R is used when the primitive lattice actually is rhombohedral. In the hexagonal system the symbol P (formerly C) is used for the primitive hexagonal lattice. The symbol P is also employed in the trigonal system for those cases classified in this system for which the true primitive lattice is hexagonal. (The symbol H was formerly sometimes employed in both these systems to represent a hexagonal cell of volume three times that of the primitive cell P and with the secondary axes rotated 30° in the basal plane from the positions in the primitive case. This symbol did not, of course, represent any new type of lattice and is no longer used.)

The 14 Bravais lattices are summarised in Table II, which gives the Hermann-Mauguin symmetry symbol of each of the seven types. The table also gives the number and type of symmetry axes present in each lattice, which were used by Bravais as the basis of his classification

TABLE II. THE 14 BRAVAIS LATTICES

System	Sym-metry	Symmetry axes					Sym-metry planes, m	Bravais lattices
		2	3	4	6	Tot.		
Triclinic	$\bar{1}$	0	0	0	0	0	0	P
Monoclinic	$2/m$	1	0	0	0	1	1	P, $[AC]$
Orthorhombic	mmm	3	0	0	0	3	3	P, $[ABC]$, F, I
Tetragonal	$4/mmm$	4	0	1	0	5	5	P, I
Trigonal (rhom-bohedral)	$\bar{3}m$	3	1	0	0	4	3	R
Hexagonal	$6/mmm$	6	0	0	1	7	7	$\left.\right\} P$
Cubic	$m3m$	6	4	3	0	13	9	P, F, I

5. GENERALISED STRUCTURE THEORY

The work of Bravais concludes one important chapter in structure theory, and provides the essential basis for all further development. Since the time of Haüy and earlier it has been accepted that crystals are built on a lattice structure; it follows, therefore, that every crystal structure must conform to one or other of the 14 Bravais lattices. This is in accord with observation. As Bravais points out, his purely geometrical classification of lattices "corresponds accurately to the classification which a patient and prolonged investigation has demonstrated among the various crystalline systems."

There is, however, a difficulty or at least a need for some extension or greater generalisation of this theory. We have seen that investigation of the crystal as an idealised geometrical figure leads to the discovery of 32 possible classes of crystalline symmetry, which can be referred to seven systems. The Bravais lattices can also be classified in the same seven systems, and each lattice displays the symmetry of the highest crystal class in each system. This is natural, because the lattice points in Bravais's investigation are regarded as geometrical points or small perfect spheres which each possess an unlimited degree of inherent symmetry and can therefore conform to all the symmetry operations that may be present. If we replace the lattice point by an object without any inherent symmetry, then the over-all symmetry of the structure is destroyed, and the cube, for example, in spite of its equal axes and angles, is reduced to the symmetry of the lowest triclinic class (C_1). Bravais went on to discuss the modification of the lattice symmetry involved by the introduction of various symmetrical figures in place of the lattice points. In this way the symmetry of the 32 classes can be explained, but this part of his structure theory is now mainly of historical interest. It involves certain elements of inherent "molecular" symmetry, which may indeed be present in certain cases, and must be present if we regard the whole lattice unit as identical with the "molecule." This aspect of Bravais's work was, however, soon to be replaced by a more generalised treatment, which can be regarded to some extent as independent of the lattice concept, or rather as including the lattice concept within its framework.

The general problem concerns all the possible ways of symmetrical repetition in space so that every point has the remaining points arranged about it in an identical manner. This, as we have seen, is a property of the lattice, but the lattice concept involves the additional requirement of the regular spacing of the rows and columns of lattice points. Camille Jordan[5] provided the mathematical methods for dealing with the problem in a perfectly general way by means of infinite groups of coincidence movements, and Leonhard Sohncke[6] employed Jordan's method to derive his 65 "regular point systems." These give all the possible systems of repetition, in accordance with the above definition, which are applicable to infinite assemblages of identical particles, and so they must include the Bravais lattices as special cases. From the point of view of the coincidence movements, or the symmetry operations, which generate the Sohncke systems, it is seen that the Bravais lattice points occupy special positions in so far as they lie on or coincide with certain of the symmetry elements present in the system. The points in the Sohncke system occupy general positions, and while achieving the property that no single point can be distinguished from any other by virtue of its position, they are more general than the Bravais lattice points; it is, however, possible to describe the Sohncke systems in terms of a number of interpenetrating congruent lattices.

If the Sohncke systems are regarded as collections of points

[5] Camille Jordan, "Mémoire sur les groupes de mouvements," *Ann. matematica pura ed applicata* (Milan), ser. 2, 1869, 2, 167–215, 322–345.

[6] Leonhard Sohncke, *Entwickelung einer Theorie der Krystallstruktur*, Leipzig, 1879.

or spherical particles, there is the danger that in some cases they will still involve certain additional symmetry elements not demanded by the fundamental generating operations. This can be overcome if we represent the points by unsymmetrical but identical objects. The systems then give all the possible distinct systems of symmetrical repetition of such identical objects and represent a complete analysis of the problem.

It is still not possible, however, to account for all the kinds of symmetry found in crystals on the basis of these 65 systems, because they fail to explain the enantiomorphous similarities exhibited in certain crystal shapes. The final step of extending the treatment to include this enantiomorphous similarity was achieved independently by E. Fedorow,[7] Arthur Schoenflies,[8] and William Barlow,[9] who, by introducing symmetry operations of the second sort, derived the 230 space groups. This step gave 165 systems additional to those enumerated by Sohncke and completed the geometrical theory of crystal structure. Regarded as infinite assemblages of particles, the properties of these arrangements are now such that every general point has exactly similar surroundings, but this similarity can either be identity or mirror image resemblance. If the general point is represented by an unsymmetrical molecule, then it follows that in the 165 new systems two kinds of molecule must be present, bearing mirror image resemblance to each other. This requirement was held by some to be improbable, but the "molecule" in this respect need not have any immediate chemical significance. With this extension of the theory it is found that all the 32 crystal classes can be accounted for exactly.

6. THE 230 SPACE GROUPS

The difficulty which beset the development of the structure theories briefly outlined above appears to have resulted largely from interwoven speculation regarding the nature of the molecule in crystals, about which no direct evidence was available until the next epoch in crystallography began with von Laue's discovery of X-ray diffraction in 1912. The whole position becomes

[7] E. Fedorow, *Trans. Russian Min. Soc.*, 1885, **21**, 1–279; 1888, **25**, 1–52; *Symmetry of Regular Systems of Figures*, 1890 (in Russian).
[8] Arthur Schoenflies, *Krystallsysteme und Krystallstruktur*, Leipzig, 1891.
[9] William Barlow, *Z. Kryst. Mineral.*, 1894, **23**, 1–63; 1895, **25**, 86.

greatly simplified if we ignore all speculations of this kind and
focus attention on the purely geometrical aspect of the problem.
This really concerns the finding of all the possible self-consistent
sets of symmetry operations which are applicable to a discontinuous but infinitely extended medium.

In a previous section we have discussed the idea of symmetry
as applied to a geometrical figure. The figure is said to possess
symmetry if it can be brought into self-coincidence by means
of certain operations such as rotations, reflections, or combinations of these, which are specified by the symbols 1, 2, 3 $\cdots n$
and $\bar{1}, \bar{2}, \bar{3} \cdots n$. The collection of operations present in any
given body or figure must form a self-consistent set, or a group
in the mathematical sense. The operations so far considered
have always been such as to leave one point of the figure unmoved,
and the groups described have been point groups.

We have now to consider the symmetry properties of structures
which are ultimately built up on regular lattices, indefinitely
extended in every direction. As the lattice consists of a regular
array of identical units, typified by the lattice points, it is clear
that if it contains an element such as an axis of symmetry, then
it must contain an infinite number of parallel axes.

Further, the structure can now be brought into self-coincidence
in a new way, namely by *translations* along any of the lattice
directions. The lattice units are by definition identical, so such
movements, over the repeat distance, leave the structure unchanged. Two successive reflections in a plane, or n successive
operations of an n-fold axis, bring a figure back to its original
position; but a lattice may be brought back to its original position
or to a position one translation removed, and in each case self-
coincidence is achieved. This alternative method of reaching
self-coincidence in a lattice structure means that a new class of
symmetry operations is applicable to such structures, in addition
to those which apply to finite geometrical figures. These new
operations are obtained by combining reflections and rotations
with translations along the lattice directions.

The operation obtained by combining a rotation axis with a
translation is called a *screw axis*. If the axis is 2-fold, the translation must be *half* the primitive translation of the lattice in the
direction of the axis, in order that complete rotation of 2π will
restore the initial aspect (one primitive translation further on).

41

The symbol for this operation is 2_1. If the axis is 3-fold, the translation must be one-third or two-thirds, and the symbols are 3_1 or 3_2. In general, the symbol p_q means a rotation of $2\pi/p$ combined with a translation of q/p in the direction of the axis. The multiplicity of these screw axes is, of course, governed by the general symmetry limitations which we have already discovered, so the complete set of these operations which are applicable to lattice structures may be written as

$$2_1 \qquad 3_1 \ 3_2 \qquad 4_1 \ 4_2 \ 4_3 \qquad 6_1 \ 6_2 \ 6_3 \ 6_4 \ 6_5.$$

It is readily seen that the "pure" rotation axes, 2, 3, 4, and 6, are special cases of these generalised operations.

As the name implies, most of these axes have a definite sense of screw. On drawing them out[10] one can see that if successive operations of the axis 3_1 are taken to give a right-hand screw, then successive operations of the axis 3_2 will give a left-hand screw. The axes 4_1 and 4_3 stand in the same enantiomorphic relationship, as do 6_1 and 6_5, and 6_2 and 6_4. The other types, 2_1, 4_2, and 6_3, where the suffix is exactly half the principal number, exhibit no sense of screw.

When a plane of symmetry is combined with a translation, the resulting operation is called a *glide plane*, and as reflection is a 2-fold operation, a half translation is always involved. The terminology employed indicates the direction of the translation, a, b, and c denoting glide planes with translations of $a/2$, $b/2$, or $c/2$, and n a diagonal translation of $(b+c)/2$, $(c+a)/2$, or $(a+b)/2$. In the case of centred nets a quarter diagonal translation may be involved, and this case is denoted by the symbol d. It will be clear that these various symbols for glide planes do not represent fundamentally different operations, but that they are related to the choice of reference axes in the lattice.

We have now enumerated all the kinds of symmetry operation that are applicable to infinitely extended structures based on the lattice principle. The self-consistent sets of these operations constitute groups of movements which are infinite, but discontinuous, and these are called *space groups*. The investigation of all the different space groups that are theoretically possible is a lengthy mathematical operation, and, as we have seen in the

[10] See, for example, F. C. Phillips, *Introduction to Crystallography*, London: Longmans, 1946.

last section, it was completed between the years 1885 and 1894 independently and almost simultaneously by the three investigators, Fedorow, Schoenflies, and Barlow, who showed that 230 different space groups are possible.

The method of development of the space groups employed by Schoenflies[8] is based on the fact that each space group is "isomorphous" with one of the 32 point groups, i.e., the operations of the space group are derived from those of the point group by multiplying by a translation. The planes and axes of the space group have the same directions, and the angles of rotation of the axes have the same values, as in the point group. The translation groups available are those represented by the Bravais lattices of appropriate symmetry. The problem of deriving the space groups can thus be systematised, but to establish that each case is unique and that no further possibilities exist is a long and arduous task.

Table III gives a list of the 230 space groups arranged under the appropriate point groups or crystal classes. The symbols for the space groups employed in the main body of the table are those of the shortened Hermann-Mauguin notation,[11,12] which is now generally employed. The capital letter (P, C, I, etc.) denotes the Bravais lattice, and is followed by the symbols for the fundamental symmetry elements present, which lie along special directions in the crystal. These directions are chosen as follows: in the monoclinic system the 2-fold axis (b); in the orthorhombic system the three perpendicular axes (a, b, and c); in the tetragonal system the principal 4-fold axis, the edge of the square base and its diagonal, at 45°; in the trigonal system the principal 3-fold axis and the edge of the triequiangular net (and its perpendicular in the same plane if the primitive lattice is hexagonal); in the hexagonal system the principal 6-fold axis, the edge of the hexagonal base and its perpendicular; in the cubic system, the cube edges, cube diagonals and face diagonals.

The fundamental symmetry elements named in the symbol are only the minimum necessary to define the space group. Their presence generally involves the existence of a number of further

[11] *Internationale Tabellen zur Bestimmung von Kristallstrukturen*, Berlin, 1935; *International Tables for X-ray Crystallography*, Birmingham, Eng.: Kynoch Press, 1952.

[12] C. Hermann, *Z. Krist.*, 1928, **68**, 257; 69, 226.

TABLE III. THE 230 SPACE GROUPS

(The eleven classes of distinct Laue symmetry are separated by double rulings)

System	Point groups Schfl.	H.-M.	Space groups						
Triclinic	C_1	1	$P1$						
Triclinic	C_i	$\bar{1}$	$P\bar{1}$						
Monoclinic	$C_2^{(1-3)}$	2	$P2$	$P2_1$	$C2$				
Monoclinic	$C_s^{(1-4)}$	m	Pm	Pc	Cm	Cc			
Monoclinic	$C_{2h}^{(1-6)}$	$2/m$	$P2/m$	$P2_1/m$	$C2/m$	$P2/c$	$P2_1/c$	$C2/c$	
Orthorhombic	$D_2^{(1-9)}$	222	$P222$ $I222$	$P222_1$ $I2_12_12_1$	$P2_12_12$	$P2_12_12_1$	$C222_1$	$C222$	$F222$
Orthorhombic	$C_{2v}^{(1-22)}$	$mm2$	$Pmm2$ $Pba2$ $Abm2$ $Ima2$	$Pmc2_1$ $Pna2_1$ $Ama2$	$Pcc2$ $Pnn2$ $Aba2$	$Pma2$ $Cmm2$ $Fmm2$	$Pca2_1$ $Cmc2_1$ $Fdd2$	$Pnc2$ $Ccc2$ $Imm2$	$Pmn2_1$ $Amm2$ $Iba2$
Orthorhombic	$D_{2h}^{(1-28)}$	mmm	$Pmmm$ $Pcca$ $Pbca$ $Ccca$	$Pnnn$ $Pbam$ $Pnma$ $Fmmm$	$Pccm$ $Pccn$ $Cmcm$ $Fddd$	$Pban$ $Pbcm$ $Cmca$ $Immm$	$Pmma$ $Pnnm$ $Cmmm$ $Ibam$	$Pnna$ $Pmmn$ $Cccm$ $Ibca$	$Pmna$ $Pbcn$ $Cmma$ $Imma$
Tetragonal	$C_4^{(1-6)}$	4	$P4$	$P4_1$	$P4_2$	$P4_3$	$I4$	$I4_1$	
Tetragonal	$S_4^{(1-2)}$	$\bar{4}$	$P\bar{4}$	$I\bar{4}$					
Tetragonal	$C_{4h}^{(1-6)}$	$4/m$	$P4/m$	$P4_2/m$	$P4/n$	$P4_2/n$	$I4/m$	$I4_1/a$	
Tetragonal	$D_4^{(1-10)}$	422	$P422$ $P4_12_12$	$P42_12$ $I422$	$P4_122$ $I4_122$	$P4_12_12$	$P4_222$	$P4_22_12$	$P4_322$
Tetragonal	$C_{4v}^{(1-12)}$	$4mm$	$P4mm$ $P4_2bc$	$P4bm$ $I4mm$	$P4_2cm$ $I4cm$	$P4_2nm$ $I4_1md$	$P4cc$ $I4_1cd$	$P4nc$	$P4_2mc$
Tetragonal	$D_{2d}^{(1-12)}$	$\bar{4}2m$	$P\bar{4}2m$ $P\bar{4}n2$	$P\bar{4}2c$ $I\bar{4}m2$	$P\bar{4}2_1m$ $I\bar{4}c2$	$P\bar{4}2_1c$ $I\bar{4}2m$	$P\bar{4}m2$ $I\bar{4}2d$	$P\bar{4}c2$	$P\bar{4}b2$
Tetragonal	$D_{4h}^{(1-20)}$	$4/mmm$	$P4/mmm$ $P4/ncc$ $P4_2/nmc$	$P4/mcc$ $P4_2/mmc$ $P4_2/ncm$	$P4/nbm$ $P4_2/mcm$ $I4/mmm$	$P4/nnc$ $P4_2/nbc$ $I4/mcm$	$P4/mbm$ $P4_2/nnm$ $I4_1/amd$	$P4/mnc$ $P4_2/mbc$ $I4_1/acd$	$P4/nmm$ $P4_2/mnm$
Trigonal	$C_3^{(1-4)}$	3	$P3$	$P3_1$	$P3_2$	$R3$			
Trigonal	$C_{3i}^{(1-2)}$	$\bar{3}$	$P\bar{3}$	$R\bar{3}$					
Trigonal	$D_3^{(1-7)}$	32	$P312$	$P321$	$P3_112$	$P3_121$	$P3_212$	$P3_221$	$R32$
Trigonal	$C_{3v}^{(1-6)}$	$3m$	$P3m1$	$P31m$	$P3c1$	$P31c$	$R3m$	$R3c$	
Trigonal	$D_{3d}^{(1-6)}$	$\bar{3}m$	$P\bar{3}1m$	$P\bar{3}1c$	$P\bar{3}m1$	$P\bar{3}c1$	$R\bar{3}m$	$R\bar{3}c$	
Hexagonal	$C_6^{(1-6)}$	6	$P6$	$P6_1$	$P6_5$	$P6_2$	$P6_4$	$P6_3$	
Hexagonal	$C_{3h}^{(1)}$	$\bar{6}$	$P\bar{6}$						
Hexagonal	$C_{6h}^{(1-2)}$	$6/m$	$P6/m$	$P6_3/m$					
Hexagonal	$D_6^{(1-6)}$	622	$P622$	$P6_122$	$P6_522$	$P6_222$	$P6_422$	$P6_322$	
Hexagonal	$C_{6v}^{(1-4)}$	$6mm$	$P6mm$	$P6cc$	$P6_3cm$	$P6_3mc$			
Hexagonal	$D_{3h}^{(1-4)}$	$\bar{6}m2$	$P\bar{6}m2$	$P\bar{6}c2$	$P\bar{6}2m$	$P\bar{6}2c$			
Hexagonal	$D_{6h}^{(1-4)}$	$6/mmm$	$P6/mmm$	$P6/mcc$	$P6_3/mcm$	$P6_3/mmc$			
Cubic	$T^{(1-5)}$	23	$P23$	$F23$	$I23$	$P2_13$	$I2_13$		
Cubic	$T_h^{(1-7)}$	$m3$	$Pm3$	$Pn3$	$Fm3$	$Fd3$	$Im3$	$Pa3$	$Ia3$
Cubic	$O^{(1-8)}$	432	$P432$ $I4_132$	$P4_232$	$F432$	$F4_132$	$I432$	$P4_332$	$P4_132$
Cubic	$T_d^{(1-6)}$	$\bar{4}3m$	$P\bar{4}3m$	$F\bar{4}3m$	$I\bar{4}3m$	$P\bar{4}3n$	$F\bar{4}3c$	$I\bar{4}3d$	
Cubic	$O_h^{(1-10)}$	$m3m$	$Pm3m$ $Fd3c$	$Pn3n$ $Im3m$	$Pm3n$ $Ia3d$	$Pn3m$	$Fm3m$	$Fm3c$	$Fd3m$

symmetry elements, which do not therefore require to be enumerated in the symbol.   For example, in the class *mmm*, the three mutually perpendicular reflection planes involve three mutually perpendicular 2-fold axes as well, and the full symbol for the space group *Pmmm* would be *P2/m, 2/m, 2/m*, the stroke indicating that the reflection plane is perpendicular to the axis named.   But as the axes in this case are necessarily involved, they are omitted.   It is, however, often necessary to specify a plane of symmetry perpendicular to a principal axis, as in *P4/mmm*, where a more complete symbol would be *P4/m, 2/m, 2/m*.   In groups of high symmetry like this quite a number of additional symmetry elements such as screw axes and glide planes may also be involved.

This particular symbolism for the space groups is more useful for many purposes than the older Schoenflies notation, given on the left of the table, where the space group is denoted by a simple numerical index attached to the Schoenflies symbol for the point group.   One advantage of the Hermann-Mauguin system, in addition to giving the fundamental symmetry elements present, is that it specifies the choice of the arbitrary reference axes where this is necessary.   Thus, in the monoclinic system, the *a* and *c* axes may be chosen arbitrarily, while in the orthorhombic system there is no immediately accessible symmetry criterion for naming the *a*, *b*, and *c* axes in any specified order.   If glide planes are present, however, the translations involved provide in many cases a means for naming the reference axes in a unique manner, and the symbols employed in the table follow certain conventions in this respect.   If the reference axes are named in a different way, then the Hermann-Mauguin symbol is different, although it refers to the same space group.   Thus, in the monoclinic system, the space group *Pc* may be described as *Pa* or *Pn* for a different choice of the monoclinic reference axes *a* and *c*.   In the orthorhombic system, a larger number of different orientations is possible, and some care is needed in making the transformations. If the reference axes *abc* are renamed as *cab*, *bca*, *acb*, *bca*, or *cba*, then the symmetry plane symbols *m*, *n*, *a*, *b*, or *c* must be rearranged accordingly, *and* the symbols *a*, *b*, and *c* for the glide planes must also be renamed.   For example, *Pbcn* becomes *Pnca*, *Pbna*, *Pcnb*, *Pcan*, and *Pnab* for the above transformations, and *Pcca* becomes *Pbaa*, *Pbcb*, *Pbab*, *Pccb*, and *Pcaa*.   On this system we see, therefore, that the space group symbol indicates in a precise manner the way in which the crystal was oriented when the reference axes were named.

It is not necessary to dwell on the description of the 230 space groups because they have been adequately tabulated and classified in a number of recent works.[11,10]

7. THE CRYSTAL MOLECULE

Although the best way to study a space group is to confine attention to the self-consistent collection of symmetry operations or to regard it as a scaffolding of symmetry elements arranged in space, it is necessary to conclude with some reference to the molecular structure from which the crystal is built. Writing a report on the structure of crystals in 1901, Miers and Barlow say:

> Until we know more about the units of which the crystal really consists, there will necessarily be speculation as to whether the units are situated at the most general sorts of homologous points in a given type, or whether they are symmetrical bodies situated at the singular points; whether they are all of the same sort or of more than one sort.[13]

For any given space group these different possibilities are easily enumerated. With the discovery of X-ray diffraction in 1912, the means of distinguishing them became available for the first time, and the importance of space group theory in X-ray crystal analysis is that it immediately presents all the different possibilities that have to be distinguished. A few examples will illustrate how inherent molecular symmetry may be inferred from a knowledge of the space group and the number of molecules present in the unit cell. For this purpose some monoclinic space groups are illustrated graphically in Fig. 25. Some of these space groups are of frequent occurrence in the organic crystal structures with which this book is mainly concerned.

The diagrams follow the symbolism of the *International Tables*,[11] but for these simple groups it is convenient to show the general point positions and the symmetry elements on one diagram, as in the tables of Astbury and Yardley.[14] The unit cell

---

[13] H. A. Meirs and W. Barlow, *Brit. Assoc. Advancement Sci., Rept.*, 1901, pp. 297–337.

[14] W. T. Astbury and K. Yardley (K. Lonsdale), *Trans. Roy. Soc. (London)*, 1924, **A224**, 221.

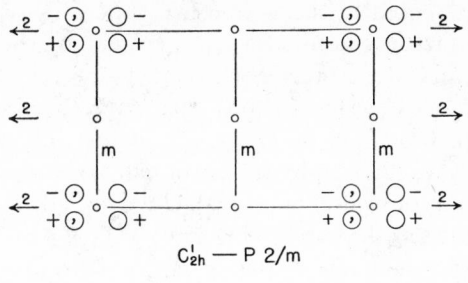

$C_{2h}^1 — P\,2/m$

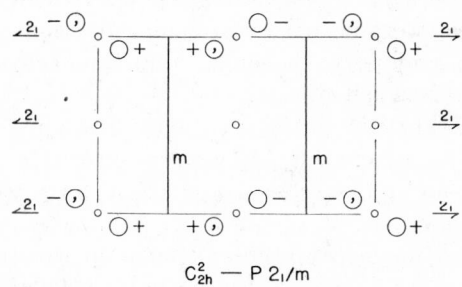

$C_{2h}^2 — P\,2_1/m$

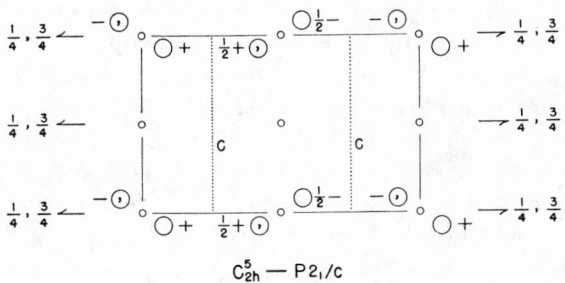

$C_{2h}^5 — P\,2_1/c$

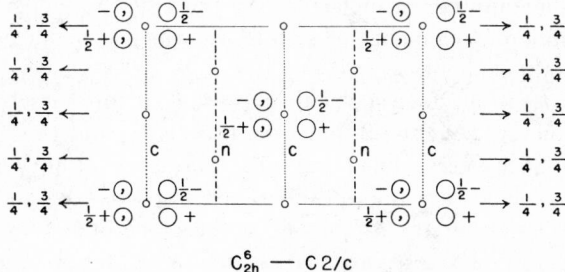

$C_{2h}^6 — C\,2/c$

Fig. 25. Some monoclinic space groups.

boundary is denoted by a fine line, with the origin at the upper left-hand corner, y co-ordinate axis (b axis) to the right, $x(a)$ downwards, and $z(c)$ towards the reader, normal to the paper. Reflection planes (m) normal to the paper are denoted by heavy lines, glide planes (a, c, or n) by dotted lines of various sorts, and rotation axes (2 or 2_1) by arrows or half arrows. The general point position is indicated by a small circle placed at an arbitrary position (x, y) near the corner of the unit cell, and at an arbitrary height (z) from the plane of the drawing. This last co-ordinate is denoted by writing $+$ or $-$ near the circle, to denote $+z$ or $-z$. Other heights are indicated by attaching fractions to the point positions or symmetry elements. Centres of symmetry are denoted by very small circles.

If the general point position is taken to represent an asymmetric molecule at an arbitrary position in the unit cell, then we see that the multiplying effect of the symmetry elements produces four molecules, all of which are required to build up the full symmetry of the structure (or eight in the case of $C2/c$, which is based on an end-face-centred lattice). Further, as these space groups involve operations of the second sort (reflections), two of the molecules must be enantiomorphs of the other two. (This is denoted by placing a comma in the circles concerned.) As this involves an inherently different molecular structure, it may appear to be physically improbable. It must be remembered, however, that a large number of the more common molecules possess at least a plane of symmetry, and so are identical with their mirror images. An inherent structural difference is only involved for the truly asymmetric molecule.

If four molecules are known to be present in these unit cells (eight in $C2/c$), they may be placed in the most general positions. If a smaller number of molecules are present, they must possess certain inherent elements of symmetry which coincide with the space group elements.

In $P2/m$, if only one molecule is present, it must itself possess the symmetry $2/m$, the 2-fold axis, the plane, and the resulting centre of symmetry coinciding with the space group elements. If two molecules are present, there are two possibilities. They may possess either the 2-fold axis *or* the plane of symmetry, the 2-fold axis involving a variable parameter y, and the plane two variable parameters, x and z.

48

In $P2_1/m$ the possibilities are more restricted. Separate individual molecules cannot make use of the space group screw axis (or a glide plane) as an inherent symmetry element. If they did, the molecules could have no boundary within the crystal. If two molecules are found to be present, then the only possibilities are the plane of symmetry with two parameters, x and z, or a centre of symmetry coinciding with one of the space group centres.

In $P2_1/c$, which involves a combination of glide planes and screw axes, the possibilities are further reduced, and, if two molecules only are present in such a unit cell, they must possess centres of symmetry coinciding with a set of the space group centres.

The space group $C2/c$ is based on an end-face-centred lattice, and a pure rotation axis is here involved as well as a screw axis. If four molecules are present, there are two possibilities: either centres of symmetry coinciding with a set of the space group centres *or* a 2-fold axis of symmetry along the space group axis, with variable y.

If the molecular weight and the density of a substance are known, it is now possible by a few simple X-ray measurements to determine the number of molecules in the unit cell, and so deduce the molecular symmetry (as revealed by the crystal) with certainty. The main difficulty usually lies in a certain determination of the crystal class, which cannot always be carried out directly by X-ray methods.

Such determinations of molecular symmetry are positive. They tell us that the crystal is utilizing this symmetry in creating its indefinitely extended lattice structure. But if a negative result is obtained, we cannot say with like certainty that the molecule does not contain any inherent symmetry. The negative result really only means that the crystal is not making use of any molecular symmetry in building up its lattice structure. Certain elements of molecular symmetry may still be present which do not happen to coincide with the required space group elements.

The treatment we have given above is applicable if we are considering discrete molecules as building units in the crystal, as we have to do in practically the whole field of organic chemistry. If, however, it is necessary to consider structures built from single atoms, or from ions not segregated into individual molecules, this treatment is not so suitable.   It is better then to consider the space groups analytically by tabulating the co-ordinates of equivalent points, for the general positions and the various special positions.   This was first done exhaustively for all the space groups by Wyckoff,[15] and the results will also be found in the *International Tables*.[11]   If atoms or ions of different kinds are present in different numbers, then they can often be assigned uniquely to special positions, or to more general positions involving one or two variable parameters.

[15] R. W. G. Wyckoff, *The Analytical Expression of the Results of the Theory of Space-Groups*, Washington: Carnegie Inst. Washington, 1922; also P. Niggli, *Geometrische Krystallographie des Diskontinuums*, Leipzig, 1919.

IV

The Intimate Structure of
Crystals; X-Ray Diffraction

1. THE DISCOVERY OF X-RAY DIFFRACTION

A new epoch in crystallography began in 1912 with von Laue's discovery of the diffraction of X-rays by crystals. At that time it was not known whether X-rays were electromagnetic waves or whether they were corpuscular in nature, but the polarization of X-rays had been demonstrated and Sommerfeld had shown that, if they consisted of waves, then the wave-length must be of the order of 10^{-8} cm. (1 Å). A reasonably accurate estimate of Avogadro's number was also available at this time, and so it was known from the observed densities of crystals that the average distance between atoms in the solid state must also be of the order of 1 Å. These considerations led to von Laue's suggestion that the regularly spaced repeating units in a crystal, which in the case of an elementary substance or a simple salt would most probably consist of single atoms or ions, might act as a three-dimensional diffraction grating for X-rays.

The experiment was immediately tried by Friedrich and Knipping,[1] and photographs were soon obtained showing that diffraction actually occurred. The first successful X-ray photograph, the forerunner of all the now well-known Laue patterns, is shown in Fig. 26. This discovery was truly one of the great events in science, for on the one hand it established the nature of X-rays and initiated the study of X-ray spectroscopy, while on the other it provided the means for exploring the structure of crystal-

---

[1] W. Friedrich, P. Knipping, and M. von Laue, *Sitzber. math.-physik. Klasse Bayer. Akad. Wiss. München*, 1912, p. 303.

line matter on an atomic scale and in a degree of detail hitherto unimagined. These consequences, however, did not follow immediately, nor were the first results easy to interpret. A good account of the difficulties that beset the first investigations, in relation to the state of knowledge at the time, has recently been given by K. Lonsdale.[2]

Although the results given by the photographs could not at first be fully interpreted, von Laue set out quite fully the conditions for diffraction by a crystal lattice, regarded as a three-dimensional grating. If we consider first the diffraction effects to be expected from a single row of the lattice of period a (Fig. 27), the condition to be fulfilled is that the path difference between waves scattered from successive points should equal a whole number of wavelengths, $n\lambda$. Reinforcement then occurs with the production of a diffracted beam. If the incident beam makes an angle α_0 with the row, and the diffracted beam an angle α, this condition is

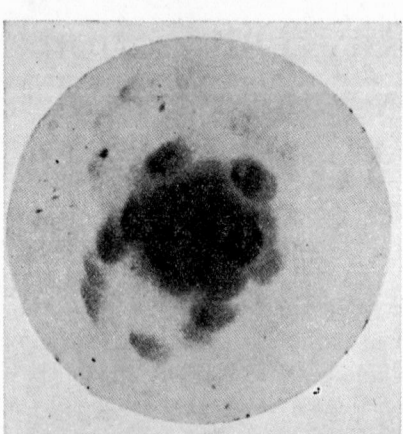

Fig. 26. The first X-ray diffraction photograph.

$$a\ (\cos \alpha_0 - \cos \alpha) = n\lambda.$$

For a lattice, with the incident beam making angles $\alpha_0, \beta_0, \gamma_0$ with the three lattice directions a, b, c, the direction of the diffracted beam is given by the angles α, β, γ when the three Laue equations,

$$a\ (\cos \alpha_0 - \cos \alpha) = n_1\lambda,$$
$$b\ (\cos \beta_0 - \cos \beta) = n_2\lambda, \qquad \cdots (4.1)$$
$$c\ (\cos \gamma_0 - \cos \gamma) = n_3\lambda,$$

are simultaneously satisfied. The triple set of integers, $n_1 n_2 n_3$,

[2] K. Lonsdale, *Crystals and X-Rays*, London: G. Bell and Sons, 1948.

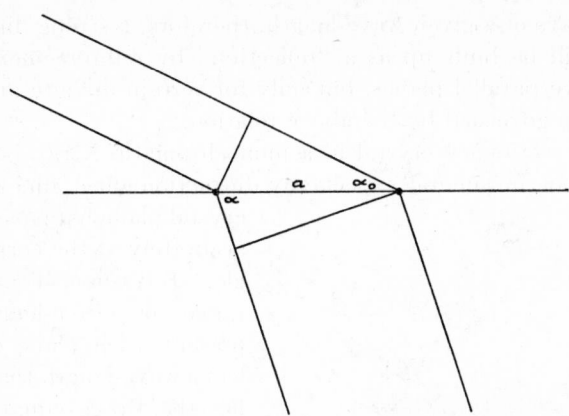

Fig. 27. Diffraction from a single row.

then denotes the order of the spectrum.   In the Laue experiment the crystal is stationary and the direction of the incident beam is fixed, but diffraction effects are obtained because radiation of appropriate λ can be selected by the crystal from the continuous range.

2. THE BRAGG LAW

The deduction of the crystal structure from the general Laue conditions is difficult and was not at first successful.   A great advance and simplification was made when the work was taken up in England by W. L. Bragg and W. H. Bragg, and later by Moseley and Darwin.   The success of W. L. Bragg's treatment of the problem was based on his introduction of the idea of reflection of X-rays from crystal planes, and his realisation that in the Laue experiment the X-ray beam was heterogeneous, consisting of a continuous range of wave-lengths.[3]

When an X-ray beam AB of wave-length λ is incident on a crystal plane at an angle θ (Fig. 28), a reflected beam BC will be formed by Huygens' principle.   Reinforcement by X-rays reflected from the next parallel crystal plane, at distance d, will occur when the path difference $DE+EF$ is equal to a whole number of wave-lengths, or when

$$n\lambda = 2d \sin \theta. \qquad \cdots (4.2)$$

[3] W. H. Bragg, *Nature*, 1912, **90**, 219, 360, 572; W. L. Bragg, *ibid.*, 1912, **90**, 410; *Proc. Cambridge Phil. Soc.*, 1913, **17**, 43.

For X-rays of a given wave-length, therefore, a strong diffracted beam will be built up as a "reflection" by reinforcement from successive parallel planes, but only for certain definite angles of incidence governed by the above relation.

For a stationary crystal in a monochromatic X-ray beam we should not in general expect any diffraction effect unless some

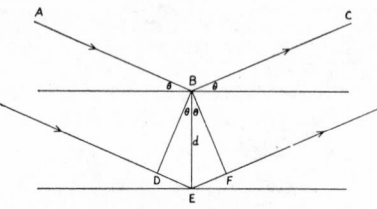

Fig. 28. Bragg reflection condition.

crystal plane happened to lie accurately at the correct angle. But when a continuous range of wave-lengths is present, each plane can select a wave-length that satisfies the Bragg equation for its particular orientation, and so give rise to a diffracted beam. A group of planes parallel to a certain line, or zone axis, in the crystal, inclined at an angle θ to the incident beam (Fig. 29), will give rise to a series of "reflected" beams, each inclined at the angle θ to the zone axis and so forming a cone of rays. The intersection of this cone of reflections with a screen or photographic plate placed normal to the incident beam gives an

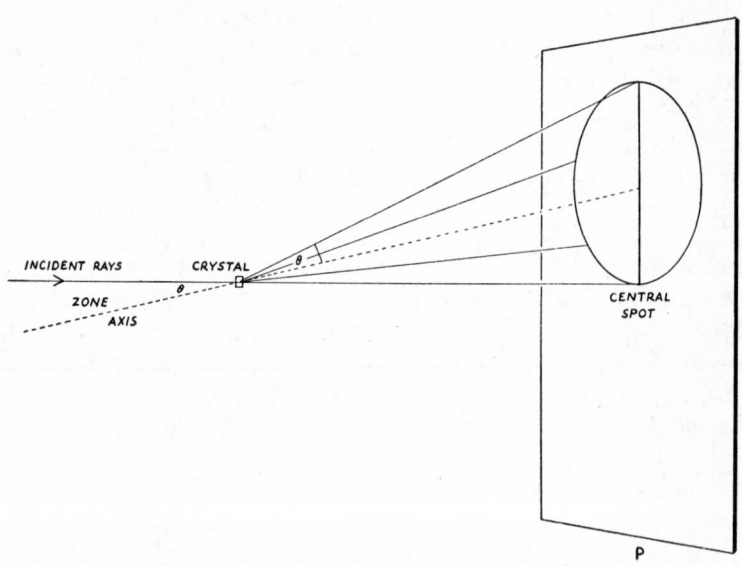

Fig. 29. Formation of ellipse in Laue diagram.

ellipse, one end of the major axis coinciding with the central spot
where the undeviated beam strikes the plate.   This simple construction, due to W. L. Bragg, is capable of explaining all the Laue
patterns, where the diffracted spots are always found to lie on a
series of intersecting ellipses.   If the incident beam passes along
one of the principal crystal axes, a complicated pattern of high
symmetry may be formed (Fig. 30), but it is usually easy to trace
out the courses of the intersecting ellipses.   That the crystal
planes are acting like mirrors in reflecting the X-rays according
to this construction is easily shown by moving the crystal slightly,
when the spots are found to change position on the screen accordingly.

By means of this inter-
pretation, W. L. Bragg was
able to carry out the first
crystal analysis and deduce
the structures of potassium
and   sodium   chloride.[4]
These fundamental results
were confirmed by the use
of   homogeneous   X-rays
and   the   ionisation   spectrometer, methods which
soon proved to be much
more powerful in the analysis of complex crystal
structures.   The   Laue
method, however, has con-
tinued to be useful in many

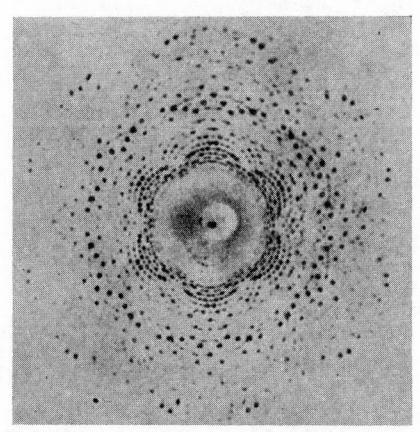

Fig. 30. Laue photograph of high
symmetry (kaliophilite, after Bannis-
ter).

cases; the further development of this method and its application
to crystal structure determination has been very fully described
by Wyckoff.[5]

The equivalence of the Laue and Bragg equations is shown in
Fig. 31, for a beam incident at an angle α on a row of particles of
period a.   The traces of the equivalent Bragg reflection planes,
spacing d, are indicated by the dotted lines.   The Laue condition for diffraction is

[4] W. L. Bragg, *Proc. Roy. Soc.* (London), 1913, **A89**, 248.
[5] R. W. G. Wyckoff, *The Structure of Crystals*, 2d ed., New York:
Chemical Catalog Co., 1931.

$$a \, (\cos \alpha_0 - \cos \alpha) = n\lambda$$

or

$$2a \sin \frac{\alpha + \alpha_0}{2} \sin \frac{\alpha - \alpha_0}{2} = n\lambda.$$

But

$$2\theta = \alpha - \alpha_0 \quad \text{and} \quad d = a \sin \frac{\alpha + \alpha_0}{2}.$$

Hence,

$$2d \sin \theta = n\lambda.$$

In the general case of diffraction by a crystal lattice, the triple set of integers, $n_1 n_2 n_3$, which define the order of the spectrum,

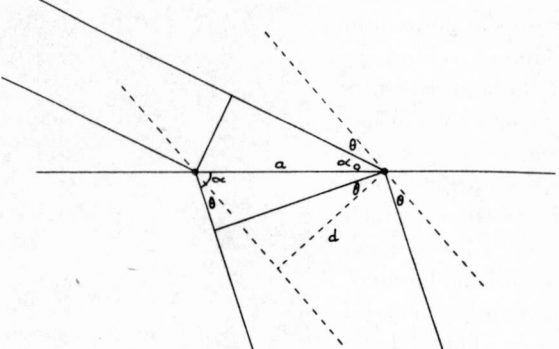

Fig. 31. Equivalence of Laue and Bragg equations.

are proportional to the Miller indices (hkl) of the crystal plane which is parallel to the Bragg reflection plane. For the first order spectrum these integers simply give the Miller indices of the crystal lattice plane, but for the second, third, and higher orders they must be divided by the common factors, 2, 3, etc. However, it is customary now to make no sharp distinction between the Bragg reflection plane and the actual lattice plane, or rather to extend the use of the Miller indices to define reflections of any order. The indices defining the Bragg reflections are usually written without parentheses, so that, for example, 111, 222, 333, and so on, refer to the first, second, third, and higher orders of reflection from the (111) crystal lattice plane. We also refer to the "spacing" of the (222) "plane" as being half that of the (111), and so on.

3. CRYSTAL ANALYSIS BY HOMOGENEOUS X-RAYS; ROTATION PHOTOGRAPHS

The introduction of the ionisation spectrometer by W. H. Bragg,[6] by means of which both the position and the intensity of X-ray crystal reflections could be accurately measured, led to immediate and rapid advances in the analysis of crystal structures and in X-ray spectroscopy.   The discovery of the characteristic line spectrum of X-rays was due to the use of this instrument. As the structures of sodium and potassium chloride were now known with certainty, the absolute values of the spacings of the reflecting planes could be calculated accurately from a knowledge of the density and Avogadro's number.   For a cubic crystal we have

$$a^3 = \sum A/N\rho \qquad \cdots (4.3)$$

where a^3 gives the volume of the unit cell, $\sum A$ the sum of the chemical atomic weights contained in it, N the Avogadro number, and ρ the density of the crystal.   These crystals, and others whose structures were soon determined, could therefore be used to evaluate accurately the wave-lengths of the peaks of characteristic monochromatic radiation which were found superimposed on the background of continuous radiation emitted from anticathodes of platinum, palladium, rhodium, copper, nickel, and other elements.   The further development of X-ray spectroscopy and the work of Moseley soon followed.

If we now assume a source of homogeneous X-rays of known wave-length, the process of crystal analysis is greatly simplified. The crystal, set up at X (Fig. 32) in the monochromatic beam, will not in general give rise to any reflection.   But if it is slowly rotated about some zone axis, which we suppose perpendicular to the direction of the incident beam, one lattice plane after another will come into a reflecting position, satisfying the relation $n\lambda = 2d \sin \theta$, and the reflected beam will flash out, to be recorded in the ionisation chamber or Geiger counter set at the angle 2θ,

[6] W. H. Bragg and W. L. Bragg, *Proc. Roy. Soc.* (London), 1913, **A88**, 428.

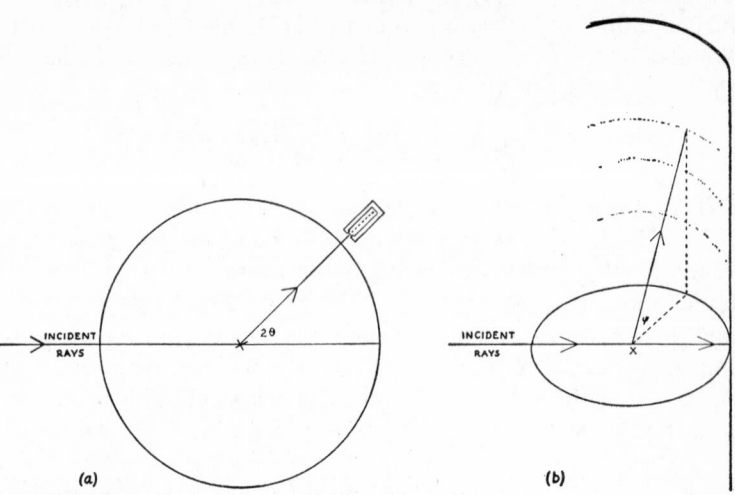

Fig. 32. Rotating crystal method. (a) Ionisation spectrometer. (b) Photographic film.

or on a photographic film surrounding the crystal. These reflections take place over a certain narrow range of angle, generally from a few minutes to about a degree of arc, depending on the state of perfection of the crystal and the parallelism of the X-ray beam. With the ionisation spectrometer, curves of the type shown in Fig. 33 are obtained from average crystals.

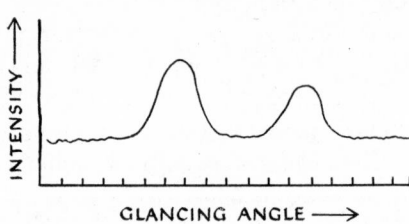

Fig. 33. Ionisation spectrometer curves.

When the rotating crystal is surrounded by a film contained in a cylindrical camera (Fig. 32), a *rotation photograph* is obtained which has the characteristic appearance shown in Fig. 34, the reflections being grouped on straight lines known as *layer lines*.

We suppose the crystal to rotate about a prominent zone axis, say a. Then the $(0kl)$ planes, parallel to the rotation axis, clearly give rise to the equatorial layer line of reflections. The first and subsequent layer lines are readily accounted for in terms of first and higher order diffractions from a row of points, period a, with incident beam normal to the row. The Laue equation

$$a\ (\cos \alpha_0 - \cos \alpha) = n\lambda$$

becomes

$$a \cos \alpha_0 = n\lambda$$

or

$$a \sin \psi = n\lambda \qquad \cdots (4.4)$$

where ψ is the angle subtended from the crystal to the nth layer line, which contains all the reflections whose indices are (nkl).

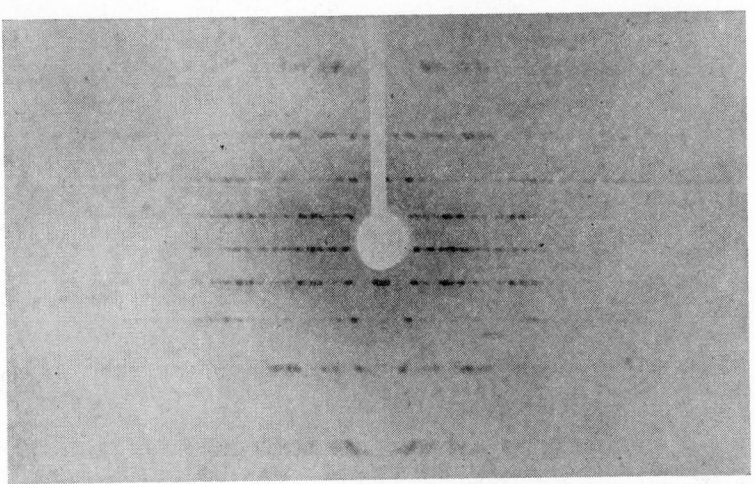

Fig. 34. Rotation photograph. Pyrene crystal rotated about c axis. Cu $K\alpha$ radiation.

The fundamental periodicity along the row of the lattice about which the crystal is rotated can thus be immediately determined from a simple measurement of the distance between the layer lines on the rotation photograph. If a set of such photographs is taken for the crystal rotating about its different axial directions, the dimensions of the unit cell are obtained. The importance of rotation photographs lies in the fact that they give an unequivocal determination of the true lattice, and this cannot easily be done by mere measurements of spacings. If the cell is body centred, for example, a spacing measurement of the (100) plane will not give a but $a/2$, because there is an identical structure half-way along a. A rotation photograph, however, will give a measure of the true periodicity and determine a directly. When, as in this case, the cell chosen is not primitive, further rotation photographs about the body diagonal, or the face diagonals, will reveal the subdivision of the lattice in these directions.

Rotation photographs of any but the most simple types of crystal show large numbers of spots crowded together with frequent overlap on the different layer lines, and the task of indexing these various reflections presents some difficulty. One index is, of course, determined by the layer line on which the reflection lies. If the same reflection could be identified on rotation photographs taken about the a, b, and c axes in turn, then it could easily be indexed. However, this is not in general possible unless the photographs are very simple or the reflection is of outstanding intensity, in which case a particular reflection may be picked out and checked against its constant spacing or θ value.

In addition to the horizontal layer lines, the spots on a rotation photograph are arranged on vertical curves, not usually so easily distinguished, known as *row lines*. The reflections on the row lines have two indices constant, while the third varies with the layer line, e.g., $0k_1l_1$, $1k_1l_1$, $2k_1l_1$, etc. The row lines form a set of closed curves, and charts showing the course of these curves have been calculated by Bernal[7] and are of great value in the interpretation of rotation photographs.

4. OSCILLATION AND MOVING FILM PHOTOGRAPHS

If single crystals are available, however, there is no need to attempt a detailed analysis from rotation photographs alone. Methods can easily be devised to record the angular setting of the crystal, with respect to some chosen reference line, at which each reflection occurs. With this additional information the assignment of the correct indices to each reflection presents no difficulty if the work is done systematically. When reflections are recorded by Geiger counter or ionisation spectrometer, the angular setting of the crystal at each reflection can be read off directly. Photographically, two principal methods, employing the *oscillation* photograph or the *moving film* photograph, are generally employed to achieve the same result.

In the former method the crystal is set about some principal axis and oscillated to and fro by means of a cam or other mecha-

[7] J. D. Bernal, *Proc. Roy. Soc.* (London), 1926, **A113**, 117.

nism through some small angle, say, 5°, 10°, or 15°. The small number of planes which now come into a reflecting position produce spots which are well separated and easily identified. They can be assigned to their correct layer lines, and the course of the row lines is also more clearly visible, especially if the crystal possesses certain symmetry.   Simple inspection of such photographs often provides some very useful information regarding symmetry and the nature of the lattice.   Fig. 35a shows a diagram of such a photograph recording a small number of reflections and displaying no particular symmetry.   In Fig. 35b the upper and lower halves of the photograph are symmetrical with respect to the equatorial or zero layer line, and this at once indicates the presence of a symmetry plane in the Bravais lattice normal to the rotation axis.   In Fig. 35c the row lines are not continuous but exhibit a characteristic staggered effect, due to the absence of alternate reflections in the series $0k_1l_1$, $1k_1l_1$, $2k_1l_1$, $3k_1l_1$, $\cdots$. This indicates a "general halving." For example, reflections may only appear when the sum of the indices is even, because the lattice is body centred.   The appearance of the photograph shows this in a very direct manner.

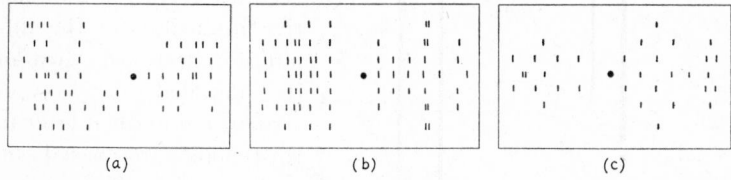

Fig. 35. Oscillation photographs. (a) Asymmetric. (b) Symmetry plane normal to rotation axis. (c) Centred lattice.

Moving film methods provide a still more useful means of analysis and are now generally employed in most single crystal work.   The principle is simple, the film being continuously translated in some direction by a movement synchronised to the rotation of the crystal, so that the angular setting of the crystal at the time of each reflection can be read off directly from the position of the spot on the film.   Generally, as in the original Weissenberg method,[8] the film holder is moved to and fro, or up and down, on a track parallel to the rotation axis, and the layer lines are recorded one at a time, the unwanted reflections being

[8] K. Weissenberg, *Z. Physik*, 1924, **23**, 229.

excluded by suitably placed screens (Fig. 36). Various modifications of this method have been devised for special purposes.[9,10,11] Some typical photographs are shown in Fig. 37, where the reflections are seen to lie spread out on certain characteristic curves. For each reflection, one co-ordinate on the photograph gives the angular setting of the crystal, and the other gives the spacing. There is thus no difficulty in identifying and indexing all the reflections, if the unit cell dimensions are known. A very full account of the interpretation of various types of Weissenberg photographs has been given by M. J. Buerger.[12]

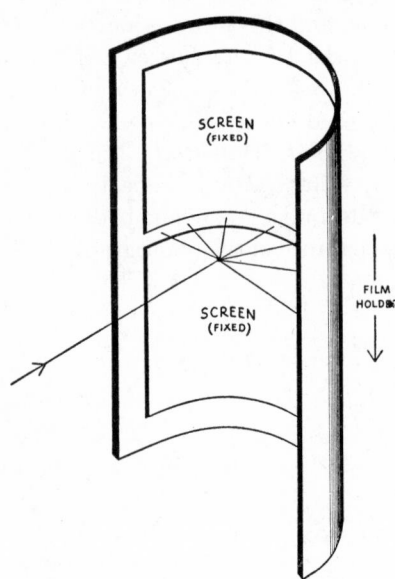

Fig. 36. Weissenberg arrangement for taking moving film photographs.

In addition to providing a convenient means of indexing the reflections and so leading to a determination of the geometry and symmetry of crystals, the moving film photograph, if taken with suitable precautions, is of the utmost importance in providing an accurate record of the relative intensities of the different reflections. Most of the detailed analyses of structure described later in this book are based on such records. The moving film method is particularly adapted for this work, for two reasons. In the first place, each equatorial layer line, containing all the reflections in one zone, can be recorded completely on one film in the course of one exposure, and so the true relative intensities are available for direct measurement, without the troublesome correlations which are necessary and which often lead to inaccuracies

[9] E. Sauter, *Z. Krist.*, 1933, **85**, 156.

[10] J. M. Robertson, *Phil. Mag.*, 1934, **18**, 729.

[11] M. J. Buerger, *Z. Krist.*, 1936, **94**, 87.

[12] M. J. Buerger, *X-Ray Crystallography*, New York: J. Wiley and Sons, 1942.

when the reflections are recorded separately on oscillation photographs, or by means of ionisation spectrometer or Geiger counter measurements. The latter methods may give more accurate individual measurements, but they are certainly more troublesome to apply to a complete zone, which may contain several hundred different reflections, unless extremely elaborate and carefully stabilized apparatus is employed.

In the second place, if the screens are suitably adjusted, the photographic film in the neighbourhood of each spot need only be exposed to X-radiation during the short time that the reflection is taking place. In this way the background of general scattering is reduced to a minimum, and this is a limiting factor in the recording of very weak reflections. In oscillation photographs the background will generally be at least five times, and in complete rotation photographs more than a hundred times, as great.

The moving film photographs shown in Fig. 37 record the equatorial layer lines ($h0l$) for some of the phthalocyanine crystals, and they contain in principle all the information required to construct the electron density maps given in Figs. 108–111 (pp. 265–270). The range of intensities covered in these photographs is many thousands to one, and many of the weaker reflections are, of course, invisible on the reproductions. In order to measure the complete range of intensities it is in practice necessary to take several photographs with different exposure times, or else reduce the strong reflections by allowing the X-ray beams to pass through a pack of superimposed films, with absorbing screens between if necessary.[13]

Some more elaborate methods of single crystal photography have recently been devised which give very direct pictures of crystal geometry in terms of the reciprocal lattice (see next section). In the Sauter method[9] the film moves at right angles to the layer line by rotation instead of by translation as in the Weissenberg method. In the de Jong and Bouman method[14] the movements of crystal and film are further modified to produce completely undistorted pictures of various levels in the reciprocal

[13] J. J. de Lange, J. M. Robertson, and I. Woodward, *Proc. Roy. Soc.* (London), 1939, **A171**, 398, 404; J. M. Robertson, *J. Sci. Instruments*, 1943, **20**, 175.

[14] W. F. de Jong and J. Bouman, *Z. Krist.*, 1938, **98**, 456; **100**, 275; *Physica*, 1938, **5**, 220; W. F. de Jong, J. Bouman, and J. J. de Lange, *ibid.*, 1938, **5**, 188.

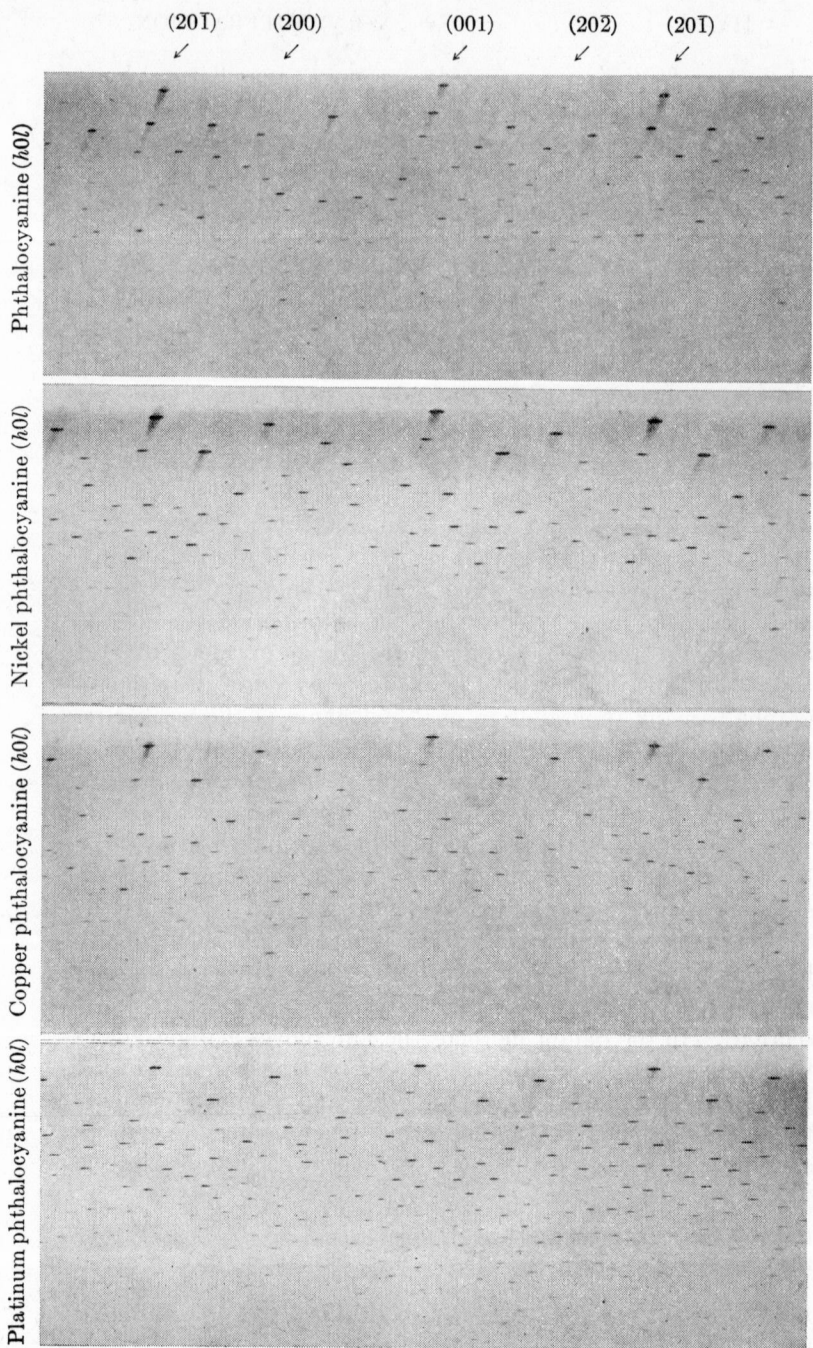

Fig. 37. Moving film photographs of equatorial layer lines of phthalo-cyanine crystals.

lattice, from which the crystallographic constants may be measured and the reflections indexed directly. The range, however, is somewhat limited. These methods and improved forms of the apparatus have been described very fully by M. J. Buerger.[15]

5. THE RECIPROCAL LATTICE

Although no attempt is made here to describe the detailed analytical and graphical methods employed in the full analysis of single crystal photographs, it is necessary to refer briefly to the concept of the reciprocal lattice. This construction is fundamental in X-ray crystallography; and in related fields of X-ray optics, when distorted crystals and deviations from strict periodicity are considered, its use is indispensable. The general concept arose early in crystallography, and is dealt with in Bravais's work under the title of polar lattices, but its development and application to X-ray problems has been made chiefly by Ewald,[16] Bernal,[7] and Buerger.[17]

To attempt to visualize a large number of planes of varying orientation in a crystal is difficult and unduly laborious. It is easier to think of the normals to the planes rather than the planes themselves, and if in addition the interplanar spacings can be represented, in the reciprocal form implied by the Bragg law,

$$\frac{n\lambda}{d} = 2 \sin \theta,$$

we have a representation which should be ideal for dealing with X-ray diffraction problems.

The construction can be described simply in the following way. Take a unit parallelopiped in the real (direct) lattice and imagine all the possible planes drawn in. Choose one point as origin, draw the normal from the origin to each plane, and produce. Represent each plane by a point on the normal at a distance ρ from the origin such that

$$\rho d = K^2 \qquad \qquad \cdots (4.5)$$

---

[15] M. J. Buerger, *The Photography of the Reciprocal Lattice*, Am. Soc. X-Ray and Electron Diffraction Monograph, no. 1, U.S.A., 1944.

[16] P. P. Ewald, *Z. Krist.*, 1921, **56**, 148.

[17] M. J. Buerger, *Z. Krist.*, 1935, **91**, 276.

where d is the interplanar spacing and K a constant. It is then found that the array of points formed in this way, representing all the planes, forms a new lattice which is a replica of the original lattice. This is called the *reciprocal lattice,* and it will be noted that each point of this lattice is the reciprocal polar of a plane in the old lattice in a sphere of radius K.

This construction is illustrated in Fig. 38, where a unit parallelogram (heavy lines) is chosen to represent the direct lattice, and some of the "planes" are drawn in. The normals to these planes are represented by thin lines, and the points of the reciprocal lattice which correspond to these planes by open circles. We have seen that second and higher order X-ray reflections from a given crystal plane can be appropriately represented as emanating from fictitious crystal planes of submultiple spacing, e.g., we refer to the second order reflection from the (100) plane as the 200 reflection, etc. There are, of course, an infinite series of these planes of submultiple spacing. All such orders of a particular plane have a common normal from the origin O, but the corresponding reciprocal lattice points are regularly spaced out along each normal in accordance with (4.5).

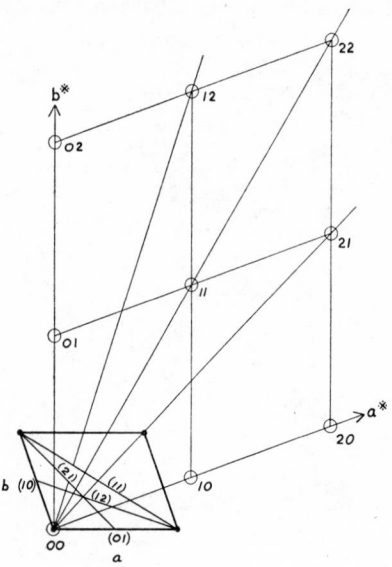

Fig. 38. Construction of a reciprocal lattice.

If we imagine Fig. 38 extended to three dimensions, there is no difficulty in seeing that the indices (hkl) of any general plane in the direct lattice are simply the co-ordinates of the corresponding point in the reciprocal lattice. We also see that a plane *net* of points in the reciprocal lattice, such as shown in Fig. 38, corresponds to a *zone* of planes in the crystal.

In describing the behaviour of crystals with homogeneous

X-rays it is convenient to make the constant K^2 in (4.5) equal to the wave-length λ, so that

$$\rho = \frac{K^2}{d} = \frac{\lambda}{d} = 2 \sin \theta. \qquad \cdots (4.6)$$

If the axes of the reciprocal lattice are labelled a^*, b^*, as in Fig. 38, it is clear that the interaxial angle is the supplement of that of the direct lattice, and the axial lengths are the reciprocals of the pinacoidal spacings. In the general 3-dimensional case

$$\left.\begin{aligned}
a^* &= \frac{\lambda}{d_{100}} = \frac{\lambda bc}{V} \sin \alpha \\[6pt]
b^* &= \frac{\lambda}{d_{010}} = \frac{\lambda ca}{V} \sin \beta \\[6pt]
c^* &= \frac{\lambda}{d_{001}} = \frac{\lambda ab}{V} \sin \gamma \\[6pt]
\cos \alpha^* &= \frac{\cos \beta \cos \gamma - \cos \alpha}{\sin \beta \sin \gamma} \\[6pt]
\cos \beta^* &= \frac{\cos \alpha \cos \gamma - \cos \beta}{\sin \alpha \sin \gamma} \\[6pt]
\cos \gamma^* &= \frac{\cos \alpha \cos \beta - \cos \gamma}{\sin \alpha \sin \beta}
\end{aligned}\right\} \qquad \cdots (4.7)$$

where V is the volume of the unit cell. For the general case of a triclinic cell

$$V = abc \,(1 - \cos^2 \alpha - \cos^2 \beta - \cos^2 \gamma + 2 \cos \alpha \cos \beta \cos \gamma)^{1/2}.$$

$$\cdots (4.8)$$

For cells of higher symmetry these expressions simplify, and for a rectangular cell

$$a^* = \frac{\lambda}{a}, \qquad b^* = \frac{\lambda}{b}, \qquad c^* = \frac{\lambda}{c} \cdot$$

In any X-ray analysis of crystal structure one of the first steps, after the cell constants have been determined from rotation photographs, is to plot out the reciprocal lattice in the form of nets, corresponding to the various zones of reflections, as in Fig. 41. If these nets are scaled in accordance with (4.6), it is clear that a sphere described about the centre of the reciprocal lattice, of radius $\rho = 2$ (which makes $\sin \theta = 1$), contains within it all the reflections that can be observed with the wave-length employed. This sphere is therefore called the *limiting sphere*. This simple construction in itself is most useful as a guide to the possible reflecting planes. If the nets are plotted carefully on a reasonable scale, giving the limiting sphere a radius of about 20 cm., the reciprocal spacings of the various planes can usually be obtained with sufficient accuracy for most purposes by direct measurement.

6. INTERPRETATION OF SINGLE CRYSTAL PHOTOGRAPHS

The greatest application of the reciprocal lattice in crystal analysis, however, lies in the simple geometrical interpretation of diffraction phenomena which it affords. In reciprocal space the incident wave will be represented by a point travelling along the normal to the wave-front. We take such an incident ray AOB (Fig. 39), passing through the reciprocal lattice origin O, with parallel rays passing over every reciprocal lattice point. As we have seen, all points within the limiting sphere ABC are capable of reflection for some direction of the incident radiation. It is required to find those points that will reflect for the direction AOB. The condition to be satisfied is expressed by the Bragg law

$$\sin \theta = \frac{\lambda}{2d},$$

which by (4.6) may be written

$$\sin \theta = \frac{\rho(P)}{2}$$

for any reciprocal lattice point P at a distance ρ from the origin.

This condition is satisfied for any point lying on the circle AOP because

$$\sin OAP = \frac{OP}{OA} = \frac{\rho(P)}{2},$$

OA being the radius of the limiting sphere and of length equal to

68

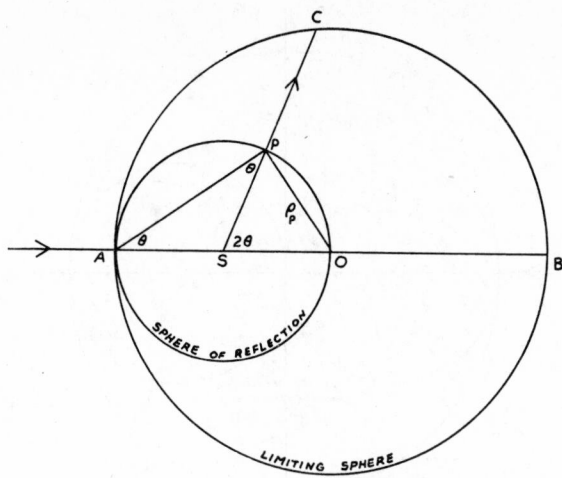

Fig. 39. Diffraction in reciprocal space.

2.   The direction of the reflected ray from the point P, which makes an angle 2θ with the direction of the incident ray, is given by SPC, S being the centre of the circle AOP.

In three dimensions the circle AOP becomes a sphere, and we obtain the very simple geometrical picture of X-ray reflection in the reciprocal lattice, that every point on the surface of this sphere, and no other point, is capable of reflecting radiation incident in the direction of its diameter AO.   The sphere AOP is thus appropriately called the *sphere of reflection.*

In the rotating crystal method, the crystal is generally turned about an axis perpendicular to the incident rays.   It is generally most convenient to regard the reciprocal lattice as fixed, with the incident ray rotating in the equatorial plane normal to this axis and carrying the sphere of reflection with it.   Each reciprocal lattice point will give rise to a reflection as it is cut by the rotating sphere, and all the points lying within the tore indicated in Fig. 40 will reflect twice during each complete rotation, as they are cut by the two sides of the moving sphere.   In the complete reciprocal lattice each point is accompanied by its inverse point, so that in general each crystal plane gives rise to four separate reflections.

When we confine attention to the equatorial layer line, it is only necessary to draw the equatorial net plane of the reciprocal

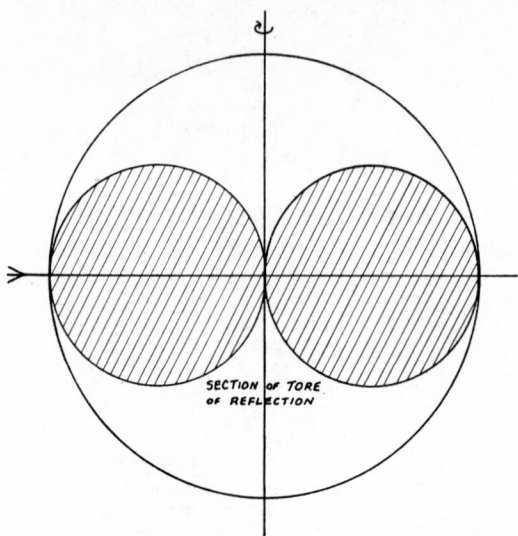

Fig. 40. Section through limiting sphere and tore of reflection.

lattice, as in Fig. 41, and consider the rotation of the equatorial circle of the sphere of reflection. The reflections that can occur on a 15° oscillation photograph of a certain crystal, for example, are those which lie on the shaded areas in Fig. 41, between the circles drawn for the two extreme positions of the incident beam. With such a simple graphical construction it is possible to index quickly and without ambiguity all the reflections on oscillation photographs. In the case of a moving film photograph, the spacing and exact reflecting angle of each plane, with reference to some fixed origin line, can be read off and tabulated. These are the two co-ordinates that determine the position of a reflection on the moving film photograph, and, given the dimensions and type of instrument, a chart can easily be prepared showing the expected positions of the reflections with their indices.

With regard to the interpretation of the higher layer lines on rotation photographs, a similar construction can be used, but in this case it is necessary to rotate some smaller circle, which is the appropriate section of the sphere of reflection, over that net plane of the reciprocal lattice which corresponds to the layer line being investigated.

It will be clear that with these powerful graphical methods and

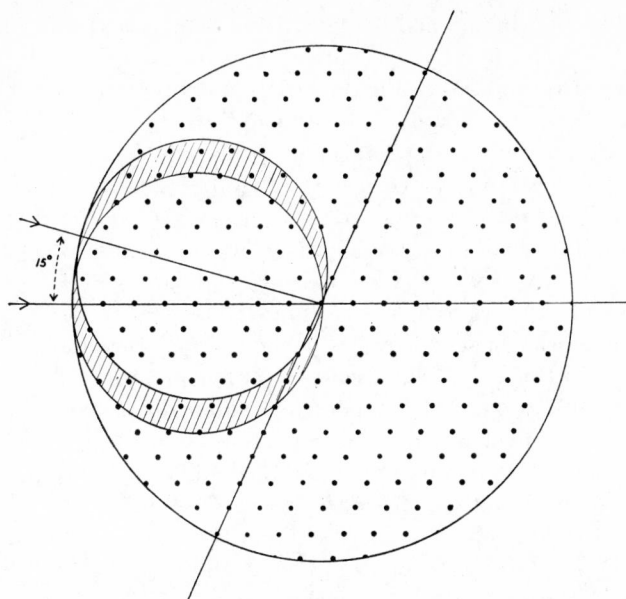

Fig. 41. Equatorial net plane of reciprocal lattice.

constructions, the process of indexing the reflections which occur on the different types of single crystal photograph can be reduced to a systematic routine, once the cell dimensions of the crystal have been accurately determined. A full analytical account of the reciprocal lattice, as well as charts to facilitate the graphical constructions, has been provided in the comprehensive investigation of J. D. Bernal.[7]

In the above discussion it has been assumed throughout that we are dealing with ideal crystals and that the reflecting power can be regarded as concentrated at the reciprocal lattice points. This treatment is sufficient for the main purposes of this book, but for completeness it should be mentioned here that some of the most important applications of the reciprocal lattice concept refer to cases where the crystal departs in some way from the ideal structure we have assumed.

If the crystal is very small, less than about 10^{-4} cm., the reflecting power can no longer be regarded as concentrated at points in the reciprocal lattice, but becomes spread out over certain small volumes, and the spots or lines on the X-ray photographs become broadened.[18]   The effect is analogous to the limited resolution afforded by an optical grating containing a small number of rulings, and so making only a limited number of component wavelets available to build up the diffraction maxima. Even if the crystal is not small, but is distorted in some way, the reflections may take place over a range of angle, and the result can again be treated as due to a spreading of the reflecting power in certain directions in the reciprocal lattice.   These effects have been dealt with very fully by Wilson[19] and James.[20]

Another important factor causing the crystal to depart from ideal regularity of structure lies in the thermal movements of the atoms, and this factor is, of course, always present, but to a different degree at different temperatures and in different substances. The general effect is a decrease of intensity in the Bragg reflection, but this is accompanied by a cloudlike spreading of the reflecting power around the reciprocal lattice points.   Under certain experimental conditions this gives rise to large, diffuse reflections on the photographic plate or film, and from a study of the shape and intensity of these reflections a great deal of important information can often be deduced concerning the crystal and molecular structure.   This matter has been studied very fully by K. Lonsdale.[21]

7. DETERMINATION OF SPACE GROUP

We have seen in an earlier section that by means of rotation photographs, which measure the true periodicity along any row in the lattice, every crystal can be assigned to its correct Bravais lattice without ambiguity.   However, when all the crystal reflections have been mapped out and correctly indexed by the methods outlined above, this purely geometrical analysis of the crystal structure can be carried much further.

In the first place the Bravais lattice can be confirmed in a more general way.   We have seen that the reciprocal lattice is a replica of the direct crystal lattice.   Consequently, the particular form of this lattice, whether primitive, face centred or body centred,

[18] A. L. Patterson, *Phys. Rev.*, 1939, **56**, 978.

[19] A. J. C. Wilson, *X-Ray Optics*, London, Methuen and Co., 1949.

[20] R. W. James, *The Optical Principles of the Diffraction of X-Rays* ("The Crystalline State," vol. II), London: G. Bell and Sons, 1949.

[21] K. Lonsdale, *Reports on Progress in Physics*, 1942–1943, **9**, 256.

is immediately obvious as soon as all the reciprocal lattice points have been recorded from an analysis of the photographs. Expressed analytically, the face-centring or body-centring of a lattice results in certain sets of the general (hkl) reflections being systematically absent from the photographs, which are then said to exhibit certain "general halvings." If the lattice is body centred, all reflections are absent when $h+k+l$ is odd. If the (001) face is centred, the reflections are absent when $h+k$ is odd. This is because these odd index planes, like the one shown in Fig. 42, are all exactly interleaved with identical scattering units, resulting in complete cancellation of the scattered waves.

In a very similar way those generalised elements of symmetry required by space group theory, the glide plane and the screw axis, may be detected, but attention is now concentrated on particular zones and on particular orders of reflections instead of on the general reflections. If a

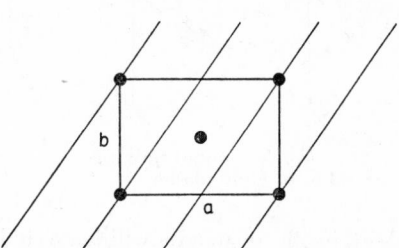

Fig. 42. General halving due to centred lattice.

glide plane n is present in a crystal structure, a mirror image of the asymmetric unit of structure occurs at the centre of the plane, as indicated by the dotted arrow in Fig. 43. If the corner arrows are regarded as pointing up, the central arrow will be pointing down, or vice versa. When viewed from some general point, these two bits of structure will appear different, and there will be no particular or predictable effect on the general reflections. But when such a structure is viewed in projection normal to the glide plane of symmetry, as in Fig. 43, the reflection unit of structure in the centre is identical with the corner units. Therefore, as far as the distribution of scattering matter across those planes normal to the glide plane is concerned, there is an identical unit at the centre, and the situation is the same as in Fig. 42. If the glide plane is parallel to (001), the particular planes concerned are ($hk0$), and these reflections will be absent when $h+k$ is odd. Such zonal halvings affect only the equatorial nets in the reciprocal lattice. If an a glide plane is involved, the halving occurs when h is odd; if b, when k is odd.

In the case of a screw axis, we find a rotated unit of structure in some region along the axis, and the only reflecting planes which can perceive an identical distribution of scattering matter in this case are the orders of that crystal plane which is perpendicular to the screw axis. Some of these orders will then be absent, thus giving rise to an axial halving or other subdivision of the spectra. A 2_1 axis, parallel to a, as in Fig. 44, will cause the odd orders of 100 to disappear. A 4_2 or a 6_3 axis will have the same effect.

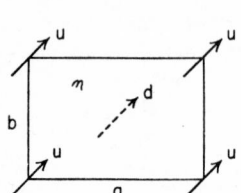

Fig. 43. Zonal halving due to glide plane.

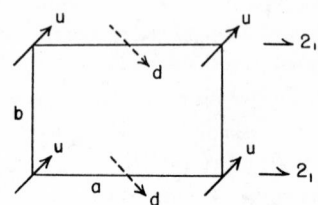

Fig. 44. Axial halving due to screw axis, 2_1.

A 3_1, 3_2, 6_2, or 6_4 axis will leave only every third order, a 4_1 or 4_3 axis will leave only every fourth order, causing a quartering, while a 6_1 or 6_5 axis will leave only every sixth order.

When all the systematically absent reflections have been determined for the various zones and axial planes, the space group or possible space groups to which the crystal can belong may be deduced in a straightforward manner. Tables have been prepared to facilitate this process (see Chapter II, Ref. 9). It will be found, however, that the determination is not in every case unique. We shall now illustrate this for the case of the space groups represented in Fig. 25 (p. 47), and then discuss the cause of these limitations and the methods by which they can be removed.

The first space group illustrated, $C_{2h}^1 - P2/m$, contains no glide plane or screw axis, and consequently gives rise to no systematically absent spectra. This is also true, however, of the space groups $C_2^1 - P2$ and $C_s^1 - Pm$ in the lower classes of the monoclinic system, and these space groups cannot be differentiated directly by their X-ray spectra. The most probable crystal class may be decided from morphological considerations, but unless a large number of very well-formed specimens of the crystal are available, such evidence may not be reliable. In the next space group illustrated in Fig. 25, $C_{2h}^2 - P2_1/m$, the screw axis gives rise to an axial halving, $(0k0)$ being absent when k is odd. This determination is unique, but care should be exercised if only a few orders of this reflection are accessible, in case the absences are accidental and not due to the operation of a true screw axis. In the case of $C_{2h}^5 - P2_1/c$, the same axial $(0k0)$ halving applies, and in addition there is the zonal halving of $(h0l)$ when l is odd, due to the glide plane c. Again this is a unique determination, and one of very great value because of the extremely frequent occurrence of this space group. The same cautionary remarks as have been made in the case of $C_{2h}^2 - P2_1/m$ also apply here, however, because the uniqueness of the determination depends upon the $(0k0)$ halving. The last space group illustrated in Fig. 25, $C_{2h}^6 - C2/c$, is based on a centred monoclinic lattice and so displays a general halving of the (hkl) planes when $(h+k)$ is odd. In addition, the $(h0l)$ planes are absent when h is odd and when l is odd (referred to as "all $(h0l)$ halved"), and, of course, $(0k0)$ is absent when k is odd. These halvings are not unique, however, and are exactly the same as apply to the space group of the lower class, $C_s^4 - Cc$.

It will be noted that the difficulty in determining the space group is due to inability to determine the crystal class uniquely by means of the X-ray spectra. Laue diagrams display the symmetry of a crystal beautifully and in a very direct manner, especially when taken with the incident radiation directed along the principal axis. (Fig. 30, p. 55). By this means, or by a complete examination of the spectra by the methods already described, it is possible to distinguish altogether the 11 different classes of crystal symmetry which are indicated on the space group table (Table III, p. 44). To determine the space groups without ambiguity from the absent spectra it would be necessary in the first place to distinguish the full 32 possible crystal classes.

The difficulty of distinguishing all the crystal classes by means of the X-ray spectra is due to the fact that X-rays cannot in general detect a polar structure. The waves scattered by the various atoms that contribute to the structure are combined, and we observe only the resultant intensity. The particular sequence of atoms across a crystal plane could be replaced by the inverse sequence without affecting the resultant intensity. This is sometimes expressed by saying that the X-rays add a centre of symmetry to the structure. This situation was realised very early in the history of X-ray analysis by G. Friedel[22] and is known as Friedel's law. It can be stated generally in the form that X-ray diffraction phenomena are invariant under an inversion of the crystal with respect to the incident beam. The inversion of the crystal, of course, corresponds to changing the signs of the Miller indices, and the law states that the reflection (hkl) has always the same intensity as the reflection $(\bar{h}\bar{k}\bar{l})$. The law is generally true except for the special cases that arise when the incident wave-length is very close to an absorption edge, when intensity changes in very weak reflections may be observed.[23]

The polar nature of a crystal can often be detected by other physical methods, particularly by the observation of piezo- and pyro-electric effects. A simple experiment in this case is sometimes sufficient to give a positive result, indicating polarity. However, it is not generally safe to accept a negative result of this nature as conclusive evidence that the crystal possesses the higher symmetry.

It is sometimes possible to make the decision between alternative space group possibilities by these methods, or if not to proceed on the assumption of the lower symmetry. Any higher symmetry will necessarily come to light as the atomic positions are revealed by the subsequent analysis, described in the following chapters. However, two more generalised methods by which nearly all the space groups can be distinguished may be noted here. They are both based on a detailed study of the *intensities* of all the X-ray spectra.

In the first method a statistical investigation[24] of the intensities of the reflections is carried out, and it can be shown that a different distribution is to be expected depending on whether the structure factors (see Chapter V) are real or whether they are complex quantities. This in turn depends on the presence or absence of a centre of symmetry. In more detailed studies, Wilson[25] and

---

[22] G. Friedel, *Compt. rend.*, 1913, **157**, 1533.

[23] W. H. Zachariasen, *Theory of X-Ray Diffraction in Crystals*, New York: John Wiley and Sons, 1945; D. Coster, K. S. Knol, and J. A. Prins, *Z. Physik.*, 1930, **63**, 345; I. M. Geib and K. Lark-Horovitz, *Phys. Rev.*, 1932, **42**, 908. See also Chapter VI, p. 150.

[24] A. J. C. Wilson, *Acta Cryst.*, 1949, **2**, 318.

[25] A. J. C. Wilson, *Acta Cryst.*, 1950, **3**, 258.

Rogers[26] have shown how the average intensities of certain groups of reflections are characteristic of the symmetry elements in the space group. The 230 space groups contain 11 enantiomorphous pairs whose members cannot, of course, be distinguished by any ordinary intensity study. Of the remaining 219 space groups, it is claimed that 215 can be uniquely identified by means of these statistical methods.

The other general method of differentiating the space groups depends upon a study of the vector representation of the structure by the Patterson method (see Chapter VI). Characteristic concentrations of points appear in these representations, reflection symmetries corresponding to linear concentrations and rotational symmetries to planar concentrations. Each of these concentrations corresponds to a Harker section. Tables giving a complete list of these characteristics have been prepared by M. J. Buerger,[27] and these, together with a few other special features[28] for particular cases, enable all the space groups to be distinguished, except the eleven enantiomorphous pairs.

8. POWDER PHOTOGRAPHS AND FIBRE PHOTOGRAPHS

Although in this book we are concerned almost wholly with single crystal experiments, mention should be made of two other types of X-ray photography which are of wide application in many important fields where single complete crystals are not available. These are the powder method, first devised by Debye and Scherrer[29] and by Hull,[30] and the method of the fibre photograph.

It will often happen that a specimen of material, such as a metal, which has none of the outward characteristics of a crystal, really consists of a mass of minute crystals that are oriented in all possible directions in a random manner. When a monochromatic X-ray beam passes through such a specimen, or in general through any finely powdered crystalline material, there are always a number of the minute single crystals correctly and exactly oriented to give every possible Bragg reflection.

[26] D. Rogers, *Acta Cryst.*, 1950, **3**, 455.
[27] M. J. Buerger, *Acta Cryst.*, 1950, **3**, 456.
[28] M. J. Buerger, *Proc. Nat. Acad. Sci. U. S.*, 1950, **36**, 324.
[29] P. Debye and P. Scherrer, *Physik. Z.*, 1916, **17**, 277.
[30] A. W. Hull, *Phys. Rev.*, 1917, **10**, 661.

In principle the experimental arrangement is extremely simple; it is illustrated in Fig. 45. The monochromatic beam passes through the powdered specimen, contained in a very small thin-walled tube or adhering to a hair or fine glass fibre, and the diffracted rays are registered on a photographic plate or circular film. As the component crystals lie in every possible orientation, each reflection gives rise to a cone of rays which interesect the photographic plate in a circle. The diameter of this circle gives a measure of 2θ, or the spacing d, so that only spacings and intensities can be recorded by the powder method. Because of this, the full plate or circular film is generally replaced by a narrow strip of film AB (Fig. 45), which is usually contained in a circular camera. If suitable precautions are taken and the collimating system is carefully designed, the spacings and intensities of the powder "lines" can be measured from such a film with great accuracy.

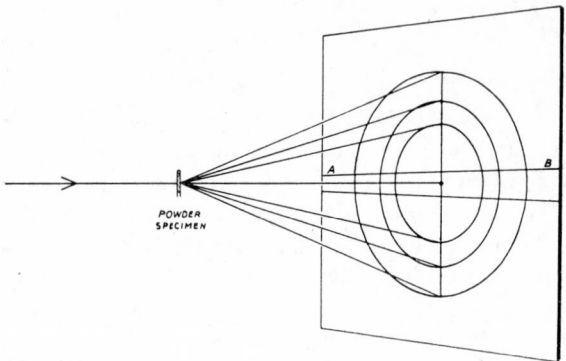

Fig. 45. Formation of powder diagram.

No information regarding the orientation of the reflecting planes in the crystal can be obtained, and so identification of the lines must depend only on spacing measurements. In lattices of high symmetry this is possible, and indices can easily be assigned to the reflections. In the cubic system, for example, the lattice is defined by one parameter a, and the spacings are given by

$$d = \frac{a}{(h^2 + k^2 + l^2)^{1/2}} \cdot$$

Hexagonal, rhombohedral, and tetragonal lattices are defined by two parameters, and the powder lines can be identified without much difficulty by graphical methods. In the lower symmetry systems identification of the lines becomes extremely difficult unless other data regarding the cell dimensions are available.[31]

Even in systems of high symmetry, however, it is clear that the reflections from many distinct crystallographic planes will coincide on the same powder line. In the cubic system the number of co-operating planes obtained by permuting the indices with their signs is 48 for the general plane. With lower symmetry, the number of co-operating planes is less, but accidental concurrences increase. These limitations often prevent any unequivocal determination of space group or complete structure analysis from powder photographic data alone.

The usefulness of the powder method, however, is very great, especially in the investigation of metals and alloys and in all cases where it is important to observe structural changes after treatment of the specimen. Particle size studies, depending on the observation of line broadening, are carried out by this method. On suitably calibrated cameras spacing measurements of extreme accuracy, to 1 part in 30,000 or better, can be obtained by using the back reflections (θ near 90°) where the resolving power is high. Another important application is in the identification of unknown substances. The powder diagram, with its many hundreds of lines of varying intensity, is completely characteristic of the crystalline structure and serves as a unique means of identification. This application, and many graphical methods for the interpretation of powder diagrams, have been described very fully by Bunn.[32]

In the powder photographic method we are dealing with diffraction from complete three-dimensional crystals, although the individual specimens are small and packed together in a random manner. A large class of natural substances, however, can be described as crystalline and oriented in one dimension, but crystalline only over limited areas, which are not mutually oriented, in the other two dimensions. Cellulose, wool, hair, and the mineral

[31] V. Vand, *Acta Cryst.*, 1948. 1, 109, 290; R. Hesse, *ibid.*, 1948, 1, 200; H. Lipson, *ibid.*, 1949, 2, 43; A. J. Stosick, *ibid.*, 1949, 2, 271.

[32] C. W. Bunn, *Chemical Crystallography*, Oxford, Clarendon Press, 1945.

asbestos, for example, are built from polymerised or chainlike molecules, and in these substances the long axes of the constituent molecules generally have a common direction, but the structure may not display much other regularity. Such substances are generally of a fibrous nature, and they give certain characteristic X-ray pictures known as *fibre photographs*.

Small groups of these long molecules generally crystallise together, and in a complete fibre of the substance there are many of these small crystallites, or micelles, with one axis in common but in random lateral orientation.. The effect of passing a monochromatic X-ray beam across a *stationary* fibre of this kind is therefore the same as passing the beam through a single crystal *rotating* about some principal axis. In each case all possible lateral orientations of the crystal are presented to the X-rays, and the result is a *rotation photograph* showing diffraction spots segregated on various layer lines. The interpretation of the fibre photograph is the same as the interpretation of the rotation photograph, but, as only one photograph is generally available, the indexing of the reflections is difficult and can only proceed by trial from postulated unit cell dimensions. The periodicity in the fibre direction, of course, can be determined immediately from the separation of the layer lines.

Further difficulties arise from the fact that the spots on the photographs are often not sharp, owing to the small size of the component crystallites, and sometimes to large deviations from a common direction. If these deviations are very great, arcs are formed and the photograph begins to resemble a powder diagram. A more completely crystalline condition is often attained by stretching the material, and excellent fibre diagrams are obtained from stretched rubber or polyisobutylene.[32] Another kind of interesting deviation from the typical fibre diagram is obtained when the sideways periodicity is reduced. If there were no sideways periodicity at all, and the diffraction occurred from disconnected rows of points having only common direction and period, the result would be continuous layer lines without separate spots. An approximation to this condition is obtained with some photographs from the mineral asbestos.

V

The Structure Factor

and Calculation

of Electron Density

1. GENERAL CONSIDERATIONS

In the previous chapter we have given some consideration to the analysis of X-ray diffraction patterns based mainly upon the observed *positions* of the diffracted beams. From this information it is possible to determine the nature of the lattice and the size of the unit cell, and in many cases to decide upon the space group to which the crystal belongs. Chemical analysis and density then specify the number and kind of atoms in the unit cell, and, if the structure is a simple one, it may be found that it is completely determined by the symmetry requirements alone. In general, however, this is not so, and several different atoms may be present in the minimum structural asymmetric unit. In the case of organic crystals the number of these atoms often corresponds to a complete chemical molecule; occasionally, if an element of molecular symmetry coincides with a space group element, the asymmetric unit will consist of some fraction of the molecule; more rarely, a small group of molecules may together form the asymmetric unit. In all these cases the precise atomic positions can only be found by a further, more detailed analysis, which depends upon a study of the relative *intensities* of the various diffracted beams.

In the derivation of the Bragg equation (Fig. 28, p. 54) we considered the X-rays to be reflected from parallel crystal planes, at a distance d apart. It was assumed that the scattering centres lay *on* these planes, so that each scattered wavelet was in phase and co-operated with those scattered from successive parallel planes to build up a strong diffracted beam. If all the scattering centres actually lay exactly on these planes, we should expect each diffracted beam to be of a uniform maximum intensity. Inspection of any single crystal photograph shows that this is not the case. The intensities of the various spots are found to vary over a wide range, and generally in an abrupt and discontinuous manner.

This is exactly what we should expect if a limited number of discrete scattering centres (atoms) are situated, not exactly on the various crystal planes, but distributed over the unit cell. It will be possible to find a few crystal planes that pass through or near to most of these scattering centres, and these planes will give rise to reflections of high intensity. Generally, however, some of the centres will be distributed in the space between successive planes, the waves scattered from these centres will be out of phase with the remainder, and the resultant intensity will be reduced. The particular distribution of the scattering matter across the various planes will be the chief factor which determines the relative intensities of the X-ray reflections.

The problem before us is to determine this distribution of scattering matter from the observed intensities of the X-ray reflections. There is no direct solution to this problem in the general case, but powerful methods of attack have been developed, which are considered here and in the following chapter. It is usually necessary, in the first instance, to assume that the scattering matter is concentrated into separate atoms, of the number and kind known from chemical considerations to be present in the unit cell. We therefore begin by considering how the amplitude of the scattered wave is modified if a number of atoms, regarded for the moment as point-scattering sources, are distributed between the successive crystal planes.

2. THE GEOMETRICAL STRUCTURE FACTOR

If a group of atoms, P_0, P_1, P_2, $\cdots P_n$ (Fig. 46) lies within the unit cell, we may choose one of these (P_0) arbitrarily as a lattice point and as an origin for co-ordinates. Draw in the traces of a Bragg reflection plane through this atom and the one representing the next lattice point (P_0'). We suppose these lattice points to lie along the a axis of the crystal, so that the path difference of the wavelets scattered by P_0 and P_0' is

$2a \sin \theta = \lambda$ for the first order reflection, or, generally, $2a \sin \theta = h\lambda$ for the $(h00)$ reflection.[1] In angular measure, the phase difference between these two wavelets (Fig. 46b) is 2π, or $2\pi h$, and the result is complete reinforcement to give a wave of double the amplitude. For the wave scattered in this direction from the atom P_1, situated at a distance x_1 across the plane, the path difference compared to the wave from the standard atom P_0 is

$2x_1 \sin \theta = \dfrac{h x_1 \lambda}{a}$ and the phase difference in angular measure is

$\dfrac{2\pi h x_1}{a}$. Similarly, the wave scattered from P_2 has a phase dif-

ference of $\dfrac{2\pi h x_2}{a}$ compared to the standard wave from P_0, and

so on.

Each of these waves is characterised by an amplitude, f_0, f_1, $f_2 \cdots f_n$, which depends upon the scattering power of the atom

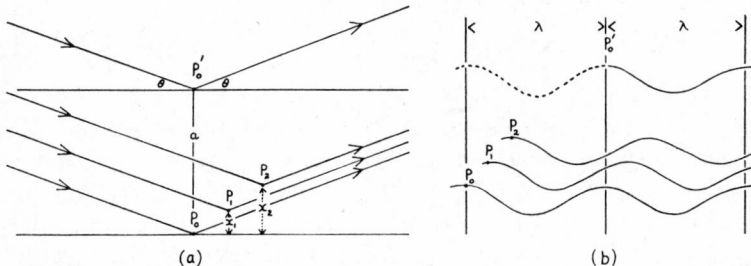

(a)   (b)

Fig. 46 (a) and (b). A series of atoms contributing to the structure amplitude, F.

for this angle of reflection, and a phase contant as given above. (Equality of phase change on scattering is assumed for all the atoms.) The amplitude $|F|$, known as the *structure amplitude*, which results from combining the waves scattered by the whole contents of the unit cell in the direction required for this reflection is easily found by combining the vectors representing the component waves, as in Fig. 47. Expressed analytically,

[1] It is assumed that $d_{100} = a$ and that the axes are orthogonal. The formulas 5.1 to 5.6, however, apply to the general case for crystal axes inclined at any angle.

$$F(h00) = \sum_{0}^{n} f_j e^{2\pi i h x_j / a}. \qquad \cdots (5.1)$$

For the general plane (hkl) the phase change from the origin to the point $x_1 y_1 z_1$ is $2\pi(hx_1/a + ky_1/b + lz_1 c)$, and the resultant vector is

$$F(hkl) = \sum_{0}^{n} f_j e^{2\pi i (hx_j/a + ky_j/b + lz_j/c)}. \qquad \cdots (5.2)$$

This complex resultant, characterised both by an amplitude $|F|$ and a phase constant α, is known as the *structure factor*. It can be evaluated by means of the expressions

$$|F(hkl)| = \sqrt{A^2 + B^2}, \qquad \cdots (5.3)$$

$$\alpha(hkl) = \tan^{-1} \frac{B}{A}, \qquad \cdots (5.4)$$

where

$$A = \sum_{0}^{n} f_j \cos 2\pi(hx_j/a + ky_j/b + lz_j/c) \qquad \cdots (5.5)$$

$$B = \sum_{0}^{n} f_j \sin 2\pi(hx_j/a + ky_j/b + lz_j/c). \qquad \cdots (5.6)$$

In these expressions the summations are taken over all the atoms in the unit cell. If the space group is known, it is usually convenient in the first instance to carry the summations over the co-ordinates of the equivalent point positions, and this often results in a simplified expression. In particular, if a centre of symmetry is present and is chosen as origin for the co-ordinates, every vector of phase angle $2\pi x_j/a$ (Fig. 47) must be accompanied by another vector of phase angle $-2\pi x_j/a$. The resultant

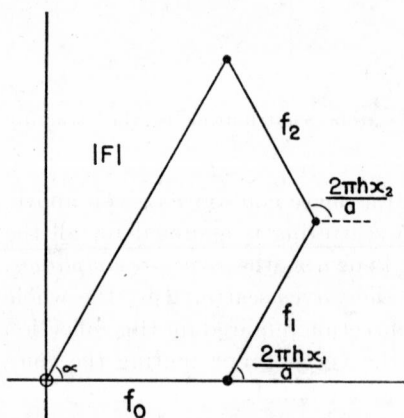

Fig. 47. Vector composition of F.

84

therefore lies on the horizontal axis, and can be obtained by summing the cosine terms alone, as in (5.5). The possible phase angles of the resultant structure factor are thus limited to 0 or π, depending on whether the expression for A in (5.5) is positive or negative. The expression for B (5.6) must always be zero.

As an example, we may sum the expressions (5.5) and (5.6) over the co-ordinates of the equivalent points for the space group $C_{2h}^5 - P2_1/c$ which are

$$x, y, z; -x, -y, -z; x, \frac{b}{2}-y, \frac{c}{2}+z; -x, \frac{b}{2}+y, \frac{c}{2}-z.$$

$$\cdots (5.7)$$

We then obtain

$$A = 4 \sum f_j \cos 2\pi \left(hx/a + lz/c + \frac{k+l}{4} \right) \cdot \cos\ 2\pi \left(ky/b - \frac{k+l}{4} \right)$$

$$\cdots (5.8)$$

$$B = 0,$$

where the summations have now to be carried over the smaller number of atoms in the asymmetric unit. For particular combinations of the indices these expressions simplify further so that when $k+l$ is even

$$A = 4 \sum f_j \cos 2\pi(hx/a + lz/c) \cdot \cos 2\pi ky/b, \quad \cdots (5.9)$$

and when $k+l$ is odd

$$A = -4 \sum f_j \sin 2\pi(hx/a + lz/c) \cdot \sin 2\pi ky/b. \quad \cdots (5.10)$$

We note that the last expression is zero when k is zero, or when h and l are zero, so that the ($h0l$) reflections are absent when l is odd, and ($0k0$) when k is odd. These are, of course, the halvings which characterise this space group.

The general expression for the structure factor has been summed over the equivalent points for each of the 230 space groups, and the results have been arranged in convenient form by K. Lonsdale.[2]

If the positions of the atoms in the asymmetric unit are known or can be guessed, there is now no difficulty in calculating the relative structure amplitudes for all the X-ray reflections, pro-

[2] K. Lonsdale, *Structure Factor Tables*, London: G. Bell and Sons, 1936.

vided that some estimate of the atomic scattering factors, f_j, can be made. These factors will depend primarily on the atomic numbers of the atoms concerned, and on the angle of reflection, θ. To compare the results with the observed X-ray intensities it is also necessary to know how the actual crystal intensities depend upon the structure amplitudes. These matters are discussed in the next section, but it may be noted here that for very small crystals the intensities depend upon the squares of the structure amplitudes. It is therefore possible to test any postulated structure very thoroughly against the observed intensities.

It is clear, however, that the observed intensities can provide no information concerning the relative phase constants of the structure factors, which are complex quantities (5.2). This is the fundamental difficulty in X-ray analysis, which prevents any immediate or direct solution to the problem of finding the atomic arrangement. Often some process involving trial and error at some stage must be adopted. A knowledge of the chemical structure combined with the observation of a few outstanding intensities may suggest probable positions for the atoms, and these can be immediately tested by calculation of all the resultant structure amplitudes.

One outstanding feature of the X-ray method of analysis is that, with a few important exceptions, when the true structure is once discovered it need never be in doubt. For most crystals the number of observed reflections very greatly exceeds the number of atoms in the unit cell. In fact, in the case of organic crystals of average density, the number of possible distinct reflections which lie within the limiting sphere for copper radiation ($\lambda = 1.54$ Å) is about 50 per atom (not counting hydrogen) and this number can easily be increased by using a radiation of shorter wave-length. As the position of each atom is defined by only three co-ordinates, and each structure factor provides a relation between all the co-ordinates, it is clear that the structure is excessively over-determined. Owing to this wealth of data, a structure, once discovered, can also be refined to a high degree of accuracy by various more elaborate methods of analysis, which are outlined in later sections.

3. SCATTERING BY ELECTRONS AND ATOMS

Crystal structures may be tested and solved by comparing calculated structure factors in a rough and ready way with the relative intensities of the spots observed on single crystal photo-graphs, as indicated in the previous section. Absolute intensity measurements, however, which relate the actual intensity of the primary incident beam to that of the various reflected beams, introduce problems of great complexity both on the experimental side and in the theoretical treatment of the results. Such measurements are of vital importance in establishing a sound theoretical basis for the whole subject, and also in aiding the solution of complex crystal structures. It is a remarkable fact that as early as 1914 not only were methods evolved by W. H. Bragg[3] of defining and measuring the true integrated intensities of crystal reflections, but a very complete theoretical treatment of the absolute reflection intensities to be expected from real crystals had been given by Darwin,[4] and similar formulas were later deduced independently by Ewald.[5] Since that time the general theory of X-ray diffraction in crystals has received intensive study. Very complete accounts are available in the works of Compton,[6] Zachariasen,[7] and James.[8] In the present section it is only possible to summarize a few of the results that are of immediate importance in structure analysis.

The units ultimately responsible for scattering X-rays are the electrons. Scattering by a free electron can be calculated by classical electromagnetic theory.[9] If an incident wave of amplitude A_0 is scattered by a single electron, it can be shown that the amplitude A of the scattered wave at a distance r (large compared to the wave length) is given by

$$A = A_0 \frac{e^2}{mrc^2} \qquad \cdots (5.11)$$

where e and m are the charge and mass of the electron and c is

[3] W. H. Bragg, *Phil. Mag.*, 1914, **27**, 881.

[4] C. G. Darwin, *Phil. Mag.*, 1914, **27**, 315, 675.

[5] P. P. Ewald, *Ann. Physik.*, 1916, **49**, 1, 117; 1917, **54**, 519, 557.

[6] A. H. Compton, *X-Rays and Electrons*, New York: Van Nostrand and Co., 1926; A. H. Compton and S. K. Allison, *X-Rays in Theory and Experiment*, New York: Van Nostrand and Co., 1935.

[7] W. H. Zachariasen, *Theory of X-Ray Diffraction in Crystals*, New York: J. Wiley and Sons, 1945.

[8] R. W. James, *The Optical Principles of the Diffraction of X-Rays*, London: G. Bell and Sons, 1948.

[9] J. J. Thomson, *Conduction of Electricity through Gases*, Cambridge: Cambridge University Press, 1928–1933.

the velocity of light. It is assumed that the electric vector is perpendicular to the plane containing the incident and scattered rays. When it lies in this plane, the amplitude depends on the scattering angle 2θ (the angle between the incident and scattered rays) and is given by

$$A = A_0 \frac{e^2}{mrc^2} \cos 2\theta. \qquad \cdots (5.12)$$

Although actual electrons do not behave like the free classical electron assumed above, these formulas may be taken to define a convenient unit in terms of which the coherent scattered radiation can be measured. If instead of giving the electron a definite position it is represented by a probability function, then the scattering can be regarded as due to an electron cloud, the amount of scattering being proportional to the density of the cloud at each point. For an extended distribution the resultant amplitude will be decreased by interference, as in the case of the distribution of point-scattering sources considered in the last section.

For an atom containing z electrons the scattered amplitude will approach zA for very small scattering angles, but in general it must be written as fA where f is the *atomic scattering factor* for the particular atom. f is a function of the scattering angle, and obviously depends upon the distribution of electrons in the atom. These distributions have been calculated theoretically by Hartree[10] and others from a number of atoms. As the free atom is spherically symmetrical, they are generally expressed as radial distribution functions, $U(r)$, $U(r)dr$ being the number of electrons contained between shells of radii r and $r+dr$. When this function is known, the atomic scattering factor may be calculated from the integral

$$f_0 = \int_0^\infty 4\pi r^2 U(r) \frac{\sin(4\pi r \sin \theta/\lambda)}{4\pi r \sin \theta/\lambda} \, dr. \qquad \cdots (5.13)$$

Atomic scattering factors derived in this way for various atoms have been calculated by James and Brindley,[11] Thomas,[12] Fermi,[13]

---

[10] D. R. Hartree, *Proc. Cambridge Phil. Soc.*, 1928, **24**, 89, 111; R. Mc-Weeny, *Acta Cryst.*, 1941, 4, 513; 1952, **5**, 463.

[11] R. W. James and G. W. Brindley, *Phil. Mag.*, 1932, 12, 81.

[12] L. H. Thomas, *Proc. Cambridge Phil. Soc.*, 1927, **23**, 542.

[13] E. Fermi, *Z. Physik.*, 1928, **48**, 73.

and Pauling and Sherman,[14] and the results are conveniently arranged in the *International Tables* (see Chapter II, Ref. 9).

In these theoretical scattering factors the atoms are assumed to be at rest, but thermal movements have an important effect in all practical cases. The general result is to spread the electron distribution and so decrease the intensities of the spectra. If f_0 is the factor calculated for the atom at rest, then the temperature corrected factor f is given by

$$f = f_0 e^{-B(\sin \theta / \lambda)^2}. \qquad \cdots (5.14)$$

A theoretical treatment of the subject and methods of evaluating B from the fundamental constants and heat capacity data have been given by Debye[15] and Waller.[16] A simple relation exists between this factor and the mean square displacement of the atom at right angles to the reflecting plane, $\bar{\mu}^2$, viz.,

$$B = 8\pi^2 \bar{\mu}^2. \qquad \cdots (5.15)$$

The mean atomic oscillations at different temperatures can thus be estimated in a direct way from the intensities of the X-ray reflections. In structure analysis, empirical or semi-empirical f-curves are frequently employed, which include an approximate temperature factor for the general type of crystal structure being investigated.

4. SCATTERING BY CRYSTALS

From a knowledge of the appropriate f-values and the positions of all the atoms in the unit cell, the structure amplitudes, $|F(hkl)|$, can now be readily evaluated for all the reflections, by the methods given in Section 2. The relation of these structure amplitudes to the absolute values of the intensities of the reflections given by the whole crystal, however, involves calculations of considerable difficulty, in which the state of perfection of the crystal specimen plays a large part.

A very small crystal element will reflect radiation over a certain angular range on either side of the angle θ_0 given by the Bragg equation (4.2), and it is necessary in the first place to measure and make use of this total reflected energy in the calculations. In

[14] L. Pauling and J. Sherman, *Z. Krist.*, 1932, **81**, 1.
[15] P. Debye, *Ann. Physik.*, 1914, **43**, 49.
[16] I. Waller, *Z. Physik.*, 1923, **17**, 398; *Ann. Physik.*, 1927, **83**, 153.

the method evolved by W. H. Bragg[3] the crystal is turned through the reflecting position with angular velocity ω, and the total reflected energy, E, received in the ionization spectrometer or on the photographic plate is recorded. This is inversely proportional to ω. If I_0 is the intensity of the incident beam (energy per second per square centimetre at the crystal) the ratio $E\omega/I_0$ then measures what is called the *integrated reflection*, and this ratio is found to be characteristic of a given crystal plane. The integrated reflection can also be defined as

$$E\omega/I_0 = \int_{\theta_0-\epsilon}^{\theta_0+\epsilon} R(\theta)d\theta/I_0 \qquad \cdots (5.16)$$

where $R(\theta)$ is the energy reflected per second at the angle θ, and the limits $\pm\epsilon$ are taken great enough to include all the reflected radiation.

For a very small crystal block of volume δV, in which absorption of the X-ray beam is not appreciable, it can be shown[4] that the integrated reflection is proportional to the volume δV and is independent of the shape of the crystal block. The following relation can be derived:

$$\frac{E\omega}{I_0} = \left[N \; \frac{e^2}{mc^2} \; F(hkl) \right]^2 \lambda^3 \frac{1 + \cos^2 2\theta}{2 \sin 2\theta} \, \delta V. \quad \cdots (5.17)$$

In this expression N is the number of unit cells per unit volume of the crystal, or the reciprocal of the volume V of the unit cell. The factor e^2/mc^2 has been discussed in (5.11) and (5.12). The incident radiation is now assumed to be unpolarized, with its electric vector distributed uniformly in all directions perpendicular to the beam direction, and the *polarization factor*, $(1+\cos^2 2\theta)/2$, provides the necessary averaging.

Expression 5.17 should only hold for minute, almost submicroscopic crystals, such as might be realised in an extremely fine powder. If ordinary macroscopic crystals were perfect structures throughout their volume, the X-ray reflections would be extremely sharp and take place over a few seconds of arc only. Over most of this range they would reflect all the radiation incident upon them, and quite a different intensity formula would apply (5.18 below). It is found, however, that expression 5.17 is in fact applicable, with modifications to allow for the absorption of the X-ray beam, to many comparatively large single crystals. It must be inferred that in such crystals the perfect continuity of the lattice structure is only maintained over very small regions, and that some break or slight change in orientation then occurs. As Darwin[4] has pointed out, the crystal behaves like a conglomerate or a *mosaic* of small blocks, "each block a perfect crystal, but the adjacent blocks not accurately fitted together." Recent direct evidence from the electron microscope[17] has shown how some of these discontinuities may arise, and in fact they appear to be an essential condition of crystal growth for many substances.

In the case of most organic crystals, expression 5.17 can be applied directly if extremely small specimens, weighing perhaps 0.1 mg. or less, are completely immersed in the X-ray beam. Some correction for absorption should still be applied, but this is generally small. Caution is necessary, however, because the crystal may not always conform to the ideal mosaic type, and the regions of perfect crystal structure may be too large. In such cases, and particularly for the strong X-ray reflections, extinction of the incident X-ray beam by reflection from the surface layers is much more important than ordinary absorption in the crystal. This effect is called *primary extinction.* Repeated immersion of the crystal in liquid air may succeed in reducing it to a less "perfect" condition, but it is difficult to devise any systematic corrections for this effect. *Secondary extinction* occurs even in good mosaic crystals and is due to the screening of the lower blocks in the crystal by the upper blocks, which can reflect away an appreciable amount of the incident radiation in the case of very strong reflections. The effective absorption coefficient is thus increased when the crystal is set at the reflecting angle. Systematic corrections for this effect can be applied.

For other experimental conditions, such as reflections from an extended crystal face, or from a powder, expression 5.17 must be suitably modified. The appropriate formulas, together with absorption and extinction correction data, are listed in the *International Tables.*

The case of the really *perfect crystal,* as contrasted to the ideal mosaic crystal type discussed above, is comparatively rare.

[17] I. M. Dawson and V. Vand, *Proc. Roy. Soc.* (London), 1951, **A206,** 555.

Some types of diamond and freshly cleared calcite surfaces may, however, approximate to this type. The theory of reflection in this case has also been given by Darwin[4] and Ewald,[5] who have shown that reflection should be total over a certain very narrow range of glancing angle. The integrated reflection now depends upon the first power of the structure factor, and is given by the formula

$$\frac{E\omega}{I_0} = \frac{8}{3\pi} N \frac{e^2}{mc^2} F(hkl)\lambda^2 \frac{1 + |\cos 2\theta|}{2 \sin 2\theta} . \quad \cdots (5.18)$$

It must be remembered that the angular range over which the reflection is total is extremely minute, and hence the integrated reflection in this case is actually smaller than for the mosaic crystal.

Expression 5.18 is seldom used in crystal analysis, but by means of (5.17) for the ideal mosaic crystal, and accurate intensity measurements, it is possible to obtain observed F values on an absolute scale. This is generally done by comparison with some standard crystal on which absolute measurements have already been made.[18,19] A set of absolute F values obtained in this manner is of the very greatest assistance in the determination of crystal structures. If only relative F values are available, the initial stages of structure determination, by trial and error, are generally much more difficult. Fictitious agreements may be obtained if only a limited number of relative F values are studied in the case of complex structures, but such false starts are far less likely if the absolute scale is known. When electron density distributions are calculated by the methods outlined in the following sections, an absolute scale enables the number of electrons associated with each atom or part of the molecule to be estimated with some accuracy.

5. GENERALISED DEFINITION OF THE STRUCTURE FACTOR

The expression for the structure factor given in equation 5.2 is obtained by assuming that all the scattering matter in the unit cell is concentrated into a number of spherically symmetrical atoms situated at known points $(x_jy_jz_j)$. It is further assumed that the electron distribution function, $U(r)$, in each of these

[18] B. W. Robinson, *Proc. Roy. Soc.* (London), 1933, A142, 422.
[19] J. M. Robertson, *Phil. Mag.*, 1934, 18, 729.

atoms is known, and that the theoretical atomic scattering factors, f_0, can therefore be calculated by equation 5.13. It is also necessary that some estimate of the temperature factor for each atom should be available, so that the effective scattering factors may be obtained by equation 5.14.

In an actual crystal structure it will seldom be possible to give a perfectly precise expression of the electron distribution in these various terms. Expecially in the case of low-symmetry molecular crystals, the thermal movements of the atoms may be anisotropic; and apart from this, the presence of bonding electrons must imply certain departures from spherical symmetry in the atoms themselves. It is clearly desirable to give a more general definition of the structure factor, which will dispense with the idea of separate atoms and electrons at definite situations in the crystal. This more general definition is also necessary in order to develop a more powerful method of structure analysis, based on the use of Fourier series. .

Let ρ (xyz) be the density of scattering matter at any point x, y, z in the unit cell. This density will be expressed in electronic units, so that $\rho(xyz)dxdydz$ will give the number of electrons in the volume element $dxdydz$. For the general case, where the crystal axes, a, b, c are inclined at any angles, the number of

electrons in the volume element will be $\rho(xyz)\dfrac{V}{abc}\,dxdydz$, where

V is the volume of the unit cell. The scattering matter in each volume element of this continuous distribution will make a contribution to the resultant amplitude for the whole unit cell, and for the general plane (hkl) the phase change from the origin to the point x, y, z is $2\pi(hx/a + ky/b + lz/c)$. The resultant vector is therefore

$$F(hkl) = \frac{V}{abc} \int_0^a \int_0^b \int_0^c \rho(xyz)e^{2\pi i(hx/a+ky/b+lz/c)}dxdydz.$$
$$\cdots (5.19)$$

As ρ (xyz) is expressed in electronic units, we see that the structure amplitude $F(hkl)$ is simply the ratio of the amplitude actually received from the whole contents of a given unit cell to that which would be received if the contents of the unit cell were replaced by a single classical electron.

6. EXPRESSION OF ELECTRON DENSITY BY FOURIER SERIES

A crystal structure will be completely solved if we can evaluate the density function, $\rho(xyz)$, in equation 5.19, throughout the whole unit cell. As a first step in the attempt to do this we represent the function by means of a Fourier series in the general form,

$$\rho(xyz) = \sum \sum_{-\infty}^{\infty} \sum A(pqr)e^{2\pi i(px/a+qy/b+rz/c)}, \quad \cdots (5.20)$$

p, q, and r being integers and $A(pqr)$ the at present unknown coefficient of the general term.

The essentially periodic nature of all crystals, which arises from the lattice postulate, suggests expansion in the form of a Fourier series as the natural representation of any physical property in the crystal. The idea that the matter composing a crystal structure might be represented in such a way was first suggested by W. H. Bragg[20] in 1915, who realised that each X-ray reflection must correspond to one of the component sinusoidal distributions of density in the medium. Bragg's treatment was based on the analogy with the optical methods of Abbe and A. B. Porter for the case of image formation in the microscope. At this time there were many difficult and uncertain factors affecting the treatment of X-ray intensities, and the Fourier series method was not immediately put into a useable quantitative form. Later developments were due Ewald,[21] Shearer,[22] Duane,[23] Havighurst,[24] and Compton,[6] the latter making the important step of showing exactly how the Fourier series coefficients are related to the X-ray intensities. In 1929 W. L. Bragg[25] extended the treatment further and made the first practical use of a double Fourier series treatment.

In order to evaluate the series 5.20 and so obtain the electron density at any point in the crystal it is necessary to calculate the coefficients, $A(pqr)$. To do this we substitute this series for

[20] W. H. Bragg, *Trans. Roy. Soc.* (London), 1915, **A215**, 253.

[21] P. P. Ewald, *Z. Krist.*, 1921, **56**, 129.

[22] G. Shearer, *Proc. Roy. Soc.* (London), 1925, **A108**, 655.

[23] W. Duane, *Proc. Nat. Acad. Sci. U. S.*, 1925, **11**, 489.

[24] R. J. Havighurst, *Proc. Nat. Acad. Sci. U. S.*, 1925, **11**, 502.

[25] W. L. Bragg, *Proc. Roy. Soc.* (London), 1929, **A123,** 537.

$\rho(xyz)$ in the general expression for the structure factor (5.19) and obtain

$$F(hkl) = \frac{V}{abc} \int_0^a \int_0^b \int_0^c \left[\sum \sum_{-\infty}^{\infty} \sum A(pqr)e^{2\pi i(px/a + qy/b + rz/c)} \right]$$
$$\cdot e^{2\pi i(hx/a + ky/b + lz/c)} dx dy dz.$$

On integrating, every term is zero except that for which $p = -h$, $q = -k$, and $r = -l$, which gives

$$F(hkl) = \frac{V}{abc} \int_0^a \int_0^b \int_0^c A(\bar{h}\bar{k}\bar{l}) dx dy dz = VA(\bar{h}\bar{k}\bar{l})$$

$$A(\bar{h}\bar{k}\bar{l}) = \frac{F(hkl)}{V}. \qquad \cdots (5.21)$$

The Fourier series which represents the electron density distribution at every point in the crystal may therefore be written

$$\rho(xyz) = \sum \sum_{-\infty}^{\infty} \sum \frac{F(hkl)}{V} e^{-2\pi i(hx/a + ky/b + lz/c)}. \qquad \cdots (5.22)$$

The zero term of the series is $F(000)/V$, and this is a constant which may be obtained from (5.19),

$$F(000) = \frac{V}{abc} \int_0^a \int_0^b \int_0^c \rho(xyz) dx dy dz = Z \qquad \cdots (5.23)$$

where Z is the total number of electrons in the unit cell.

We therefore obtain the very important and very beautiful result that, when the electron density distribution in the crystal is represented in this manner by a triple Fourier series, the coefficients of the various terms in the series are given by the absolute values of the structure factors for the corresponding crystal planes, divided by the volume of the unit cell. The individual terms in this Fourier series may be visualised as corresponding to the various reciprocal lattice points, if these latter are weighted according to the values of the structure factors with which they are associated, the origin point (000) being given a weight corresponding to the total number of electrons in the unit cell.

In general these weights will decrease as we pass outwards from the origin point, and for sufficiently high values of $\sin \theta/\lambda$ they will become negligible. This is clear from the general form obtained for the atomic scattering curves (5.13).   As the coefficients finally become negligible, it follows that the Fourier series for the electron density function converges.   This convergence will become more rapid for a diffuse distribution of electrons, as will occur when the temperature factor (5.14) assumes large values.

It must be remembered that $F(hkl)$ in (5.22) is a complex quantity, representing not only the amplitude but also the phase of the diffracted beams, with reference to some chosen origin in the crystal.   It may generally be assumed that the phase change on scattering is the same for all parts of the unit cell, that $F(hkl)$ and $F(\bar{h}\bar{k}\bar{l})$ are conjugate, and $\rho(xyz)$ is everywhere real.   The amplitudes of $F(hkl)$ and $F(\bar{h}\bar{k}\bar{l})$ are then equal in accordance with Friedel's law (p. 76).   If there is a phase change on scattering at different atoms in the unit cell, as happens in the special case when the incident wave-length is near one of the atomic absorption edges, then Friedel's law breaks down and $\rho(xyz)$ becomes complex in the above expressions.   We can neglect this special case at present, although it is of great importance as a means of absolute phase determination in certain circumstances referred to later in Chapter VI.

7. FOURIER SYNTHESIS

It is convenient to write the Fourier series (5.22) in the form

$$\rho(xyz) = \sum \sum_{-\infty}^{\infty} \sum \frac{|F(hkl)|}{V} \cos[2\pi hx/a + 2\pi ky/b + 2\pi lz/c - \alpha(hkl)] \quad \cdots (5.24)$$

where $\alpha(hkl)$ represents the phase constant associated with the amplitude $|F(hkl)|$ and has the same meaning as in equation 5.4.   As already pointed out in Section 2, the observed intensities enable us to calculate $|F(hkl)|$ but provide no information concerning the relative values of the phase constants, $\alpha(hkl)$.   This limitation prevents any immediate or direct application of the series to the solution of crystal structures except in special cases (see Chapter VI), but the value of the method as a means of fully representing and refining the results of crystal analysis remains.

The method is most readily applied when the crystal structure contains a centre of symmetry and the origin for co-ordinates is chosen at this point. The possible phase angles are then limited to 0 or π. This is immediately obvious when we remember that each term in the series 5.24 represents a component sinusoidal distribution of electron density covering the whole structure and completely accounting for the intensity of reflection from one crystal plane. If the structure as a whole has a centre of symmetry at the origin, then this symmetry must apply to each of the component distributions which define the structure, i.e., either a trough or a peak of each of the component waves must coincide with the origin. The phase angle, $\alpha(hkl)$, can be regarded as measuring the displacement of the peak of the wave from the origin. In the centro-symmetrical case this displacement is restricted to 0 or π in angular measure. These two possibilities can be covered by making the signs of the coefficients either positive or negative, and so for a centrosymmetrical structure the series may be written:

$$\rho(xyz) = \sum \sum_{-\infty}^{\infty} \sum \pm \frac{F(hkl)}{V} \cos 2\pi(hx/a + ky/b + lz/c).$$
$$\cdots (5.25)$$

Now if the positions of the atoms in a structure are approximately known, from chemical considerations combined with a number of trial calculations or by application of some of the more advanced methods described in the next chapter, then the signs of the more important coefficients may be determined with reasonable certainty from the calculations of the geometrical structure factor (equations 5.2 to 5.6). Application of the series 5.25 will then provide a picture in three dimensions showing high concentrations of electron density (electrons per $\mathring{A}^3$) in the regions of the atomic sites. This picture of the density distribution will be defective, because the series employed will have been incomplete, with many weaker terms of uncertain sign omitted. But the distribution obtained will of necessity account *exactly* for all the coefficients in the series used to generate it. It will be the picture of a structure which would yield the precise F values employed, and make all the other F values not employed exactly zero. Such a structure will generally contain regions of negative electron density, without physical meaning, and will also contain atoms that are somewhat distorted from the expected spherical symmetry. Nevertheless, when the most

probable atomic positions are estimated from the main concentrations in this distribution, it will usually be found, if the initial
trial structure has not been too wide of the mark, that a considerable refinement of the atomic parameters has been achieved.
Recalculation of the structure factors from these revised atomic
positions will generally lead to improved agreements with the
measured values and enable the signs of further weak terms in the
series to be included in the next synthesis, which will yield a still
more precise picture of the actual structure.   This process of
successive approximation to the true structure is of the greatest
value, especially in the case of structures containing a large
number of parameters where the independent variation of all the
parameters by trial and error methods is quite impossible.

The computation of a series of the form (5.25) is, however,
extremely laborious.   It is often used in the final stages of analysis, when great accuracy is required (Figs. 83, 84, p. 186)
or when the structure is so complex that other methods are of
little value (Fig. 119, p. 284).   In a great many cases a *double
Fourier series* may be employed, confined to two co-ordinates
and expressing the density per unit area (electrons per Å²) in any
projection of the structure.   The computational work is now
much more manageable, since the series takes the form

$$\rho(xy) = \sum_{-\infty}^{\infty}\sum \frac{F(hk0)}{A} \cos[2\pi hx/a + 2\pi ky/b - \alpha(hk0)].$$

$$\cdots (5.26)$$

This formula may be derived from equation 5.24 by integrating
over z (see equation 5.28). A is now the area of the face upon
which the projection is made.   This form of the series was first
employed by W. L. Bragg[25] to represent the diopside structure.
It is used extensively in the later parts of this book, and is particularly useful in the case of structures involving planar aromatic
molecules, where some projection can often be found which reveals
the separate atoms clearly.   Another important feature of this
form of analysis is that the structure factors employed are
obtained from the reflections belonging to a single zone in the
crystal.   We have seen in the previous chapter how the reflections
from any one zone form an equatorial layer line, which can be
most conveniently and accurately recorded in one operation on a
moving-film Weissenberg type of apparatus.   All the required

intensities are then automatically on the same scale, and trouble-some correlations are avoided. When the series is finally evaluated, the result gives a projection of the structure along the axis about which the crystal was rotated. By combining the results from two or more projections obtained in this way, the complete structure can often be established with very great accuracy.

The most simple form of the Fourier series is confined to *one dimension*:

$$\rho(x) = \sum_{-\infty}^{\infty} \frac{F(h00)}{d_{100}} \cos\left[2\pi hx/a - \alpha(h00)\right], \quad \cdots (5.27)$$

and this gives the density per unit length (electrons per Å) averaged for sheets parallel to any given plane. The coefficients,

$\dfrac{F(h00)}{d_{100}}$, are now the structure factors obtained for the successive

orders of reflection, $(h00)$, of that plane, divided by the spacing, d_{100}, of the fundamental plane. Although this form of the series has important applications to simple structures and is very easy to apply, it unfortunately gives little help even with moderately complex distributions.

When any of the expressions 5.24 to 5.27 are integrated over a definite region of the structure, say over a single atom or part of a molecule, the result gives directly the number of electrons in that region, provided that the coefficients have been evaluated on an absolute scale. The possibility of making such an *electron count* is a valuable feature of the Fourier method. If the integration is carried over the whole unit cell, the result, of course, gives the total number of electrons in the unit cell, Z, because the effect of integrating each individual term over a complete period is zero, except in the case of the term $F(000)$.

From what has been said it will be clear that the triple Fourier series representation (5.24) is by far the most powerful, but its use is limited by the forbidding amount of computational work involved. If the series is evaluated only on certain planes to give a *section* of the structure at height z, say, the computational work is reduced, but many such sections may be required to give a reasonable definition of the whole structure. Again, the series may be evaluated only along certain *lines*, generally chosen to pass as nearly as possible through the atomic centres. The variable in the summation is now reduced to one co-ordinate, but in general a large number of separate line syntheses will be required to define the atomic positions fully.

Two methods have been introduced by Booth[26] which have advantages somewhat intermediate between those of the standard methods already described. The first is known as the method of *bounded projections* or *section projections*, where attention is confined to the projection of that portion of the electron density distribution contained between two planes, given for example by $z = z_1$ and $z = z_2$. This portion of the structure is defined by

$$B_{z_1}^{z_2} = \int_{z_1}^{z_2} \rho(xyz)dz, \qquad \cdots (5.28)$$

and expression 5.24 becomes

$$B_{z_1}^{z_2} = \sum \sum_{-\infty}^{\infty} \sum \frac{c}{2\pi l} \frac{|F(hkl)|}{V} \{ \sin[2\pi hx/a + 2\pi ky/b + 2\pi lz_2/c$$
$$- \alpha(hkl)] - \sin[2\pi hx/a + 2\pi ky/b + 2\pi lz_1/c - \alpha(hkl)] \}. \cdots (5.29)$$

This can be regarded as a generalized projection synthesis, and when $z_2 = z_1 + c$ the expression reduces to (5.26). On the other hand, when z_2 is taken very close to z_1, the synthesis is equivalent to a normal three-dimensional synthesis evaluated for the section at height z_1, because

$$\frac{1}{\delta z_1} B_{z_1}^{z_1 + \delta z_1} \to \rho(xyz_1). \qquad \cdots (5.30)$$

Although expression 5.29 is cumbersome, it can, for computational purposes, be reduced in a manageable way and simplified for certain space groups. The method is extremely useful for separating a molecule from its companion molecule in the unit cell. In the normal Bragg projection (5.26) it is often found that the atoms of neighbouring molecules overlap to such an extent that they cannot be separately resolved, although the atoms of a single molecule would be separated for the same projection direction. It is in such cases that the Booth method is valuable,

[26] A. D. Booth, *Trans. Faraday Soc.*, 1945, **41**, 434.

if the locations of the boundary planes are carefully selected.

Booth's second method is known as the method of *projected sections,* and while this does not introduce any information about the structure not provided by the ordinary three-dimensional synthesis evaluated for a number of sections at different heights, $z_1, z_2 \cdots z_n$, it reduces the computational burden by combining the production of these separate sections into one operation. It will often happen that the groups of atoms lying at the different heights $z_1, z_2 \cdots z_n$ will have widely differing (xy) co-ordinates; in this case the sections at these different heights may be combined without interference.

For a single section at height z_1 the series is

$$\rho(xyz_1) = \frac{1}{V} \sum \sum_{-\infty}^{\infty} \sum |F(hkl)| \cos [2\pi hx/a + 2\pi ky/b + 2\pi lz_1/c - \alpha(hkl)].$$

The result of combining a number of such sections gives

$$R(x, y, \overline{z_1 \cdots z_n}) = \sum_{r=1}^{n} \rho(xyz_r) = \frac{1}{V} \sum_{r=1}^{n} \left\{ \sum \sum_{-\infty,hkl}^{\infty} \sum |F(hkl)| \right.$$
$$\left. \cdot \cos [2\pi hx/a + 2\pi ky/b + 2\pi lz_r/c - \alpha(hkl)] \right\}.$$
$$\cdots (5.31)$$

Again, the expression may be simplified for computational purposes, and it does not in fact involve much more labour than in the case of a single section. If the positions of the atoms are known well enough for the section levels to be selected so that they are free from interference, there is no reduction in accuracy.

8. COMPUTATIONAL METHODS

Although this is no place to give a detailed account of computational methods, the problem is so important in X-ray crystal analysis that some reference to the more important advances should be made. Even in the comparatively simple double Fourier summation the computational work is heavy. For example, in the projection of the platinum phthalocyanine structure shown in Fig. 111 a series containing 302 terms was evaluated at 1800 different points on the asymmetric unit (half the molecule). If no shortening devices were employed, the total number of terms of the type $\dfrac{F(h0l)}{A}$ cos $2\pi(hx/a+ky/b)$ which it would be necessary to evaluate and add for this projection alone would be well over half a million. Fortunately, a number of methods have now been devised which shorten and speed up this work.

These methods fall into two classes which are based (a) on the speeding up of normal digital operations by special devices or special machines and (b) on the design of special analogue computers, in which the density function is represented by some mechanical device or physical property such as weight, light intensity, or electrical potential.

8A. Digital operations.—Under this first category, a way in which the numerical work can be simplified may be illustrated for the most common case of two-dimensional synthesis. Writing F_{hk} for the coefficients, θ_1, θ_2 for $2\pi x/a$, $2\pi y/b$ in (5.26), and assuming for the moment that the phase constants can be included in the signs of the coefficients (centrosymmetrical case), the expression becomes

$$\rho(xy) = \sum_{-\infty}^{\infty}\sum F_{hk} \cos (h\theta_1 + k\theta_2) = \sum_{-\infty,k}^{\infty}\left[\sum_{-\infty,h}^{\infty} F_{hk} \cos h\theta_1\right]$$

$$\cdot\cos k\theta_2 - \sum_{-\infty,k}^{\infty}\left[\sum_{-\infty,h}^{\infty} F_{hk} \sin h\theta_1\right]\sin k\theta_2. \qquad \cdots (5.32)$$

The summations in the brackets may be carried out first (*preliminary summations*), and the totals obtained become the coefficients for the *final summations*, which may be written simply in the form

$$\rho(xy) = \sum_{-\infty,k}^{\infty} A \cos k\theta_2 - \sum_{-\infty,k}^{\infty} B \sin k\theta_2.$$

The number of terms in the final summation is greatly reduced by this method, but the number of different coefficients, A and B, is, of course, increased. It becomes worth while to make out a list of sine and cosine factors of all the natural numbers likely to be required, at some small interval, say 3° (or 120ths of the cell edge). In the method of Beevers and Lipson[27] strips of numbers

[27] C. A. Beevers and H. Lipson, *Phil. Mag.*, 1934, **17**, 855; *Nature*, 1936, **17**, 825; *Proc. Phys. Soc.* (London), 1936, **48**, 772; C. A. Beevers, *Acta Cryst.*, 1952, **5**, 670.

are printed on cards covering all two-figure coefficients, positive and negative, for sines and cosines, and all values of the index k from 0 to 30. About 13,000 separate cards are then required, which can be conveniently arranged in a special box. The synthesis then reduces to selecting sets of cards and adding up successive columns of already printed figures.

In other methods, three-figure coefficients are employed. This would make the number of cards in the above process excessive, and only one card is employed for each coefficient. Positive and negative numbers, sines and cosines, and variations in the value of the index k are then taken care of by a semi-mechanical sorting device[28] or by selection by means of punched stencil cards laid over the strips.[29] The synthesis again reduces to the mere adding of columns of already printed figures, which can be carried out very quickly on electrical adding machines.

The inverse operation of structure factor calculation (see Section 2) can also be systematised and carried out by similar methods. The Beevers-Lipson strips are not in general suitable for such work, but by a slight modification of the sorting scales or stencils the alternative strip methods described above can easily be adapted for accurate structure factor calculations.[30] Such methods are particularly useful when dealing with a large number of similar atoms, where a single f factor can be used. A graphical method of structure factor calculation has been devised by Bragg and Lipson[31] in which contour maps are plotted over a projection of the unit cell to show the contributions of the atoms, one at a time, to the structure factor. These contributions vary with the indices and the symmetry of the space group. The method is especially useful for space groups of complicated symmetry and for consideration of a small number of atoms.

If the structure has no centre of symmetry, but approximate values for the phase constants, α_{hk}, have been obtained, Fourier synthesis may be effected by the same numerical methods with little extra labour. The double series

[28] J. M. Robertson, *Phil. Mag.*, 1936, **21**, 176.

[29] A. L. Patterson, *Phil. Mag.*, 1936, **22**, 753; A. L. Patterson and G. Tunell, *Amer. Mineral.*, 1942, **27**, 655; J. M. Robertson, *J. Sci. Instruments*, 1948, **25**, 28.

[30] J. M. Robertson, *Nature*, 1936, **138**, 683.

[31] W. L. Bragg, *Nature*, 1936, **138**, 362; W. L. Bragg and H. Lipson, *Z. Krist.*, 1936, **95**, 323.

$$\rho(xy) = \sum_{-\infty}^{\infty} \sum F_{hk} \cos (h\theta_1 + k\theta_2 - \alpha_{hk})$$

may be written in the form

$$\rho(xy) = \sum_{-\infty,k}^{\infty} \left[\sum_{-\infty,h}^{\infty} (F_{hk}' \cos h\theta_1 + F_{hk}'' \sin h\theta_1) \right] \cos k\theta_2$$

$$- \sum_{-\infty,k}^{\infty} \left[\sum_{-\infty,h}^{\infty} (F_{hk}' \sin h\theta_1 + F_{hk}'' \cos h\theta_1) \right] \sin k\theta_2$$

$$\cdots (5.33)$$

where $F_{hk}' = F_{hk} \cos \alpha_{hk}$ and $F_{hk}'' = F_{hk} \sin \alpha_{hk}$. The procedures outlined above can then be applied as before.

When the final summations have been completed, an array of totals representing $\rho(xy)$ at some small interval, such as 60ths or 120ths of the cell side, will be obtained. It is usual to represent the results graphically over the area of the projection by drawing contour lines through points of equal density. If the interval of summation is sufficiently fine, about 0.1 to 0.15 Å, there is no difficulty in doing this accurately by graphical interpolation from the array of summation totals. The positions of the atoms and general shape of the molecule then become clearly outlined on the resulting *electron density map*. Many examples are given in the later chapters of this book.

When the Fourier series is developed for a given crystal, care is needed in noting the way in which the phase constant varies in the different quadrants of the reciprocal lattice, even though the numerical values of the structure factors may be the same. This depends on the symmetry of the particular space group to which the crystal belongs. For example, in $P2_1/c - C_{2h}^5$ the origin of co-ordinates may be taken at a centre of symmetry and then $F(hk0) = F(\bar{h}k0)$ if k is even, but $F(hk0) = -F(\bar{h}k0)$ if k is odd. Similarly, $F(0kl) = F(0k\bar{l})$ when $(k+l)$ is even, but $F(0kl) = -F(0k\bar{l})$ when $(k+l)$ is odd. These relations can be seen to follow from a consideration of the geometrical structure factor (see Section 2). The important step of working out all such relations completely for all the space groups and summing the expression for the electron density over all the quadrants so that only positive values of the indices (hkl) need be considered, has been carried out by K. Lonsdale.[2]

Although the computational methods described above enable the operations of all ordinary double Fourier syntheses to be carried out quite expeditiously, they are as a rule too laborious for triple Fourier summations, except in the most simple examples. To render such operations feasible, it is necessary to make the work still more automatic. This can be done if the information contained on the Beevers-Lipson strips, or some modified version of this information, is transferred to punched cards. The summations can then be carried out, and the results printed automatically, on any of the standard forms of punched-card computing equipment, such as the Hollerith or IBM machines. The earliest applications of this method were made by Comrie,[32] and other systems have been devised by Shaffer, Schomaker, and Pauling,[33] Cox and Jeffrey,[34] and Grems and Kasper.[35] This field is obviously one for professional computers, and for details reference should be made to the papers cited above. The methods possess the outstanding advantages of great accuracy and speed. It should be noted, however, that even when the arrays of printed totals for the function $\rho(xyz)$ are presented by the machines, a great deal of careful work is still required in order to portray the density distribution fully in the form of graphs or sections, from which the atomic positions and other features of the structure may be estimated.

As an alternative to punched-card methods, special digital machines can be devised to handle the computations directly. One machine of this type, based on the use of impulse counters and standard telephone switch-gear, has been described by MacEwan and Beevers.[36]

8B. Analogue computers.—In this second category a large number of devices and machines have been described, such as the simple optical superposition of sinusoidal alternations of

---

[32] L. J. Comrie, private papers of the Scientific Computing Service, Ltd., 23 Bedford Square, London, W.C. 1; L. J. Comrie, G. B. Hey and H. G. Hudson, *J. R. Statist. Soc. Suppl.*, 1937, **4**, 210.

[33] P. A. Shaffer, V. Schomaker, and L. Pauling, *J. Chem. Phys.*, 1946, **14**, 648, 659; J. Donohue and V. Schomaker, *Acta Cryst.*, 1949, **2**, 344.

[34] E. G. Cox and G. A. Jeffrey, *Acta Cryst.*, 1949, **2**, 341; E. G. Cox, L. Gross, and G. A. Jeffrey, *ibid.*, 1949, **2**, 351.

[35] M. D. Grems and J. S. Kasper, *Acta Cryst.*, 1949, **2**, 347.

[36] D. MacEwan and C. A. Beevers, *J. Sci. Instruments*, 1942, **19**, 150.

Fig. 48. General view of XRAC.

light and shade (Bragg,[37] Huggins[38]) or more elegant methods involving real optical interference effects (Bragg[39]).   A machine for adding suitably adjusted electrical potentials has been designed by Hägg and Laurent[40] and one for summing the harmonic terms on a balance by Vand;[41] other mechanical analogue devices have been made by MacLachlan.[42]   A very accurate form of multi-component integrator based on the planimeter principle[43] is also a promising development.   All these arrangements, however, become dwarfed into relative insignificance in comparison with the magnificent electronic analogue computer (XRAC) designed and built by R. Pepinsky[44] and his colleagues.

[37] W. L. Bragg, *Z. Krist.*, 1929, **70,** 475.

[38] M. L. Huggins, *J. Amer. Chem. Soc.*, 1941, **63,** 66; *J. Chem. Phys.*, 1944, **12,** 520; *Nature*, 1945, **155,** 18.

[39] W. L. Bragg, *Nature*, 1939, **143,** 678; 1942, **149,** 470.

[40] G. Hägg and T. Laurent, *J. Sci. Instruments*, 1946, **23,** 155; see also T. B. Rymer and C. C. Butler, *Phil. Mag.*, 1944, **35,** 606.

[41] V. Vand, *J. Sci. Instruments*, 1950, **27,** 257, 261.

[42] D. MacLachlan, Jr., and E. F. Champaygne, *J. Applied Phys.*, 1946, **17,** 1006.

[43] C. A. Beevers and J. M. Robertson, paper at the Computer and Phase Conference, Pennsylvania State College, April, 1950; *Computing Methods and the Phase Problem in X-Ray Crystal Analysis*, State College, Pa., 1952, p. 119.

[44] R. Pepinsky, *J. Applied Phys.*, 1947, **18,** 601.

Fig. 49. Electron density projection of phthalocyanine made by XRAC. (Compare Fig. 108, p. 265.)

A general view of the XRAC computer is shown in Fig. 48, and some of the contoured electron density maps which it produces are given in Figs. 49 and 50.   In principle the machine adds electrical potentials, but these are generated in wave form with controlled amplitude and phase, added, and finally presented in two-dimensions by a television scan on the screen of a cathode ray oscilloscope.   The banks of rheostats for controlling the

107

amplitude and phase of the many hundreds of terms which the machine can handle simultaneously are shown in Fig. 48. Once these terms are set on the rheostat dials the result of the summation is given instantaneously on the oscilloscope over the whole projection surface for any two-dimensional synthesis. One important feature of the machine is that instead of presenting the function by areas of light and shade on the screen, the actual

Fig. 50. Electron density projection of hexamethylbenzene made by XRAC. (Compare Fig. 77, p. 177.)

contour levels are traced out on the surface by a special electronic device, and the atomic positions as well as the whole finer detail of the structure can thereby be determined with great accuracy.

At the present time this machine can handle two-dimensional syntheses of up to 400 terms, but extension to three-dimensional work appears to be a possibility. As the machine virtually abolishes all computational labour in the application of Fourier series methods to crystal analysis, its use is likely to have the most far-reaching effects on the general technique of the subject, although the fundamental problems discussed in Chapter VI still remain.

9. REFINEMENT OF ATOMIC PARAMETERS

Although in a few very important cases discussed in the next chapter the Fourier series method can be applied directly to the solution of crystal structure problems, its most general application lies in the refinement of structures which are already partially known, and for which at least some of the phase constants have been determined. It is therefore necessary to examine the limitations of the method when applied to the precise determination of atomic positions, and also to enquire whether other methods not involving the use of Fourier series, or modifications of the Fourier series method, may not be equally suitable for this purpose.

9A. Resolving power and accuracy.—If all the coefficients and phase constants in the Fourier series could be determined with complete accuracy, and if the measurements could be extended indefinitely to smaller and smaller terms by the use of short wave-lengths, the series when summed would provide a perfect representation of the crystal structure. In practice, however, the measurements are never completely accurate, and the number of terms employed is limited by the experimental conditions, particularly by the wave-length used. These two factors combine to make the representation imperfect in all practical cases.

The most serious imperfections arise if the coefficients in the Fourier series are still large when it is terminated by the experimental conditions. Defects then appear in the form of false detail and regions of negative density which have no physical meaning. Soon after the introduction of the Fourier series method this matter was studied by Havighurst[45] and by Bragg and West.[46] It was realised that the projections obtained by double Fourier series are precisely analogous to the case of image formation by an optical instrument with a circular stop, the defects corresponding to those produced by optical diffraction. If $\theta_{max.}$ is the limiting glancing angle up to which the spectra are measured, the numerical aperture in X-ray analysis must be defined as $2 \sin\theta_{max.}$, and detail cannot be distinguished unless it is on a coarser scale than about $0.6\lambda/2 \sin\theta_{max.}$. This expression defines the *resolving power*, and atoms which are closer together than this cannot be distinguished as separate peaks on the elec-

[45] R. J. Havighurst, *Phys. Rev.*, 1927, **29**, 1.
[46] W. L. Bragg and J. West, *Phil. Mag.*, 1930, **10**, 823.

tron density maps.   For $\theta_{max.} = 90°$ and copper radiation ($\lambda = 1.54$ Å) this distance is about 0.5 Å.   As atoms are always further apart than this, no serious limitation is imposed for three-dimensional work, but in two-dimensional projections of a structure, overlaps to within this distance with lack of resolution are a common occurrence.

This limitation of resolving power does not set a corresponding limit to the *accuracy* with which the positions of atomic centres can be determined.   However, if the series is terminated while the coefficients are still large, each peak will be surrounded by diffraction "ripples," and this will have an effect in displacing the peaks from their true positions.   This source of error, due to the termination of the series after a limited number of terms, and also errors arising from experimentally inaccurate F values, have been studied very fully by Booth[47] in a number of papers. Tests have also been made on purely hypothetical structures by direct computation of the series, with and without various forms of error in the F values.[48]

Purely random errors in the experimental F values have been shown to have a remarkably small effect on the atomic positions. By tests on an average organic structure in which the F values were estimated carefully but independently by different observers using different methods, it has been shown that the positional standard deviations for the atoms due to this source of error alone is well below 0.01 Å.   The errors due to the finite limits of the summation may, however, be more serious.   For a simple system containing two carbon atoms, Booth has calculated that the probable positional error is about 0.02 Å if the series is terminated at $2 \sin \theta_{max.} = 1.5$, and about 0.005 Å if terminated at $2 \sin \theta_{max.} = 2.0$ (for copper radiation).

For the polyatomic case, no general analytical expression for the errors can be obtained, but Booth has suggested a very useful method of ascertaining the errors approximately in any given case.   The co-ordinates obtained from the results of the Fourier synthesis are used to calculate the structure factors, and

[47] A. D. Booth, *Proc. Roy. Soc.* (London), 1946, **A188,** 77, 85; **A190,** 482, 490; *Nature,* 1945, **156,** 51.

[48] J. M. Robertson and J. G. White, *Proc. Roy. Soc.* (London), 1947, **A190,** 329.

these *calculated* structure factors are then employed in a new synthesis, omitting any terms which were omitted in the original synthesis. The co-ordinates derived from this new synthesis will deviate slightly from those originally obtained; these deviations give the errors, with reversed signs, present in the original set of co-ordinates. This method of correction represents a valuable means of additional refinement if there is already good general agreement between the calculated and observed *F* values. It is, of course, difficult to apply in the absence of reliable atomic scattering curves for all the atoms. Trials with this method on average, carefully determined organic structures generally reveal average probable errors of about ±0.02 to ±0.03 Å due to this cause. A very full discussion of methods for correcting for both systematic and random errors in electron density maps, together with the application of statistical significance tests in the comparison of different bond lengths, has been given by Cox and Cruickshank.[49]

The existence of a large temperature factor in a crystal causes a rapid fall off in the intensities for small spacing planes (large θ) and hence ensures rapid convergence in the Fourier series, so reducing the errors due to the finite summation limits. The use of an artificial temperature factor in the form $e^{-B \sin^2\theta}$ has been suggested[46] to ensure convergence, but while this may be of advantage in certain extreme cases, it makes the whole structure more diffuse and so increases the errors due to overlapping. There is a danger that the errors introduced in this way may be larger than those which it is designed to eliminate. It is also clear that by artificially reducing the final coefficients to near zero, a great deal of potentially valuable information is being sacrificed.

9B. Differential synthesis.—The methods of correcting a Fourier synthesis mentioned above imply that the work has already reached a fairly advanced state of refinement, and that the co-ordinates of the atoms are known with some precision. To achieve this state it will generally have been necessary to make repeated application of the Fourier series method, introducing further terms at each stage as it becomes possible to

[49] E. G. Cox and D. W. J. Cruickshank, *Acta Cryst.*, 1948, **1**, 92; D. W. J. Cruickshank, *ibid.*, 1949, **2**, 65.

decide the phase constants of the weaker reflections. This process is most laborious, especially in the case of three-dimensional syntheses, and recently a number of new methods have been devised with the object of proceeding to the optimum atomic positions as quickly as possible, without the necessity of mapping out the entire electron distribution accurately at each stage.

Concentrating attention on the problem of finding the positions of the maxima of the function rather than on working out the general electron distribution, Booth[50] has introduced the method of *differential synthesis*. Adopting a contracted notation with $\theta = 2\pi xh/a + 2\pi yk/b + 2\pi lz/c$, equation 5.24 may be written:

$$\rho(xyz) = \frac{1}{V} \sum_3 |F| \cos (\theta - \alpha). \qquad \cdots (5.34)$$

At each point of maximum density

$$\frac{\partial \rho}{\partial x} = \frac{\partial \rho}{\partial y} = \frac{\partial \rho}{\partial z} = 0, \qquad \cdots (5.35)$$

giving three equations of the form

$$-\frac{2\pi}{aV} \sum_3 h |F| \sin (\theta - \alpha) = 0. \qquad \cdots (5.36)$$

The problem is to find the precise points where these conditions hold, which will be somewhere near the approximate peak positions already known.   Let a true maximum be given by the co-ordinates $(x+\epsilon_x,\ y+\epsilon_y,\ z+\epsilon_z)$ where ϵ_x, ϵ_y, and ϵ_z are small, and let $\epsilon_\theta = 2\pi h\epsilon_x/a + 2\pi k\epsilon_y/b + 2\pi l\epsilon_z/c$. Equation 5.36 then becomes:

$$-\frac{2\pi}{aV} \sum_3 h \{ |F| \sin (\theta - \alpha) \cos \epsilon_\theta + \cos (\theta - \alpha) \sin \epsilon_\theta \} = 0,$$

but as ϵ_θ is small this may be written:

$$-\frac{2\pi}{aV} \sum_3 h \{ |F| \sin (\theta - \alpha) + \epsilon_\theta \cos (\theta - \alpha) \} = 0. \qquad \cdots (5.37)$$

Expanding this equation we obtain:

[50] A. D. Booth, *Trans. Faraday Soc.*, 1946, **42**, 444, 617.

$$-\frac{2\pi}{aV} \sum_3 h|F| \sin(\theta - \alpha) - \frac{4\pi^2}{a^2V} \epsilon_x \sum_3 h^2|F| \cos(\theta - \alpha)$$

$$-\frac{4\pi^2}{abV} \epsilon_y \sum_3 hk|F| \cos(\theta - \alpha)$$

$$-\frac{4\pi^2}{acV} \epsilon_z \sum_3 hl|F| \cos(\theta - \pi) = 0,$$

or

$$\frac{\partial\rho}{\partial x} + \frac{\partial^2\rho}{\partial x^2}\epsilon_x + \frac{\partial^2\rho}{\partial x\partial y}\epsilon_y + \frac{\partial^2\rho}{\partial x\partial z}\epsilon_z = 0.$$

This equation and two similar ones derived from (5.35) provide a means of determining the quantities ϵ_x, ϵ_y, and ϵ_z exactly. In the convenient notation adopted by Booth, where

$$A_h = \frac{\partial\rho}{\partial x} = -\frac{2\pi}{aV} \sum_3 h|F| \sin(\theta - \alpha)$$

$$A_{hk} = \frac{\partial^2\rho}{\partial x\partial y} = -\frac{4\pi^2}{abV} \sum_3 hk|F| \cos(\theta - \alpha), \text{ etc.},$$

these simultaneous equations may be written:

$$\left. \begin{array}{l} A_{hh}\epsilon_x + A_{hk}\epsilon_y + A_{hl}\epsilon_z + A_h = 0, \\ A_{hk}\epsilon_x + A_{kk}\epsilon_y + A_{kl}\epsilon_z + A_k = 0, \\ A_{hl}\epsilon_x + A_{kl}\epsilon_y + A_{ll}\epsilon_z + A_l = 0. \end{array} \right\} \quad \cdots (5.38)$$

To solve these equations involves summing the series in the expression for A_h, A_{hk}, etc., at the original peak locations, and this requires computations similar in form to those used in structure factor calculations (see Section 8). A great simplification can be effected by assuming spherical symmetry in the density distribution in the region of the maxima. For orthogonal systems of reference axes, the coefficients with mixed suffixes, A_{hk}, A_{hl}, etc., are then zero, and we obtain

$$\epsilon_x = -A_h/A_{hh},$$

$$\epsilon_y = -A_k/A_{kk}, \quad \cdots (5.39)$$

$$\epsilon_z = -A_l/A_{ll}.$$

The second differential coefficients may be readily computed for the various atoms. For the same kind of atom it should be found to have a fairly constant value, and if this is so no great error will result from adopting a mean value for the curvature.

The formulas which apply when the reference axes are not orthogonal have been given by Booth,[51] who has also studied the application of differential synthesis to the refinement of the phase angle α at the same time as the atomic co-ordinates. Recent very accurate structural investigations utilizing the method have been made by Cox, Gillot, and Jeffrey.[52]

9C. Method of least squares.—The earliest attempt to supplement or even dispense with the method of Fourier synthesis in the final stages of crystal analysis was made by Hughes,[53] who introduced the *least squares* refinement of atomic parameters. In this method the observed and calculated structure factors for each reflection, which we shall denote by F_o and F_c respectively, are compared, and the object is to find the most probable values for the atomic parameters, i.e., those values which will make the sum of the squares of the discrepancies, $\sum(|F_o| - |F_c|)^2$, a minimum. Each structure factor supplies one observational equation for a least squares solution, but as these equations are trigonometric they are not, of course, immediately suitable for least squares treatment. But if a reasonably good approximation to the structure exists, linear equations can be developed.

As before, we suppose the true atomic positions to be given by the co-ordinates $(x+\epsilon_x, y+\epsilon_y, z+\epsilon_z)$, and let $\theta = 2\pi(hx/a+ky/b+lz/c)$ and $\epsilon_\theta = 2\pi(h\epsilon_x/a+k\epsilon_y/b+l\epsilon_z/c)$. Then

$$F_o = \sum^{N} f \cos(\theta + \epsilon_\theta), \qquad \cdots (5.40)$$

$$F_c = \sum^{N} f \cos\theta, \qquad \cdots (5.41)$$

using (5.5) for the centrosymmetrical case, the summations being taken over all the N atomic positions. Hence

$$F_o - F_c = \sum^{N} f\{\cos\theta(1 - \cos\epsilon_\theta) - \sin\theta \sin\epsilon_\theta\},$$

[51] A. D. Booth, *Fourier Technique in X-Ray Organic Structure Analysis*, Cambridge: Cambridge University Press, 1948.

[52] E. G. Cox, R. J. J. H. Gillot, and G. A. Jeffrey, *Acta Cryst.*, 1949, **2**, 356.

[53] E. W. Hughes, *J. Amer. Chem. Soc.*, 1941, **63**, 1737.

or, as ϵ_θ is small,

$$F_o - F_c = - \sum_{}^{N} f\epsilon_\theta \sin \theta$$

$$= \frac{\partial F_c}{\partial \theta} \epsilon_\theta, \qquad \cdots (5.42)$$

and in this way linear equations can be formed.

Hughes gives the equation

$$\sqrt{w} \sum_{}^{3N} \frac{\partial F}{\partial x} \epsilon_x = \sqrt{w} (F_o - F_c) \qquad \cdots (5.43)$$

where the summation is taken over all the co-ordinates of all the atoms, and w is a weighting factor depending on the reliability of the F value. The number of equations, one for each structure factor, will usually greatly exceed the number of unknowns. By the usual least squares reduction $3N$ normal equations for the $3N$ unknowns are obtained, and these can be solved by ordinary methods.

9D. Method of steepest descents.—More generalised methods for minimizing the function

$$R = \sum (|F_o| - |F_c|)^2 \qquad \cdots (5.44)$$

or certain other functions such as

$$R' = \sum (F_o{}^2 - F_c{}^2)^2 \qquad \cdots (5.45)$$

have been studied by Booth,[54] Vand,[55] and Qurashi.[56] The figure of merit represented by the function R is everywhere positive and as a function of the $3N$ atomic parameters (x_i, y_i, z_i) represents some surface in the $3N$ dimensional space defined by these co-ordinates. The object of the treatment is to bring the figure of merit to a minimum, and when this has been done the solution corresponds to the least squares solution. The steepest descent method means going "downhill" in the $3N$ dimensional space. Successive approximations are obtained by moving from a point on the surface specified by R along the normal to the R contours

[54] A. D. Booth, *Nature*, 1947, **160**, 196; 1948. **161**, 765; *Proc. Roy. Soc.* (London), 1949, **197**, 336.

[55] V. Vand, *Nature*, 1948, **161**, 600; 1949, **163**, 129; *Acta Cryst.*, 1951, **4**, 285.

[56] M. M. Qurashi, *Acta Cryst.*, 1949, **2**, 404.

to another surface of smaller R.   The shifts ϵ_{x_i}, etc., have to be proportional to the components of grad R:

$$\epsilon_{x_i} = K\frac{\partial R}{\partial x_i} = -2K\sum_{hkl}(|F_o| - |F_c|)\frac{\partial F_c}{\partial x_i}. \qquad \cdots (5.46)$$

Qurashi has examined the convergence of the method, which is shown to depend on the scales of representation chosen for the parameters.   By transformation from elliptical to circular contours he derives a formula,

$$\epsilon_{x_i} = \sum_{hkl}(F_o - F_c)\frac{\partial F_c}{\partial x_i} \bigg/ \sum_{hkl}\left(\frac{\partial F_c}{\partial x_i}\right)^2. \qquad \cdots (5.47)$$

It can be shown that this formula, which, with the inclusion of a weighting factor, may be written

$$\sum_{hkl} w\left(\frac{\partial F}{\partial x_i}\right)^2 \epsilon_{x_i} = \sum_{hkl} w(F_o - F_c)\frac{\partial F_c}{\partial x_i},$$

is equivalent to the solution obtained from the normal equations in the least squares method when these are symmetrical, only the diagonal terms of the determinant being employed.

9E. $(F_o - F_c)$ synthesis.—Results which are equivalent to those obtained by the steepest descents method can be achieved by means of ordinary Fourier synthesis if the residuals $(F_o - F_c)$ are employed as coefficients.   This *difference synthesis*, which for the centrosymmetrical case may be formulated as

$$D = \sum_3 (F_o - F_c) \cos 2\pi(hx/a + ky/b + lz/c), \qquad \cdots (5.48)$$

has been employed by several investigators[57] and studied theoretically by Booth and Cochran.[58]   If the calculated co-ordinates (x_c, y_c, z_c) are marked on the map obtained for D, the directions of steepest *ascent* at these points give the directions of the shifts

[57] G. W. Brindley and R. G. Wood, *Phil. Mag.*, 1929, **7**, 616, 619; D. Crowfoot, C. W. Bunn, B. W. Rogers-Low, and A. Turner-Jones, *The Chemistry of Penicillin*, Princeton: Princeton University Press. 1949, pp. 310, 334; C. Finbak and N. Norman, *Acta Chem. Scand.*, 1948, **2**, 813; and many later applications.
[58] A. D. Booth, *Nature*, 1948, **161**, 765; W. Cochran, *Acta Cryst.*, 1951, **4**, 408.

obtained by the steepest descents procedure, and the gradients are proportional to the magnitudes of the shifts.

This Fourier method, which can be computed by standard techniques, has many advantages. In the foregoing methods it has been generally assumed that the electron distribution can be accounted for entirely in terms of atomic positions which are approximately known, and this is very often not the case. The difference synthesis gives a complete picture of the cause of the discrepancies between F_o and F_c; neglected hydrogen atoms appear as well-defined peaks, and lack of spherical symmetry in the form of characteristic saddle points. This additional information, which is obtained during the process of refinement of the atomic parameters, is most valuable. Another valuable aspect of the method[59] is that series-termination effects are eliminated. Again, by leaving out strong terms, the observed values of which may be influenced by extinction, the disturbing effects of this phenomenon may also be excluded.

9F. Advantages and limitations.—Finally, it must be emphasised that the various methods which we have outlined above require for their success that a good approximation to the structure has already been obtained; that the quantities ϵ_x, ϵ_y, and ϵ_z really are small. In general, this first stage has to be carried out by ordinary Fourier methods. For the later stages of refinement, however, these other methods possess certain advantages.

In Fourier synthesis it is essential that all possible observed F values be included. The "termination of series" errors which arise owing to the omission of certain terms by wave-length or other experimental limitations have already been discussed. In the least squares treatment the omission of certain data does not cause the same difficulty, and the weights attached to the different F values may be varied.

Another advantage arises particularly in two-dimensional work, where the Fourier series method is severely limited by lack of resolution. The same limits do not apply in the case of the least squares refinement. Although with experience and certain subsidiary calculations the positions of unresolved peaks can be estimated fairly well in normal electron density projections, the

[59] R. Pepinsky, private communication.

least squares refinement procedure offers a more definite approach to a solution in terms of the known or assumed electron distributions incorporated in the f factors.

Although the refinement procedures mentioned possess various advantages, care is needed in discussing the merit of the structure finally deduced in such a way. The methods have certain natural limitations which are not always fully appreciated. The objective set is generally that of minimizing the quantity $\sum(|F_o| - |F_c|)$. As a result a better account of the experimental observations may be achieved in the end, *but only in terms of some preconceived structure* which generally contains atoms that are assumed to be spherically symmetrical, and, for any given element, to possess exactly equal scattering power. The particular values of the parameters which give optimum agreement with such a model are not necessarily those which are intrinsically most correct.

In the actual structure the atoms will not be exactly spherical, owing to directed valency bonds and anisotropic thermal motions. Nor will they necessarily have exactly equal scattering power, even though they are chemically identical. It is generally found that atoms lying far from the centre of the molecule, or from some central ion, appear to have a more diffuse electron distribution with lower peak values for the density than atoms near the centre. This effect is probably also a thermal one, the outer atoms executing somewhat larger vibrations than those near the centre.

A careful approach to the structure by Fourier methods and successive approximations will reveal those various features, and enable the true positions of the atoms to be estimated. Provided that any necessary finite series corrections have been made, the structure obtained in this way is likely to be accurate. It may not, however, give an extremely low value for $\sum(|F_o| - |F_c|)$ unless the F_c values take account of the hydrogen atoms and the various anisotropic factors mentioned above.

VI

The Phase Problem
and Methods of Solution

1. THE PHASE PROBLEM

The discussions in the previous chapter on the use of Fourier series and allied methods all start from the assumption that the arrangement of the atoms in the crystal is at least approximately known from some external evidence. In this chapter we are now concerned with the more fundamental problem of whether it is possible to determine a crystal structure from intensity data alone, or, if not, what is the minimum of other evidence necessary. Logically, this problem should have been discussed earlier, but it was necessary to present the Fourier series development first, because it is fundamental to the subject.

It will be clear from the earlier discussion that in general an infinite number of different electron distributions may be derived from the measured structure amplitudes in a given crystal. These are obtained by assigning various arbitrary values to the unknown phase constant associated with each amplitude and then summing the resulting Fourier series for the electron distribution. Each distribution obtained in this way will give a perfectly exact account of all the measured amplitudes. Obviously, therefore, there is no unique *mathematical* solution to the general problem, unless conditions are attached to the density function.

The question of whether there is a unique *physical* solution is different, and it is not an easy question to answer. Many of the electron distributions obtained by varying the unmeasured phase constants will contain regions of negative density, and these can be eliminated as not corresponding to any physical reality. The use of such a positivity criterion to limit the possible phase relations in an unknown structure is a mathematical problem that has recently received considerable attention. It has been shown by Harker and Kasper that the result of this condition is to establish certain sets of inequality relations involving the structure factors. Some of these inequalities are more powerful than others, and can be used to limit the possible values of the unknown phases. In the case of crystals where the number of atoms is not too great, this method can be of great value in helping to solve the structure in the initial stages, and so bringing it to a point where the Fourier refinement method can be applied. Unfortunately, the power of the inequality relations appears to diminish as the complexity of the crystal structure increases.

The same positivity criterion can be used in a very direct and practical manner on R. Pepinsky's electronic computer, XRAC, which presents an instantaneous summation of double Fourier series (see Chapter V, Section 8B). The effect of varying the phase constants can be observed directly on the screen, which presents $\rho(xy)$ as a contoured map, and the effect of false values in the phases is generally to produce regions of spurious negative density. It is, of course, possible to test systematically only a relatively small number of possible phase combinations, but the method has great possibilities as a means of refinement and successive approximation in structure determination.

A much more powerful physical criterion is available, however, if it is accepted that the electron density function is composed of a number of spherically symmetric atomic functions. If in addition the chemical criterion is utilized, that the unit cell of the structure should contain the number and kind of atoms known to be present from chemical analysis and density measurement, then it should be possible at least to formulate a direct algebraic solution of the problem. This aspect has also received considerable attention,[1] but, unfortunately, for structures involving more than a few atoms, the problem generally resolves itself into the solution of equations of very high order, and the method has not often been attempted in actual practice. Pos-

[1] H. Ott, *Z. Krist.*, 1927, **66**, 136; H. Seyfarth, *ibid.*, 1927, **67**, 131, 422, 595; M. Avrami, *Phys. Rev.*, 1938, **54**, 300; *Z. Krist.*, 1939, **100**, 381.

sibly a more hopeful approach lies in the design of special equa-
tion-solving machines.[2]

When the structure is expressed in terms of atoms, of course,
the methods of trial and error, mentioned in the previous chapters, immediately present themselves, and these can usually be
reinforced by the utilization of further known chemical facts,
such as the presence of a benzene ring or similar structural
feature in the molecule. The general effect of incorporating
chemical information is to reduce the number of parameters in the
problem from three per atom to a much smaller number. If a
reasonably exact model based on the chemical structure can be
derived, the problem of finding its position and orientation in the
unit cell of the crystal can sometimes be facilitated by the method
of Fourier transforms. An important extension of this method,
which employs optical diffraction (the "diffraction spectrometer")
to facilitate the evaluation of the Fourier transforms, has recently been developed by Lipson and his co-workers.[3]

If we assume complete ignorance of the phase relationships
and confine ourselves to the measured structure amplitudes
without reference to any physical assumption, the actual information about a structure which is contained in the X-ray patterns
is most clearly expressed by a method first developed by A. L.
Patterson in 1934. The result of this treatment shows that from
the measured structure amplitudes alone, introduced as the
quantities F^2, it is possible to produce a *vector representation* of
the crystal structure. Instead of obtaining the position of each
scattering centre, we now obtain a superposition of all the vector
distances connecting these centres. Again, if the existence of
atoms is assumed, it appears that the problem should be soluble.

[2] J. M. Robertson, *Phil. Mag.*, 1932, **13**, 413; paper at the Computer
and Phase Conference, Pennsylvania State College, April, 1950; *Computing
Methods and the Phase Problem in X-Ray Crystal Analysis*, State College,
Pa., 1952, p. 98; H. Hauptman and J. Karle, *Acta Cryst.*, 1950, **3**, 478.
    [3] A. Hettich, *Z. Krist.*, 1935, **90**, 483; P. P. Ewald, *ibid.*, 1935, **90**,
493; G. Knott, *Proc. Phys. Soc.* (London), 1940, **52**, 229; D. Wrinch,
Fourier Transforms and Structure Factors, Am. Soc. X-Ray and Electron
Diffraction Monograph, no. 2, Cambridge, Mass., 1946; A. Klug, *Acta
Cryst.*, 1950, **3**, 176; T. H. Goodwin, to be published; W. L. Bragg, *Nature*,
1939, **143**, 678; 1944, **154**, 69; C. A. Taylor, R. M. Hinde, and H. Lipson,
Acta Cryst., 1951, **4**, 261; H. Lipson and C. A. Taylor, *ibid.*, 1951, **4**, 458;
A. W. Hanson and H. Lipson, *ibid.*, 1952, **5**, 362.

But if the structure contains any large number of atoms, the interpretation of such vector maps presents problems of great difficulty.

One question of great general interest still remains. If an exact solution in terms of spherically symmetric atomic functions is found, is this solution necessarily unique? The existence of the Patterson ambiguities, described in a later paragraph, shows that in many cases there is in fact more than one solution, even in terms of atomic positions. At first sight this is a disturbing fact and may seem to throw doubt on the validity of many structure determinations. It emphasises the importance of making a final appeal to something more than the basic minimum of physical assumption, such as the mere existence of atoms. The final test, that the solution obtained should be chemically reasonable, must not be overlooked.

So far in this discussion we have assumed that it is not possible to obtain any direct information about the actual phase relationships of the different X-ray intensities. There are, however, many ways in which some direct knowledge of these phase relations may be gained, either partially or completely. Such methods, although not perhaps of universal application, are obviously by far the most important and powerful. Most of the outstanding achievements in the structure analysis of complex organic compounds in recent years have been made possible by their application in some form or other.

In general these methods are partly chemical in nature, because they depend upon the introduction at some point in the structure of an atom or a succession of atoms of known scattering power. The remainder of the structure is unknown. The unknown part can be represented by a wave contribution of unknown phase but measurable amplitude for each X-ray reflection. When the known contribution is added, the resultant amplitude depends upon and gives a measure of the unknown phase. Although this method may become difficult to apply in the case of extremely large molecules, its range of application is very much greater than that of most other methods, and the final limitation is often accuracy of measurement rather than mathematical complication.

In the following sections it is not possible to present a full account of all recent developments in these various aspects of the phase problem, but some of the more important topics are discussed.

2. RELATIONS BETWEEN STRUCTURE FACTORS

To describe the relations between structure factors discovered by Harker and Kasper[4] it is convenient to use a simplified notation, writing F_H for $F(hkl)$, F_{2H} for $F(2h, 2k, 2l)$, etc., and, as before, θ for $2\pi(hx/a+ky/b+lz/c)$. We shall be considering the geometrical structure factor (expressions 5.2 to 5.6), and in this connection an average atomic scattering factor may be employed without much loss in accuracy, provided that the atoms do not differ too widely in atomic number. It is useful to divide this average factor by the number of electrons in the atom, z, and so obtain a "unitary atomic scattering factor," $\hat{f} = f_j/z_j$. $\hat{f}$, like f is a function of $\sin \theta/\lambda$ only (see Chapter V, Section 3). The structure factor in expression 5.2 can then be written:

$$F_H = \hat{f} \sum_1^N z_j e^{i\theta j}. \qquad \cdots (6.1)$$

A "unitary structure factor" for the whole unit cell may be defined similarly, dividing by the total number of electrons in the unit cell, Z, and also by the unitary scattering factor, $\hat{f}$:

$$\hat{F}_H = F_H/Z\hat{f} = \sum_1^N \frac{z_j}{Z} e^{i\theta j} = \sum_1^N n_j e^{i\theta j}. \qquad \cdots (6.2)$$

It will be noted that the factor n_j is simply the fraction of the total electrons present on any particular atom, j. The effect of this division of the structure factors by an averaged atomic scattering factor is to produce a purely geometrical structure factor, in which the atoms have been concentrated to point-scattering sources, each with its correct relative share of the electrons. If the atoms are all of one kind and exactly spherically symmetrical, the reduction can be made exactly. In general, however, it is a good approximation and has long been used to facilitate the computation of structure factors in large organic molecules containing atoms like carbon, oxygen, and nitrogen.

To derive the inequalities of Harker and Kasper, use is made of Cauchy's Inequality, which states that

$$\left| \sum_1^N a_j b_j \right|^2 \leq \left(\sum_1^N |a_j|^2 \right) \left(\sum_1^N |b_j|^2 \right) \qquad \cdots (6.3)$$

---

[4] D. Harker and J. S. Kasper, *Acta Cryst.*, 1948, **1**, 70.

where a_j and b_j are any real or complex numbers.

If we apply this expression to (6.2), using $(\sqrt{a_j})$ $(\sqrt{a_jb_j})$ as factors of a_jb_j, we obtain

$$|\widehat{F}_H|^2 \leq \left(\sum_1^N n_j \right)\left(\sum_1^N n_j \big| e^{i\theta_j} \big|^2 \right),$$

but as $\displaystyle\sum_1^N n_j = 1,$

$$\widehat{F}_H \leq 1. \qquad \cdots (6.4)$$

This applies to the most general case where the crystal has no symmetry whatever, and merely defines the unitary structure factor. But as soon as symmetry conditions are introduced, more restrictive relations are found to apply. With a centre of symmetry present the structure factor can be written to correspond with equation 5.5:

$$\widehat{F}_H = \sum_1^N n_j \cos \theta_j, \qquad \cdots (6.5)$$

and application of Cauchy's Inequality gives

$$|\widehat{F}_H|^2 \leq \left[\sum_1^N n_j \right]\left[\sum_1^N n_j (\cos^2 \theta_j) \right]$$

$$\leq \sum_1^N n_j (\tfrac{1}{2} + \tfrac{1}{2} \cos 2\theta_j),$$

and hence

$$\widehat{F}_H^2 \leq \tfrac{1}{2} + \tfrac{1}{2}\widehat{F}_{2H}. \qquad \cdots (6.6)$$

This is the first inequality derived by Harker and Kasper, and it has an immediate application in determining some of the unknown phase constants, in this case the signs of the F values. For example, if $\widehat{F}_H^2$ is found to be greater than $\tfrac{1}{2}$, then $\widehat{F}_{2H}$ must be positive; more generally, $\widehat{F}_{2H}$ is positive if it is greater than $(1 - 2\widehat{F}_H^2)$.

A large number of other inequalities have been derived by Harker and Kasper and by Gillis[5] for the symmetry elements and combinations of symmetry elements in the different space groups.

[5] J. Gillis, *Nature*, 1947, **160**, 866; *Acta Cryst.*, 1948, **1**, 76, 174.

Relations involving odd indices can be introduced by the addition or subtraction of terms, as in the following examples:

$$(\widehat{F}_H + \widehat{F}_{H'})^2 \leqq (1 + \widehat{F}_{H+H'})(1 + \widehat{F}_{H-H'}), \qquad \cdots (6.7)$$

$$(\widehat{F}_H - \widehat{F}_{H'})^2 \leqq (1 - \widehat{F}_{H+H'})(1 - \widehat{F}_{H-H'}). \qquad \cdots (6.8)$$

For details of the various relations, some of which are very complicated, reference must be made to the original papers. The relations have been applied in practice by Gillis,[5] who has succeeded in determining the signs of about 40 structure factors for the oxalic acid dihydrate crystal, and by Kasper, Lucht, and Harker,[6] who have used them to solve the structure of decaborane.

If a crystal provides a number of high $\widehat{F}$ values, it is clear that these relations supply a powerful tool that can aid in solving the structure. Unfortunately, as the crystal structure increases in complexity, the average value of $|\widehat{F}|$ must diminish; indeed, as has been shown by Hughes[7] and Wilson,[8] the root mean square average of $|\widehat{F}|$ is $1/\sqrt{N}$, where N is the number of atoms per cell.

It has been pointed out by Hughes[7] and also by MacGillavry[9] that the constant factor in the inequality relation is always equal to the reciprocal of the symmetry number of the space group under consideration. The more simple relations may be expressed generally in the form:

$$\widehat{F}_H^2 \leqq \frac{1}{p} \left[1 + \phi(\widehat{F}) \right] \qquad \cdots (6.9)$$

where the crystal has symmetry which causes the general position to be p-fold, and $\phi(\widehat{F})$ is a function of the $\widehat{F}$'s related to $\widehat{F}_H$. As the symmetry increases, the constant term in the inequality diminishes, but the number of atoms in the unit cell increases. The power of the inequality, in relation to the number of atoms in the *asymmetric unit*, will be about the same for all symmetries. For complicated crystal structures with large numbers of atoms in the asymmetric unit, the average value of the structure factor is small, and it will only be possible to make a few phase determinations by means of inequality relations. Unfortunately it is

[5] J. S. Kasper, C. M. Lucht, and D. Harker, *Acta Cryst.*, 1950, **3**, 436.
[7] E. W. Hughes, *Acta Cryst.*, 1949, **2**, 34, 37.
[8] A. J. C. Wilson, *Acta Cryst.*, 1949, **2**, 318.
[9] C. H. MacGillavry, *Acta Cryst.*, 1950, **3**, 214.

just in the case of such complicated structures that a large number of phase determinations is necessary to be of any real value in solving the problem.

It should be mentioned that precisely the same inequality relations can be derived for the structure factors themselves, without dividing by the atomic scattering factors, f. In this case the generalised expression for the structure factor given in (5.19) is employed, and this can be substituted in the relation known as Schwarz's Inequality:

$$\left| \int fg\, dr \right|^2 \leq \left(\int |f|^2 dr \right) \left(\int |g|^2 dr \right) \qquad \cdots (6.10)$$

when the same general results are obtained. It is, however, more useful to employ the stronger inequalities obtained by using $\widehat{F}$ values, which do not diminish with increasing θ.

In an analysis of great generality, Karle and Hauptman[10] (see also Goedkoop[11] and Pepinsky and MacGillavry[12]) have shown that symmetry considerations are not basic to the development of the theory of these inequality relations. The fundamental condition is that the distribution function, $\rho(xyz)$, is everywhere positive, and this condition alone imposes restrictions on the phases and magnitudes of the structure factors. These very general relations are then developed in determinant form, and they include the Harker-Kasper relations as special cases.

If instead of merely assuming the positivity of the function $\rho(xyz)$, it is known that the distribution can be expressed in terms of discrete atoms, it should be possible to develop stronger relations. Goedkoop[11] and Hughes[7] have shown how in this case some of the inequality relations then reduce to equalities, involving rigorous equations between the F's. Such equations were developed much earlier for a special case by Banerjee.[13] Again, however, the application of this method is limited if the crystal structure is complex.

Further relationships between the signs of the structure factors in centrosymmetrical crystals have recently been discussed by

[10] J. Karle and H. Hauptman, *Acta Cryst.*, 1950, **3**, 181.
[11] J. A. Goedkoop, *Acta Cryst.*, 1950, **3**, 374.
[12] R. Pepinsky and C. H. MacGillavry, *Acta Cryst.*, 1951, **4**, 284.
[13] K. Banerjee, *Proc. Roy. Soc.* (London), 1933, **A141**, 188.

Sayre,[14] Cochran,[15] and Zachariasen.[16]   These discussions are
directed particularly to the case of more complex crystals where
the Harker-Kasper inequalities are inadequate.   Having deduced
the signs of as many of the larger structure factors as possible by
the usual relationships, these authors show that the list can be
extended by means of certain new statistical relationships which
are established between the signs.

For example, if $\widehat{F}_H$, $\widehat{F}_K$, and $\widehat{F}_{H+K}$ are "large" structure factors
(greater than about 1.5 times the root mean square value), it is
shown that the following relation between the signs (S) is probably, but not necessarily, correct:

$$S_{H+K} = S_H S_K.$$

Multiplying both sides by S_{K_i} and averaging, a statistical equivalent is obtained:

$$S_H = \overline{S(S_{K_i} S_{H+K_i})}$$

where S_H applies to a particular structure factor, and S_{K_i} and
S_{H+K_i} apply to a number of pairs of structure factors.   In this
manner groups of structure factors can be employed, and the
general principle of the method is to attach signs to all the larger
structure factors so that self-consistency with the above relation
is obtained.

This method has been extensively developed by Zachariasen[16]
and applied with outstanding success to solving the structure of
the monoclinic form of metaboric acid, which contains 27 parameters for the boron and oxygen atoms alone.   The method appears to give promise of success with even more complicated
structures.

3. VECTOR REPRESENTATION OF CRYSTAL STRUCTURE

The vector representation of crystal structure, discovered by
A. L. Patterson[17] in 1934, constitutes one of the most powerful
of modern methods of crystal analysis.   Patterson's original
result was derived as an extension to crystals of the theory of
scattering of X-rays in liquids (Warren and Gingrich[18]), but was

[14] D. Sayre, *Acta Cryst.*, 1952, **5**, 60.
[15] W. Cochran, *Acta Cryst.*, 1952, **5**, 65.
[16] W. H. Zachariasen, *Acta Cryst.*, 1952, **5**, 68.
[17] A. L. Patterson, *Phys. Rev.*, 1934, 46, 372; *Z. Krist.*, 1935, **90**, 517.
[18] B. E. Warren and N. S. Gingrich, *Phys. Rev.*, 1934. **46**, 368.

rapidly developed into an entirely new direct method of portraying crystal structures. In this approach we no longer attempt to determine the phases of the structure factors, but confine ourselves solely to a study of the information that can be obtained by utilizing the magnitudes of these factors, which are introduced as the quantities F^2.

3A. The F² series of Patterson.—In Patterson's treatment of the problem a quantity $A(uvw)$, which may be called the weighted average distribution of density of scattering matter about a point (xyz) in the crystal, is defined by the equation

$$A(uvw) = \frac{1}{V} \int_0^a \int_0^b \int_0^c \rho(xyz)\rho(x + u, y + v,$$

$$z + w)dxdydz. \quad \cdots (6.11)$$

In this expression, $\rho(x+u, y+v, z+w)$ gives the distribution about (xyz) as a function of the parameters u, v, w, and it represents a distribution similar to $\rho(xyz)$ but displaced from the point (xyz) through a distance whose components are (uvw). This distribution function is weighted by the amount of scattering matter in the volume element at (xyz), viz., $\rho(xyz)dxdydz$. If the two functions $\rho(xyz)$ and $\rho(x+u, y+v, z+w)$ are now expanded by expressing them in terms of the corresponding Fourier series (as in expression 5.20), and the integration effected, we obtain

$$A(uvw) = \frac{1}{V^2} \sum \sum_{-\infty}^{\infty} \sum F^2(hkl)e^{-2\pi i(hu/a+kv/b+lw/c)}. \quad \cdots (6.12)$$

The function $A(uvw)$ can thus be calculated in terms of the *squares* of the structure factors, i.e., in terms of quantities which can be measured directly, and which are simply related to the actual X-ray intensities by equation 5.17.

The meaning of the function $A(uvw)$ in crystal analysis becomes apparent from equation 6.11, which shows that it can only attain a large value when both $\rho(xyz)$ and $\rho(x+u, y+v, z+w)$ are large. This will occur if, for example, there are atoms both at (xyz) and at $(x+u, y+v, z+w)$, separated by the vector distance (uvw). Thus a peak in the function $A(uvw)$ at $(u_1v_1w_1)$ corresponds to an interatomic distance in the crystal defined by a vector whose components are u_1, v_1, w_1. We thus see that the picture of the crystal structure which can be obtained directly from the measured X-ray intensities is not a picture of the actual atomic positions, but a *vector map*, that is, a picture showing in superposition all the interatomic vectors of the crystal. Expressions similar to (6.12) may, of course, be written in terms of double and single Fourier series, to represent the vector distribution function in projection, or averaged along a line.

This expression of the results might appear to solve the structural problem, at least as far as finding the distribution of well-separated and distinct atoms is concerned. But the difficulties of the method are considerable. In particular, for an array of n atoms, there may be a very large number of possible vectors to consider; actually $n(n-1)/2$ which are essentially distinct, as is shown later. In the general case, each of these will give rise to a separate peak in the distribution function, but the chance of resolving so many peaks by the Fourier series method is extremely meagre, except for the most simple structures (see Chapter V, Section 9A). There will usually be many chance concurrences, and large groups of peaks will tend to coalesce.

Methods of artificially improving the resolution afforded by this synthesis have been studied by Patterson,[19] Yü,[20] and others, but they do not overcome all the difficulties. We have seen in a previous section (p. 123) that dividing the F values by an averaged atomic scattering factor has the effect of reducing the atoms to point scattering sources. If the coefficients are treated in this way, there will be a sharpening of the resolution, but as the series also ceases to be convergent, the results are disappointing. Some intermediate treatments, however, can have beneficial results.

It is clear that the use of three-dimensional syntheses, with their higher powers of resolution, are of the greatest importance in the Patterson method and enable the maximum amount of information to be extracted from the X-ray data. With the improved computational methods described in the last chapter, it becomes easier to effect such syntheses, and several very successful applications of the method to complicated structures have recently been made.[21]

[19] A. L. Patterson, *Z. Krist.*, 1935, **90**, 517.
[20] S. H. Yü, *Nature*, 1942, **149**, 638.
[21] D. P. Shoemaker, J. Donohue, V. Schomaker, and R. B. Corey, *J. Amer. Chem. Soc.*, 1950, **72**, 2328; T. R. R. McDonald and C. A. Beevers, *Acta Cryst.*, 1950, **3**, 394.

3B. The Harker synthesis.—Some important modifications of the Patterson method have been made by Harker[22] in the use of three-dimensional series simplified by the symmetry properties of the crystal and applied to the detection of certain special vectors. For example, if the crystal has a 2-fold axis of symmetry parallel to b, an atom at (xyz) is accompanied by one at the equivalent point $(-x, y, -z)$ and the vector between these atoms has the components $(2x, 0, 2z)$. The function $A(uvw)$, therefore, has a maximum at $(2u, 0, 2w)$ corresponding to this interatomic distance. To find maxima of this kind it is only necessary to evaluate $A(uvw)$ in the plane $v=0$. The series 6.12 may therefore be written, neglecting absolute values,

$$A(u0w) = \sum_{-\infty,h}^{\infty} \sum_{-\infty,k}^{\infty} \sum_{-\infty,l}^{\infty} F^2(hkl) \cos 2\pi(hu/a + lw/c)$$

$$= \sum_{-\infty,h}^{\infty} \sum_{-\infty,l}^{\infty} \left\{ \sum_{-\infty,k}^{\infty} F^2(hkl) \right\} \cos 2\pi(hu/a + lw/c).$$

$$\cdots (6.13)$$

This is effectively only a two-dimensional summation, and the numerical work is not formidable. There is an excellent chance of resolving the required vector peaks, because all the measured values of $F(hkl)$ are used, and because interatomic vectors not parallel to the plane $v=0$ are eliminated.

A plane of symmetry gives an even simpler result. If the plane is perpendicular to b, the equivalent points are (xyz) and $(x, -y, z)$ with vector components $(0, 2y, 0)$. The required maximum in $A(uvw)$ lies on the b axis, and the series becomes one-dimensional:

$$A(0v0) = \sum_{-\infty,k}^{\infty} \left\{ \sum_{-\infty,h}^{\infty} \sum_{-\infty,l}^{\infty} F^2(hkl) \right\} \cos 2\pi(kv/b). \qquad \cdots (6.14)$$

Other simplifications can be derived to take advantage of the different types of symmetry which may be present. Such methods have yielded many most useful results in crystal analysis. One difficulty that should be mentioned, however, lies in the chance concurrence of one or two co-ordinates of certain atoms not actually related by the symmetry elements.

A practical example of the special Harker vectors between symmetrically related atoms is shown later for a simple two-

[22] D. Harker, *J. Chem. Phys.*, 1936, **4**, 381.

dimensional case (Figs. 54–56, p. 137). The space group in this case is $P2_1/c$, and the projections are along the a axis. The trace of the glide planes c are shown by broken lines at $\frac{1}{4}b$ and $\frac{3}{4}b$ in Fig. 54. Equivalent points in this projection are

(1) y, z (2) $-y, -z$

(3) $\frac{1}{2} + y, \frac{1}{2} - z$ (4) $\frac{1}{2} - y, \frac{1}{2} + z,$

and the following special vectors arise:

(1)(3) $\frac{1}{2}, \frac{1}{2} - 2z$ (1)(4) $\frac{1}{2} - 2y, \frac{1}{2}$

(2)(4) $\frac{1}{2}, \frac{1}{2} + 2z$ (2)(3) $\frac{1}{2} + 2y, \frac{1}{2}.$

There are eight atoms in the unit cell, or two per asymmetric unit, and so there should be eight special vectors of the above type. This is the result shown in Figs. 55 and 56, when four peaks are seen to lie on each of the dotted lines at $y = \frac{1}{2}$ and $z = \frac{1}{2}$. In applying the Harker method, the series would be summed along these lines only, with the knowledge that these particular vector peaks would be found to lie on them.

A generalisation of the Harker synthesis which has the property of indicating possible positions for the atoms in a crystal structure (still, however, subject to certain ambiguities) has been formulated by M. J. Buerger [23] under the title of implication theory. This leads to various relations which are of assistance in the general interpretation of Patterson syntheses. Another aspect of the theory[24] leads to relations between the F's and F^2's, and so to the possibility of phase determination as in the case of the Harker-Kasper inequality relations (see section 2).

4. INTERPRETATION OF VECTOR DISTRIBUTIONS

In this section we consider first simple vector distributions obtained from atoms regarded as point scattering sources. The formation and interpretation of the sets of vectors derived from isolated arrays of atoms, assuming unlimited resolution, presents relatively simple and attractive problems. It is then easy to extend this treatment to cover periodic distributions such as occur in crystals, and to consider briefly the effect of symmetry on the distribution.

[23] M. J. Buerger, *J. Applied Phys.*, 1946, **17**, 579; *Phys. Rev.*, 1948, **73**, 927; *Proc. Indian Acad. Sci.*, 1948, **28**, 324; *Acta Cryst.*, 1948, **1**, 259.
[24] M. J. Buerger, *Proc. Nat. Acad. Sci. U. S.*, 1948, **34**, 277.

The fundamental problem in X-ray analysis can be regarded as working back from such vector sets to discover the original distribution of atoms which gave rise to them. If we really had unlimited resolution with all the vectors clearly and separately defined, this could be done, apart from certain cases where fundamental ambiguities arise.   (These are discussed in the next section.)

The solution of the actual Patterson maps obtained from crystals, however, is enormously more complicated than these simple considerations indicate. This is, of course, due to the fact that in a crystal we are dealing with atoms that are not point scattering sources, and in practice the resolution of the Fourier method is very limited. Because of these difficulties it is not at present possible to outline any systematic treatment which would be applicable to every case.

4A. Simple distributions.—We take a linear arrangement of atoms with co-ordinates $x_1, x_2 \cdots x_N$, in a crystal of axial length a. f_i represents the atomic scattering factor for atom i for the plane $(h00)$. Then the square of the structure amplitude (compare equation 5.3) is given by

$$\left| F(h00) \right|^2 = \left[\sum_{i=1}^{N} f_i \cos 2\pi h x_i / a \right]^2$$

$$+ \left[\sum_{i=1}^{N} f_i \sin 2\pi h x_i / a \right]^2. \qquad \cdots (6.15)$$

If the terms in the brackets are expanded, we obtain, with a slight rearrangement,

$$\left| F(h00) \right|^2 = \sum_{i=1}^{N} \sum_{j=1}^{N} f_i f_j \cos \frac{2\pi h}{a} (x_i - x_j). \qquad \cdots (6.16)$$

It is seen that the cosine terms contain all the interatomic distances in the structure, and that the coefficient of each term is the product of the atomic scattering factors appropriate to the pair of atoms concerned. This is really a restatement of the Patterson result in terms of atomic positions, because the vector distribution in (6.16) is a representation of the crystal structure and therefore periodic. It can thus be expressed by means of a Fourier series in the usual way, when it is found that the coefficients of the terms in this series are the squares of the successive structure amplitudes.   Extensions to two and three dimensions yield analogous expressions.

If we consider an isolated group of atoms and number them 1, $2 \cdots N$, the array of vectors between them may be written:

$$u_{11} \quad u_{12} \; \cdots \; u_{1N}$$
$$u_{21} \quad u_{22} \; \cdots \; u_{2N}$$
$$\cdot \; \cdot \; \cdot \; \cdot \; \cdot \; \cdot \; \cdot \; \cdot \qquad \cdots \; (6.17)$$
$$u_{N1} \quad u_{N2} \; \cdots \; u_{NN}.$$

There are N diagonal terms, $u_{11}, \; u_{22} \cdots u_{NN}$, which express the condition that every atom is at zero distance from itself.   The remaining $N(N-1)$ terms, $u_{12}, \; u_{21}$, etc., occur in symmetrical pairs, so that only $N(N-1)/2$ terms are essentially distinct and represent different distances.   Unfortunately, this is still a large number even if the structure contains only a moderate number

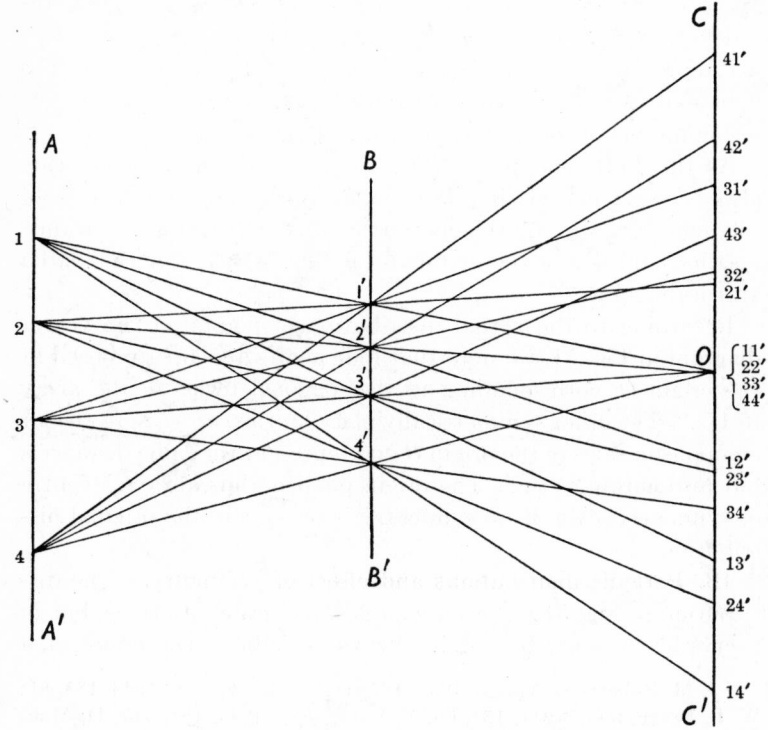

Fig. 51. Construction of vector array. Asymmetric distribution.

of atoms, and difficulties of resolution in any Fourier representation must always be very great.

A simple graphical construction is useful here, both in illustrating the problem and suggesting a possible means of solution when the Patterson peaks are not fully resolved,[25] In Fig. 51 the points 1, 2, 3, 4, on AA' represent a distribution of atoms, and these points are joined by straight lines to the point O on the parallel line CC'. The intersections on some intermediate line, BB', are marked. A line is then drawn from each point on AA' through each point on BB' and produced to meet CC'. The array of points on CC' is then the vector map of the original distribution on AA'.

This construction also holds in two dimensions, with the points distributed on planes at AA', BB', and CC'. This case can be represented by various optical devices. For example, the points distributed on a sheet at AA' can be represented by moveable apertures with corresponding opaque objects on a transparent screen at BB'. When illuminated, the shadow pattern on CC' will represent the vector map. The intensities and sizes of the light sources can be adjusted to match roughly the relative scattering powers of the atoms under investigation.

As the Patterson projection of any unknown structure can always be calculated directly from the X-ray data, the process of solving a structure by this method consists of varying the atomic positions until a good match is obtained with the calculated vector diagram.

Returning to the actual distribution of points on the vector diagram in Fig. 51, we note that four points are superimposed at the origin O, corresponding to the diagonal terms in the array (6.17). The other symmetrically placed terms, u_{21}, u_{12}, etc., occur on opposite sides of the origin O and represent the same distances, but in opposite sense. The origin point is thus always a centre of symmetry, even if no symmetry is present in the original distribution.

4B. Periodic distributions and effect of symmetry.—The distribution in Fig. 51 represents an isolated group of atoms, but in a crystal the distribution is always periodic. The effect of a

[25] J. M. Robertson, *Nature*, 1943, **152**, 411; G. Hägg, *ibid.*, 1944, **153**, 81; W. L. Bragg, *ibid.*, 1944, **154**, 69; V. Vand, *ibid.*, 1944, **154**, 545; D. MacLachlan, Jr., *Bull. Univ. of Utah*, 1950, **41**, no. 6.

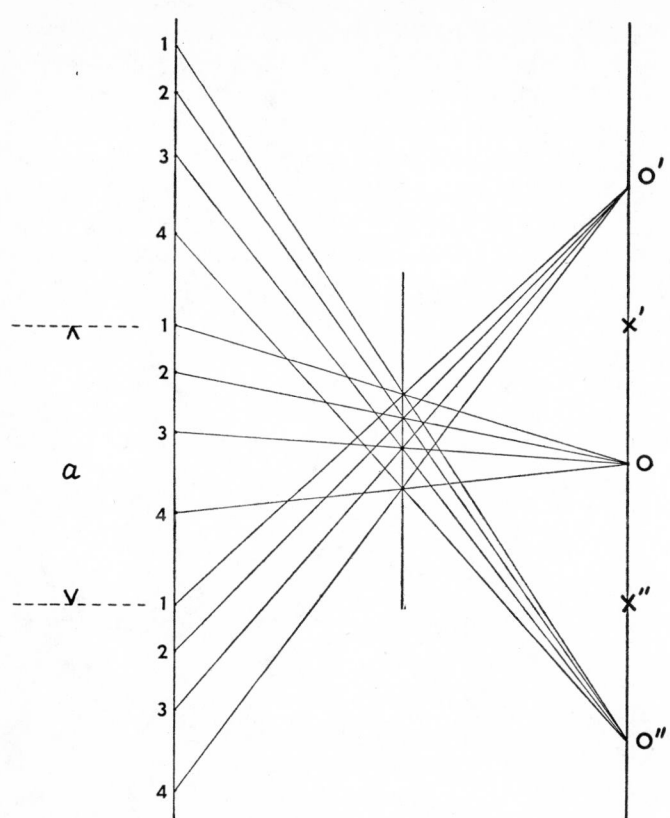

Fig. 52. Vector array from periodic distribution.

periodic distribution on our construction is illustrated in Fig. 52, where all the lines have not been drawn in. It is clear that the vector distribution itself becomes periodic, the origin point repeating at O', O'', etc. This involves further subsidiary centres of symmetry at x' and x'' as well as at the origin points, and so introduces a set of new distances which can be represented by $(a - x_i)$ where a is the primitive translation and x_i the distance. In representing an actual crystal structure by this method it will always be necessary to project the contents of at least two unit cells in order to display fully all the interatomic vectors.

The atomic distributions dealt with so far have all been asymmetric. When these distributions themselves contain

certain inherent symmetry, the resulting vector distribution
becomes modified by the superposition of certain points. For
example, in Fig. 53 a simple centrosymmetric distribution and
its vector map are shown. Of the 16 possible interactions, 4
result in points which coalesce at the origin, as before. Of the
remaining 12, 4 represent interactions across the centre of symmetry, viz., 2'2, 1'1, 11', 22', and produce single points on the
vector map. The 8 interactions that remain are superimposed
on the vector map to form 4 double points, because these vectors
are duplicated by the centre of symmetry.

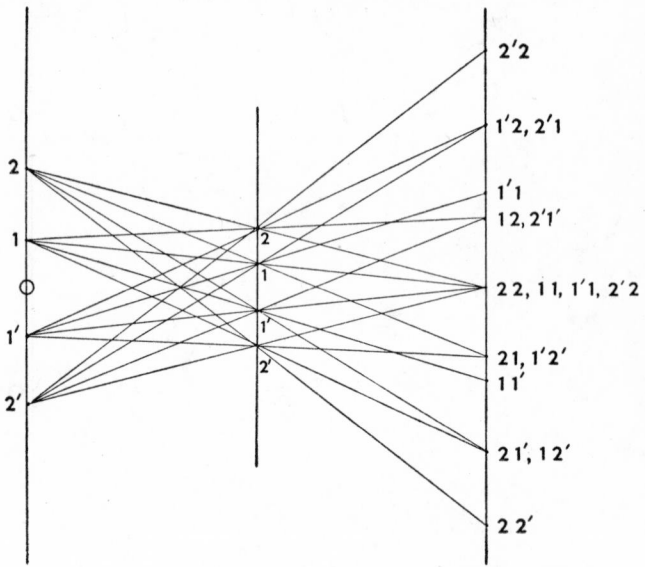

Fig. 53. Vector array from centrosymmetric distribution.

This example is better illustrated in two dimensions. Fig. 54
shows a distribution of 8 atoms in a unit cell with a centre of
symmetry, and the corresponding vector map is shown in Fig. 55.
The origin point (not shown in Fig. 55), with 8 superimposed
points, occurs at each corner. In addition, the map shows 8
single peaks corresponding to interactions across the centre of
symmetry, and 24 double peaks due to the other vectors which
occur in parallel pairs. In general, for a centrosymmetric unit
cell containing N atoms, the number of peaks due to single interactions across the symmetry centre is N, and the number of

double peaks due to the remaining equal and parallel vectors is
$N(N-1)/2$. The total number
of peaks is thus $N^2/2$.

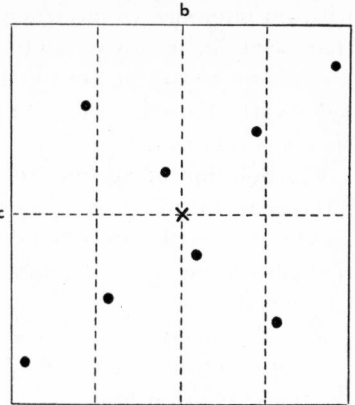

Fig. 54. Atomic distribution.

This example is taken from an
actual structure determination,[26]
and the contoured Patterson
map calculated directly from the
X-ray data is shown in Fig. 56.
The peaks of different heights
are readily distinguished, although only partial resolution is
achieved for some of the single
peaks, and there is some further
and probably spurious detail.
The actual structure contained eight bromine atoms in the unit
cell, together with a large number of carbon, nitrogen, and hydrogen atoms. The effects of all the light atoms are so small that
they can be neglected at this stage of the analysis.

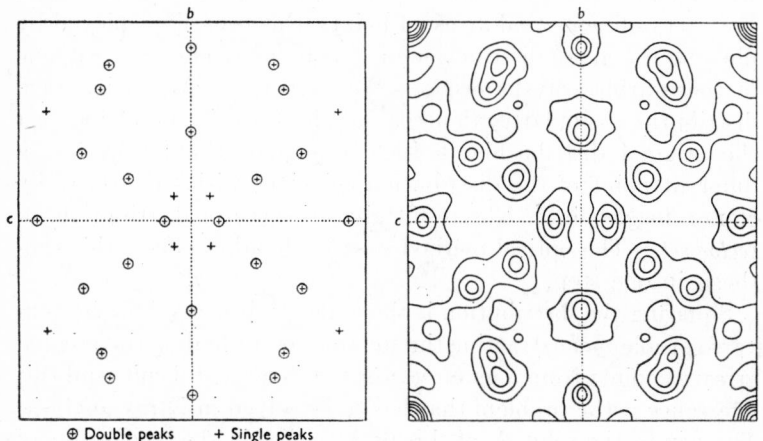

⊕ Double peaks   + Single peaks

Fig. 55 (Left). Theoretical vector map. Fig. 56 (Right). Patterson map
calculated from X-ray data.

It is a matter of some general interest that a diagram of this
kind can be produced directly from the X-ray measurements,

[26] W. P. Binnie and J. M. Robertson, *Acta Cryst.*, 1949, **2,** 116.

with no reference to any chemical data. The number of peaks that occur in it proves quite clearly that there must be eight prominent scattering centres in the unit cell, a result in agreement with that deduced from the density, unit cell volume, and chemical analysis.

4C. Solution of vector sets.—It is a matter of great general interest to examine the question of whether a structure can in fact be reconstructed from its vector representation, if we assume that all the vectors are known. This really amounts to asking the question of whether it is theoretically possible, apart from practical difficulties, to solve a crystal structure in terms of atomic positions from the X-ray data alone, without any knowledge of the phase relationships.

From what has already been said it will be clear that forming the vector array from a given distribution of points really amounts to "squaring" the distribution. This can be carried out by drawing a line from every point to every other point and translating all these lines without rotation to a common origin. Or it can be effected more readily by the method illustrated in Figs. 51–53, which emphasises the process of "squaring" the distribution. For a two-dimensional array of points, this is easily achieved by plotting the given distribution of N points and choosing any one of these arbitrarily as origin. We then take $(N-1)$ identical distributions, similarly oriented, and plot these in succession over the original distribution in such a manner that each time a different point of the distribution coincides with the arbitrarily chosen fixed point. The result gives the vector distribution, or *vector set*. The optical method described earlier effects this synthesis in one step.

Squaring the distribution is thus an easy process, but the converse problem of extracting the square root, or finding the original array of points from a given vector set, is more difficult, and this in essence is the problem that has to be solved in X-ray analysis. The key to the solution of this problem lies in the recognition of the fact that the vector set consists of N superimposed images of the original set of N points.

The problem has been studied in a very general way by D. M. Wrinch,[27] and more recently M. J. Buerger[28] has developed

[27] D. M. Wrinch, *Phil. Mag.*, 1939, **27**, 98.
[28] M. J. Buerger, *Acta Cryst.*, 1950, **3**, 87.

systematic procedures that enable the fundamental set or sets to be extracted from any given two-dimensional vector set. By means of a simple algebra based on arrays of the type of (6.17) the properties and symmetry relations of the vector set can be discussed. There is no need to describe these methods here, but it appears that if all the points representing the vector set are clearly and separately defined, the fundamental set can be extracted from it by a process of first establishing a line (two points) and finding all its images, then a triangle, quadrilateral, and so on. It is clear that this can only be done if all the points are defined with very great precision, and unfortunately this precision becomes more difficult to achieve as the number of points increases. However, the fact that the problem can be solved systematically is of very great interest and significance.

There remains the problem of trying to solve the actual Patterson map obtained from a crystal, where, if any considerable number of atoms is present, many of the vector points must coalesce and fail to resolve. Even if the separate vector peaks are not resolved, the general over-all appearance of the Patterson maps from complex structures, with their regions of high and low density, does provide information which can be utilised in giving a clue to possible atomic arrangements. Great experience is required in making such interpretations,[29] and it is difficult to provide any adequate generalisations.

Much work has recently been devoted to attempting a more systematic approach to this subject,[30] by the formation of image-seeking functions and by procedures which are parallel to those that have been developed for the solution of well-defined vector sets. If at least one peak on the vector map can be definitely assumed to represent a single vector, it is sometimes possible by means of translated overlays of such maps to isolate certain heavier resulting peaks, from which part of the original atomic distribution may be obtained.

[29] See, for example, D. Crowfoot, C. W. Bunn, B. W. Rogers-Low, and A. Turner-Jones, *The Chemistry of Penicillin*, Princeton: Princeton University Press, 1949, chap. xi; J. Boyes-Watson, E. Davidson, and M. F. Perutz, *Proc. Roy. Soc.* (London), 1947, **A191**, 83.

[30] M. J. Buerger, *Proc. Nat. Acad. Sci. U. S.*, 1950, **36**, 376, 738; D. Mac-Lachlan, Jr., *ibid.*, 1951, **37**, 115; J. Clastre and R. Gay, *Compt. rend.*, 1950, **230**, 1876; *J. Phys. radium*, 1950, **11**, 75; *Bull. soc. franç. mineral. et crist.*, 1950, p. 202.

5. THE PATTERSON AMBIGUITIES

A question of fundamental importance for crystal analysis is whether the solution of a Patterson diagram or vector set, when found, is always unique. Evidence to the contrary was first obtained by Pauling and Shappell[31] in the analysis of an actual crystal structure, and later the general problem was examined very carefully by A. L. Patterson.[32] He has shown that certain distinct distributions of points can exist which possess the same vector distances. Distributions possessing this property are called *homometric*. It is obvious that crystal structures based on homometric distributions of atoms will give rise to exactly the same X-ray diffraction patterns, and it will be impossible to distinguish such structures by any observations of the X-ray intensities alone.

Patterson has confined his investigation mainly to linear periodic arrangements of points, and such distributions can be conveniently represented, together with their vector distances, by plotting the points on the circumference of a circle and joining each point to all the remaining points. The distances can then be expressed in angular measure,

$$\theta_i = 2\pi x_i/a$$

where a is the period.

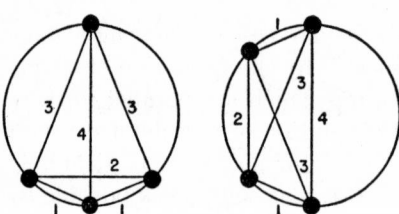

Fig. 57. A simple homometric pair.

Fig. 57 shows two distributions plotted in this way, which are obviously distinct but yet possess identical distances. This simple homometric pair can be generalised by the introduction of a variable parameter, as illustrated in Fig. 58 and so leads to an infinity of homometric pairs and corresponding ambiguities.

Patterson has examined systematically the cases which may be obtained by using r of the n vertices of an inscribed regular polygon, and out of 2664 sets examined in this way (up to $n = 16$) he has found a total of 390 homometric pairs, 7 sets of homometric

[31] L. Pauling and M. D. Shappell, *Z. Krist.*, 1930, **75**, 128.
[32] A. L. Patterson, *Nature*, 1939, **143**, 939; *Phys. Rev.*, 1944, **65**, 195.

triplets, and 3 sets of quad-
ruplets.   Many generalisations of these sets by the introduction of variable parameters can be made, and
the study can also be ex-
tended to two and three
dimensions.   There would
appear, however, to be considerable difficulty in de-

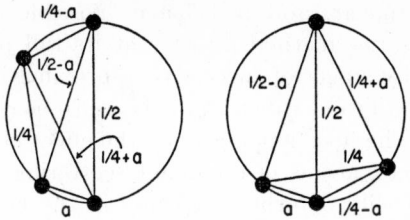

Fig. 58. Homometric pair with variable parameter.

veloping any general theory for the occurrence of homometric
sets.

With regard to the general problems of X-ray analysis and
crystal structure determination, these discoveries are rather
disturbing, in so far as they show that there are many possible
distributions of atoms which we cannot hope to distinguish by
means of X-ray intensity data alone.   They emphasise the importance of confirming the results of diffraction experiments by
an appeal to other independent lines of physical and chemical
evidence.   If alternative solutions should exist in any given case,
it seems very unlikely that they will both be chemically reasonable.

When very large numbers of atoms are present, the situation
becomes more difficult, because it may not be possible to distinguish between homometric sets on the available chemical
evidence.   Recent work by Pepinsky and Calderon[33] has further
indicated that certain fundamental ambiguities may be inherent
in the case of completely non-centrosymmetric density distributions.   This development still further complicates the difficult
question of the possibility of analysing very complex molecular
structures, such as the proteins (see Chapter XI) where the available X-ray data are in any case always less, per atom of the asymmetric unit, than for simple structures.

6. PHASE-DETERMINING METHODS

It happens that in many crystal structures it is possible to
make what amounts to a direct determination of the phase constant for some or all of the X-ray reflections.   In these cases the
difficulties which we have discussed in the preceding sections dis-

[33] R. Pepinsky and A. Calderon, paper to the Chicago meeting of the
American Crystallographic Association, October, 1951.

appear, and it becomes possible either to apply the Fourier series method directly at its full power, or at least to proceed by stages of successive approximation to the true structure.

Two distinct methods are to be considered under this heading. The first depends upon the effect of one or more "heavy "atoms, i.e., atoms of dominant scattering power, situated at known or easily determined points in the crystal structure. The second more powerful method of "isomorphous replacement" becomes available if it is possible to replace these special atoms by others of different scattering power without disturbing, or without greatly disturbing, the rest of the crystal structure. Such a substitution affects the intensities of the reflections in a way which enables deductions to be made regarding the phase constants.

These methods obviously apply only to special crystal structures that include the necessary atoms. But considered as methods for the determination of complex molecular structures, particularly organic molecular structures, they possess a much greater generality than might at first be supposed. The molecular structure in question will usually consist of some framework of "light" atoms (carbon, oxygen, nitrogen, etc.) whose spatial positions it is desired to find. In general it will contain no heavy atom, but chemical methods are often available by means of which a heavy atom can be inserted at some point without greatly disturbing the main framework of the molecular structure. If acid groupings are present, a heavy metal salt may be formed; otherwise addition or substitution products involving the heavier halogens or other elements can often be prepared. Such products will in general be crystallographically different from the parent compound, but all essential information concerning molecular structure will be equally well obtained from their analysis.

In the case of the isomorphous substitution method the conditions are more stringent, because it is necessary to be able to prepare at least two derivatives containing special atoms of different scattering power, without greatly disturbing the over-all crystalline structure. Nevertheless, this method also has considerable generality. The phenomenon of isomorphism, for example between bromine and chlorine compounds, sulphur and selenium compounds, or in the case of the salts of potassium and rubidium, lead and strontium, etc., is well known and has received much careful study.   By the exercise of some chemical ingenuity it is usually possible to prepare such isomorphous derivatives from a given molecular structure.

For the sake of clarity in the following discussion we confine attention largely to ideal cases where phase determination can be regarded as definite.   It must be realised that in practice there may be variations of these cases.   Very often an approach through the Patterson method will be desirable, and the true outlines of the molecular structure will only gradually become apparent, with many applications of trial and error calculation and successive Fourier approximation.

6A. Use of a heavy atom.—The general effect of an atom of high scattering power on the X-ray structural problem is to convert the unknown and unmeasurable phase relationships into certain intensity relationships which are susceptible to direct measurement.   In terms of the Patterson method, where F^2 values are used, the vectors between the light atoms become second order quantities, and only the vectors between the heavy atom and each of the light atoms are significant: application of the Fourier method then gives a picture of the structure direct, instead of its vector representation.   Certain ambiguities may, however, persist, depending on the crystallographic situation of the heavy atom, and it is convenient to consider the different cases separately.

The most simple case occurs when the heavy atom is situated at a centre of symmetry and makes a contribution to each X-ray reflection.   This will occur in the triclinic space group $P\bar{1}$; and for various projections in other space groups when the heavy atom positions form an effectively primitive lattice (Fig. 59a). The heavy atom can be regarded as a highly concentrated scattering source, and its contribution, F_A, to the resultant amplitude will always be positive if the origin of co-ordinates is taken at the centre of symmetry where it is situated.   The amplitude resulting from the combined effect of all the light atoms in the structure, F_O, may, however, be either positive or negative at the symmetry centre.   By addition, the total resultant amplitude, F_{AO}, will always be positive if the scattering power of the heavy atom is greater than that of all the others combined.

This last condition will not often be realised fully for every

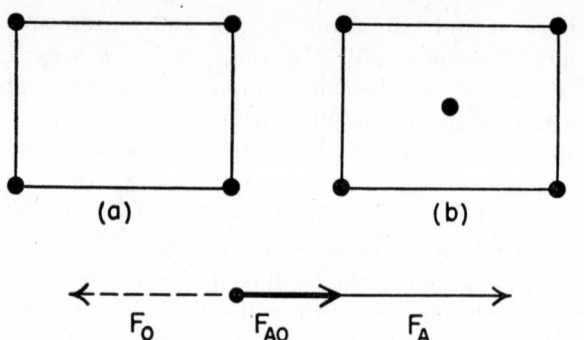

Fig. 59. Heavy atom at centre of symmetry.

X-ray reflection. It is completely realised in the case of platinum phthalocyanine[34] (p. 270) and for the majority of the reflections in the case of nickel phthalocyanine.[35] Even if the condition is only partially realised, a Fourier analysis based on the assumption of all phase signs being positive will yield some approximation to the true structure, on the basis of which it may be possible to proceed with further refinement.

There is an unfortunate but inevitable practical difficulty associated with this method, and with all its variations that are mentioned below. The major portion of the structure amplitude being due to the heavy atom contribution, all information concerning the positions of the light atoms in the structure must be derived from the smaller remaining part of the structure amplitude. To achieve the same accuracy in determinination of atomic positions as in a structure consisting of light atoms alone, it is therefore necessary to measure the intensities with much greater precision. On the other hand, the presence of the heavy atom means a higher absorption coefficient and makes the measurement of intensities more difficult. From this dilemma there is no easy escape, but a great deal can be achieved by the use of extremely small crystal specimens to minimize absorption difficulties.

It may be noted here that the isomorphous substitution method (see Section 6B) minimizes this difficulty, because after the phase

[34] J. M. Robertson and I. Woodward, *J. Chem. Soc.*, 1940, p. 36.
[35] J. M. Robertson and I. Woodward, *J. Chem. Soc.*, 1937, p. 219.

determination has been completed the final analysis can be conducted on the derivative containing the lighter atom.

The second case to be considered occurs with the heavy atom still at a centre of symmetry, but with the lattice centred, or effectively centred for the projection being studied (Fig. 59b). In this case the heavy atom makes a full contribution as before to half of the reflections (when $h+k$ is even), but for the remaining reflections (when $h+k$ is odd) the heavy atom contributions are in opposite phase and cancel each other out. Odd index reflections which do appear are due to the light atoms alone, and are not phase determined.

The best method of solution in this case is to effect a Fourier synthesis based on the phase-determined reflections alone. The result is a representation of the structure containing additional, spurious symmetry; because the reflections omitted impose exact centring on the entire structure and not merely on the positions of the heavy atoms. In projection (Fig. 59b) the effect, in combination with elements of symmetry already present such as screw axes or glide planes parallel to the sides, is to duplicate every atom in a general position. The Fourier synthesis conducted on this basis reveals the position of each light atom by a peak of half the normal height, accompanied by its mirror image in the other quadrant. To solve the structure it is then necessary to select one peak from each of the resulting pairs and reject the other, the criterion for the correct structure being that which gives calculated intensities which agree with those observed for the odd index reflections. For a small number of atoms this selection is not difficult and can be carried out systematically for every possibility. (There are 2^n possibilities for n atoms.) Usually, however, some chemically plausible features will be recognised in the structure, and these will serve as a guide in making the right selection of peaks.

A good example of a structure analysed by this method is provided by the main projection of cupric tropolone[36] (Fig. 112, p. 273). The seven carbon positions in this case can be selected in 128 ways, but, as the expected seven-atom ring is clearly visible, the correct selection can be achieved immediately and tested against the observed intensities.

[36] J. M. Robertson, *J. Chem. Soc.*, 1951, p. 1222.

The centred centro-symmetrical case described above leads immediately to the most general case, in which we have a heavy atom present and no crystallographic symmetry applicable to the structure as a whole.   Here it is convenient to choose an origin for co-ordinates at the heavy atom position, and the situation is then represented in Fig. 60.   The entirely arbitrary phase of the contribution of the light atoms, F_O, is shifted by the heavy atom contribution, F_A, to give a resultant vector, F_{AO}, of which the phase is still variable but confined within certain fairly narrow limits if the heavy atom contribution preponderates. If a Fourier synthesis is effected on the assumption that the structure is centrosymmetrical with all phases positive, the effect is clearly to introduce a spurious image of each atomic peak across the assumed centre of symmetry.

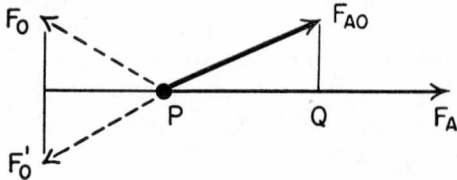

Fig. 60. Heavy atom, a non-centrosymmetrical case.

This can be seen from Fig. 60, where the addition of two symmetrically related light atom contributions, F_O and F_O', each of half weight, to the heavy atom contribution give a resultant PQ which is of approximately the same magnitude as the observed structure amplitude, F_{AO}.

The interpretation of the resulting synthesis then involves the selection of one atom from each of the resulting pairs, as described above.

The cases so far described have been simplified to illustrate the general principles of the heavy atom method.   In practice it will frequently happen that two or more heavy atoms are present in the crystal unit, and their relative positions must first be determined.   This can generally be done without much difficulty by application of the Patterson method.   The contribution of this group of atoms to each structure factor can then be calculated, and their value in phase determination assessed.   The work then generally proceeds by successive approximations as the positions of the lighter atoms become revealed.   A good example of the application of this method is found in the analysis of cholesteryl iodide by Carlisle and Crowfoot.[37] It will often happen in such cases that the situations of the heavy atoms are such that their contributions will vary from one reflection to another in such a way as to prevent a complete determination of the phases. It will then usually be necessary to combine trial and error procedures with the heavy atom method.

The extent to which the heavy atom method can be applied to phase determination in the case of a complex molecular structure (assuming a favourable situation for the heavy atom) will depend upon the value of the average structure amplitude due to the light atoms (average F_O) in relation to the heavy atom contribution, F_A. It is not necessary that F_A be larger than all the F_O's, but it should dominate a sufficient number to enable a process of successive approximation by the Fourier method to commence. It is not possible to make accurate calculations here, because it is difficult to say how many reflections must be phase determined before a refinement process can commence. The root mean square of $|F_O|$, however, is proportional to $\sqrt{N}$ where N is the number of light atoms in the unit cell. By a suitable choice of heavy atoms it may be possible to make $|F_A|$ perhaps ten times as great as $|f_O|$ (F_O for one light atom). Iodine, atomic number 53, against carbon, atomic number 6, would provide such a combination. Thus, if we had a molecule containing a hundred carbon atoms and one iodine atom, we might expect the heavy atom contribution to be about equal to the average contribution from the light atoms combined, and there would at least be a possibility that successive Fourier approximations to the structure might commence.

6B. Isomorphous replacement.—A still more powerful phase-determining method is available if it is found possible to substitute successively two different heavy atoms in a molecule without at the same time unduly disturbing the over-all crystal structure. The phase relationships can then be determined from a difference effect, and, if accurate intensity measurements can be made, it is no longer necessary that the heavy atom contribution should outweigh the contributions made by the rest of the molecule.

[37] C. H. Carlisle and D. Crowfoot, *Proc. Roy. Soc.* (London), 1945, **A184**, 64.

The most simple case arises when the replaceable atoms are situated at a symmetry centre. Let these atoms be A_1 and A_2, and represent the remainder of the molecule by O. Then if the two structures are strictly isomorphous, the contribution made by O remains constant, and we have

$$F_{A_1 O} - F_{A_2 O} = F_{A_1} - F_{A_2} = \Delta F. \qquad \cdots (6.18)$$

The quantity ΔF represents the difference in scattering power of the two replaceable atoms and is known for each reflection. This is sufficient to determine the signs of the two structure factors $F_{A_1 O}$ and $F_{A_2 O}$, if their magnitudes can be measured accurately, and if the condition of isomorphism on which the equation is based really holds exactly. Under perfect conditions a fairly small difference in scattering power between A_1 and A_2 is theoretically capable of determining all the signs, but if, as will generally be the case, exact isomorphism does not hold, then a larger difference in scattering power is necessary to avoid ambiguity. In practice the main limitation of the method does in fact arise from cases of only approximate isomorphism, where the co-ordinates of the replaceable atoms may sometimes differ appreciably.

This method of phase determination was first extensively applied to organic structures in the case of the phthalocyanines[38] (p. 262), where a large series of replaceable metal atoms is available. These include the case of complete removal of the metal atom in free phthalocyanine, where the condition of isomorphism with the other members of the series still holds. The chemical conditions pertaining to the phthalocyanines are somewhat rare, but the generality of the method is considerably greater than might at first be expected, as has already been pointed out. More recent examples illustrating the method are to be found in the work of Wiebenga and Krom[39] on camphor derivatives and of Beevers and Cochran[40] on sucrose derivatives.

It may happen that, owing to lattice centring by the replaceable atom, phase determination can only be effected for certain sets of reflections. The treatment will then be similar to that described in the case of a single heavy atom.

[38] J. M. Robertson, *J. Chem. Soc.*, 1935, p. 615; 1936, p. 1195.

[39] E. Wiebenga and C. J. Krom, *Rec. trav. chim.*, 1946, **65**, 663.

[40] C. A. Beevers and W. Cochran, *Proc. Roy. Soc.* (London), 1947, A190, 257.

When the replaceable atom site is not coincident with a centre of symmetry, equation 6.18 still holds, but the quantities are vectors with phase angle no longer restricted to 0 or π (+ or −). If an origin for co-ordinates is chosen at the replaceable atom site, the quantity ΔF is real, and again represents simply the known difference in scattering power of the two atoms concerned. The vector equation can then be solved for the phase angles of the structure factors, $\alpha_{A_1 0}$ or $\alpha_{A_2 0}$, but an ambiguity of sign remains. This will be clear from Fig. 61, which is due to Bijvoet. The phase angles are given by

$$\cos \alpha_{A_1 0} = \frac{F_{A_1 0}^2 - F_{A_2 0}^2 + \Delta F^2}{2 \Delta F \cdot F_{A_1 0}}, \qquad \cdots (6.19)$$

$$\cos \alpha_{A_2 0} = \frac{F_{A_1 0}^2 - F_{A_2 0}^2 - \Delta F^2}{2 \Delta F \cdot F_{A_2 0}}. \qquad \cdots (6.20)$$

As the signs of the phase angles remain unknown, both the positive and negative values must be employed for each term. The effect of carrying out a Fourier synthesis on this basis is clearly to introduce a spurious centre of symmetry at the origin, with duplication of every peak. The subsequent analysis is then

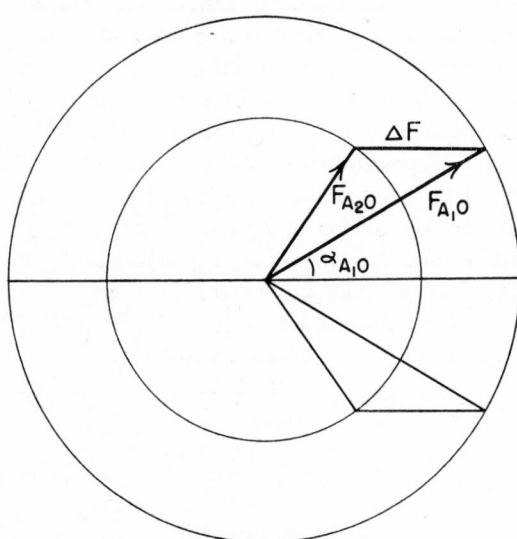

Fig. 61. Isomorphous substitution method, a non-centrosymmetrical case.

similar to the corresponding heavy atom case described in the previous section.

A beautiful example of the application of the isomorphous replacement method under these conditions is provided by the analysis of the strychnine structure by Bokhoven, Schoone, and Bijvoet.[41] The [001] projection of this structure is a noncentrosymmetrical case of the type just described. The replaceable atoms were selenium and sulphur, in the isomorphous selenate and sulphate of strychnine.

An extremely important extension of this work has recently been made by Bijvoet,[42] who has succeeded in solving the remaining ambiguity regarding the sign of the phase angle α in Fig. 61. This has been achieved by the introduction of a "phase lag" in scattering, by the use of an atom which can itself be excited if the incident radiation is of appropriate wave-length (near the absorption edge for the atom in question). Such a phase change on scattering invalidates Friedel's law (see Chapter IV, Section 7) and has been used by Coster, Knol, and Prins[43] to distinguish between 111 and $\overline{1}\overline{1}\overline{1}$ reflections of the polar zinc blende crystal. In a similar way, the sign of the phase angle for the light atom contribution (F_O) can be determined (Fig. 60) by the use of a suitable heavy atom and suitable radiation. The effect is observed by means of very small intensity differences and can usually only be applied to a few reflections. But when a structure has already been determined except for absolute configuration, one phase sign determination is sufficient to decide between the two mirror image possibilities. In this way Bijvoet and his collaborators[42] have determined the absolute configuration of D-tartaric acid by the action of Zr $K\alpha$ rays, which were used to excite the rubidium atom in sodium rubidium tartrate.[44] A small but sufficient phase change on scattering was thus introduced. The result established for the first time that the chemical convention of Emil Fischer corresponds with reality.

[41] C. Bokhoven, J. C. Schoone, and J. M. Bijvoet, *Acta Cryst.*, 1951, **4**, 275.

[42] J. M. Bijvoet, *Koninkl. Nederland. Akad. Wetenschap.*, 1949, **52**, 313; A. F. Peerdeman, A. J. van Bommel, and J. M. Bijvoet, *ibid.*, 1951, **54**, 3.

[43] D. Coster. K. S. Knol. and J. A. Prins, *Z. Physik.*, 1930, **63**, 345.

[44] C. A. Beevers and W. Hughes, *Proc. Roy. Soc.* (London), 1941, **A177**, 251.

Part Two

THE ANALYSIS OF SOME ORGANIC

MOLECULAR STRUCTURES

VII

Fundamental Structures
and Early Work
on Organic Crystals

1. FUNDAMENTAL STRUCTURES

The first X-ray diffraction experiments were carried out on a range of inorganic crystals and minerals, simplicity of chemical composition and perfection of crystalline from being the chief factors guiding the choice of substance. The work of von Laue, W. H. Bragg, and W. L. Bragg on these crystals established the fundamental principles of X-ray analysis, and during the first decade attention remained focused mainly on the inorganic field. Included in this work, however, were the crystal structures of two minerals, diamond and graphite, where the atomic arrangements are fundamental to organic chemistry.

1A. Diamond.—The diamond structure was, in fact, one of the very first to be determined,[1] and although later work[2] may have added to the accuracy of the measurements, the essential structure remains unchanged. It may be described briefly as follows. Eight carbon atoms are found in a cubic unit cell, $a = 3.56$ Å, with co-ordinates 000; $0\frac{1}{2}\frac{1}{2}$; $\frac{1}{2}0\frac{1}{2}$; $\frac{1}{2}\frac{1}{2}0$; $\frac{1}{4}\frac{1}{4}\frac{1}{4}$; $\frac{1}{4}\frac{3}{4}\frac{3}{4}$; $\frac{3}{4}\frac{1}{4}\frac{3}{4}$; $\frac{3}{4}\frac{3}{4}\frac{1}{4}$. The arrangement corresponds to that required by the space group $O_h^7 - Fd3m$, and a basal projection is shown in Fig. 62. This structure, in which the atoms are arranged on two interpenetrating face-centred lattices, is an extremely open one, each atom making close contact with only four others, which are ar-

---

[1] W. H. Bragg and W. L. Bragg, *Nature*, 1913, **91**, 557; *Proc. Roy. Soc.* (London), 1913, **A89**, 277.

[2] W. Ehrenberg, *Z. Krist.*, 1926, **63**, 320.

ranged tetrahedrally about it.   This feature of the structure is
well illustrated in the two views of a model illustrated in Figs. 63
and 64, which was set up in 1913 when the atomic arrangement
was first deduced from ionisation spectrometer measurements.
This structure clearly corresponds with the fundamental chemical
concept of the tetrahedral distribution of valency bonds about the
carbon atom, and the distance between adjacent carbon atoms,
$\frac{\sqrt{3}}{4} a = 1.54$ Å, is generally accepted as the standard carbon-
carbon single bond distance.   A number of other Group IV elements (Si, Ge, Sn) conform to the same structure type.

The diffraction pattern obtained from diamond, however, does
not correspond exactly to that which would be given by strictly
spherical atoms placed at
these positions.   In particular, a small but definite
222 reflection is observed,
which is forbidden by the
above arrangement.   This
was observed in the early
measurements and at-
tributed by Bragg[3] to a
tendency towards tetra-
hedral rather than strictly
spherical symmetry in the
distribution of scattering
matter about the carbon
atoms, an observation
which is again in accord
with chemical expectations.

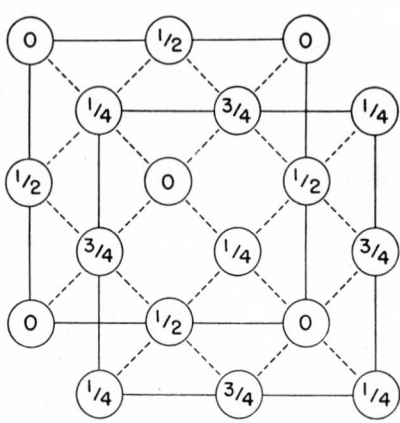

Fig. 62. Diamond structure, basal
projection, showing the two interpenetrating face-centred lattices. Figures
indicate heights above projection plane.

This departure from the
ideal structure has recently received some detailed study.   Accurate intensity measurements on a large number of reflections[4]
show distinct deviations between the observed scattering power
of the carbon atom and that predicted from the theoretical f-
curve of Hartree, especially in the region of the small angle re-

[3] W. H. Bragg, *Proc. Phys. Soc.* (London), 1921, **33**, 304.
[4] R. Brill, H. G. Grimm, C. Hermann, and C. Peters, *Ann. Physik.*,
1939(5), **34**, 393.

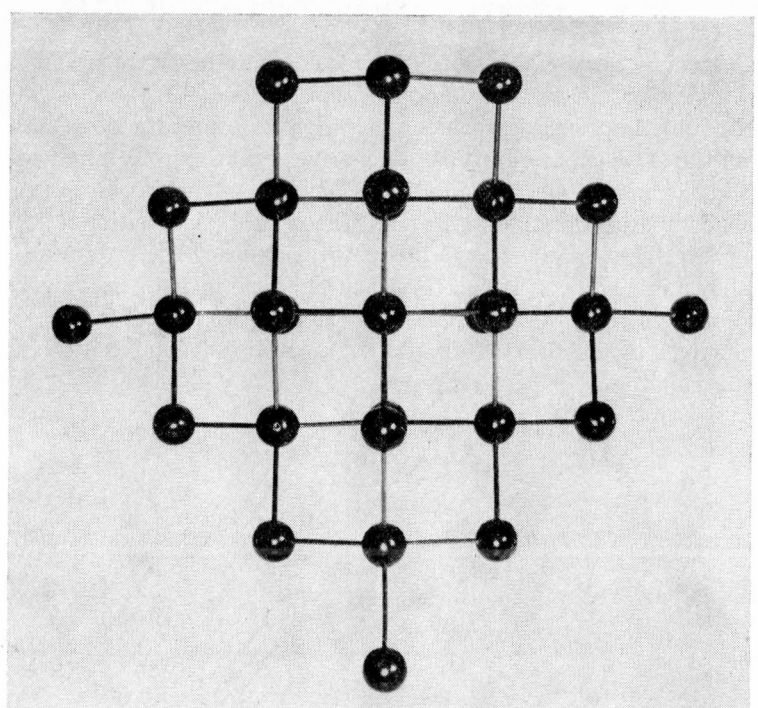

Fig. 63. Model of diamond structure, viewed as basal projection. Horizontal and vertical planes perpendicular to paper are (1ĪO) (Bragg).

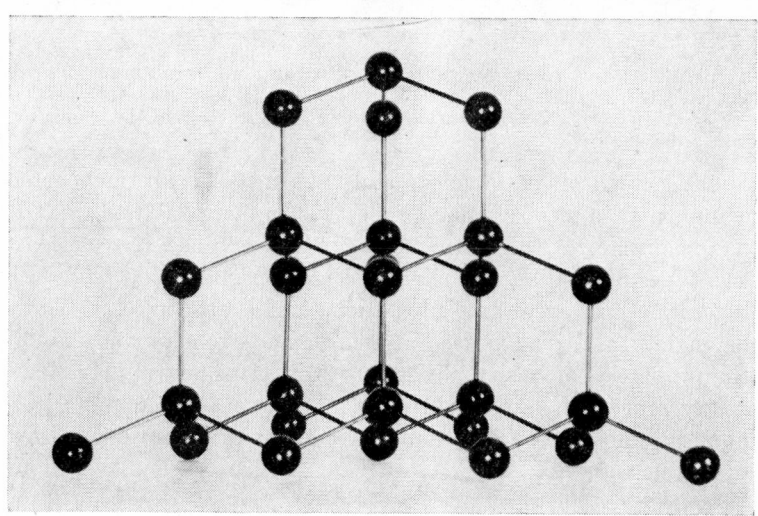

Fig. 64. Model of diamond structure with trigonal axis of cube vertical (Bragg).

flections. Brill[5] has recently expressed these deviations quantitatively by means of a difference synthesis, $(F_o - F_c)$, employing a double Fourier series which gives a projection of the deviations on the (110) plane (Fig. 65). Negative regions (shaded) indicate a deficiency in electron density, and positive regions an excess, in comparison with the Hartree theoretical distribution. The arrangement of the atoms in this projection of the structure is shown in Fig. 66. It will be clear that there is distinct evidence of an electron shift towards the lines of the chemical bonds, producing a tetrahedral deformation of the electron cloud around the carbon atom.

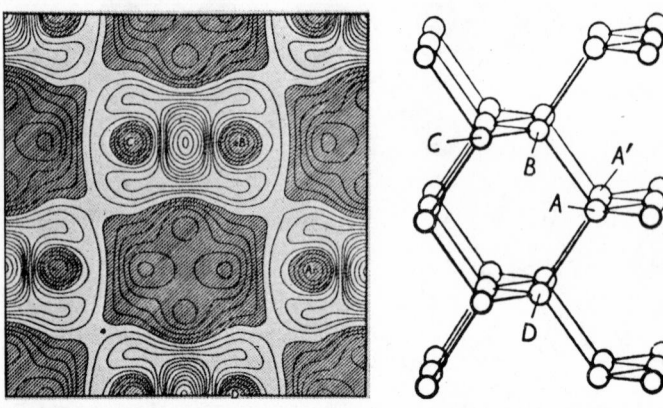

Fig. 65 (Left). Diamond. Difference synthesis projection, showing electron shift relative to Hartree spherical distribution. Negative regions are shaded (Brill).

Fig. 66 (Right). Diamond structure viewed in direction of the projection in Fig. 65.

Later work on diamond crystals has revealed an unexpected complexity in macroscopic properties and crystal texture. It has been shown[6] that specimens can be classified into at least two types depending on their physical properties, especially with regard to infra-red and ultra-violet light absorption. Those belonging to Type I possess a greater homogeneity and exhibit the properties of nearly perfect crystals, giving extremely sharp X-ray reflections. Type II diamonds have a much more pronounced mosaic structure and in certain cases reflect X-rays over a considerable range of angle. Much work has been done

[5] R. Brill, *Acta Cryst.*, 1950, **3**, 333.

[6] R. Robertson, J. J. Fox and A. E. Martin, *Trans. Roy. Soc.* (London), 1934, **A232**, 463; *Proc. Roy. Soc.* (London), 1936, **A157**, 579.

on the type of disorder present in diamond crystals, especially by observation of temperature diffuse scattering effects.[7] The method of divergent beam X-ray photography clearly reveals the difference between Type I and Type II diamonds, only the latter giving high contrast photographs owing to the absence of primary extinction. These experiments are of particular interest here, because by their means K. Lonsdale[8] has been able to re-determine the fundamental parameter in the structure with great precision. The value found for the lattice constant varies between 3.56665 and 3.56723 Å in different specimens, leading to a carbon-carbon bond length of 1.5445 ± 0.00014 Å at normal temperature.

1B. Graphite.—Just as the diamond structure proves to be the prototype for the atomic arrangement in the aliphatic compounds of organic chemistry, so the mineral graphite is found to typify the groupings present in aromatic compounds. The first X-ray studies of graphite were made soon after the discovery of the diffraction effect,[9] but largely on account of the difficulty in obtaining good single crystals of this substance details of the atomic arrangement remained in doubt until the investigations of Bernal[10] and Hassel and Mark[11] in 1924.

The system is hexagonal, with $a = 2.46, c = 6.80$ Å, $c/a = 2.76$. (Later precision measurements[12] have given $a = 2.461, c = 6.709$ Å, $c/a = 2.726$, at room temperature; there is a possiblity that the c spacing may vary slightly in different specimens.) This unit cell contains four atoms, of which two are crystallographically different from the other two, the co-ordinates being A: (000; $00\frac{1}{2}$) and B: ($\frac{1}{3}$ $\frac{2}{3}$ z; $\frac{2}{3}$, $\frac{1}{3}$, $z+\frac{1}{2}$). Bernal's investigation showed that the parameter z must be less than 1/18, and was probably

[7] K. Lonsdale and H. Smith, *Nature*, 1941, **148**, 112, 257; *Proc. Roy. Soc.* (London), 1942, **A179**, 8; K. Lonsdale, *ibid.*, 1942, **A179**, 315; C. V. Raman, *ibid.*, 1942, **A179**, 289, 302; C. V. Raman and P. Nilakantan, *Proc. Indian Acad. Sci.*, 1940, **11**, 389.

[8] K. Lonsdale, *Trans. Roy. Soc.* (London), 1947, **A240**, 219; see also D. P. Riley, *Nature*, 1944, **153**, 587.

[9] P. P. Ewald, *Sitzber. math.-physik. Klasse bayer. Akad. Wiss. München*, 1914, p. 325; P. Debye and P. Scherrer, *Physik. Z.*, 1916, **17**, 277; 1917, **18**, 291; A. W. Hull, *Phys. Rev.*, 1917, **10**, 661.

[10] J. D. Bernal, *Proc. Roy. Soc.* (London), 1924, **A106**, 749.

[11] O. Hassel and H. Mark, *Z. Physik.*, 1924, **25**, 317; C. Mauguin, *J. Phys.*, 1925, **6**, 38.

[12] W. Trzebiatowski, *Roczniki Chem.*, 1937, **17**, 73.

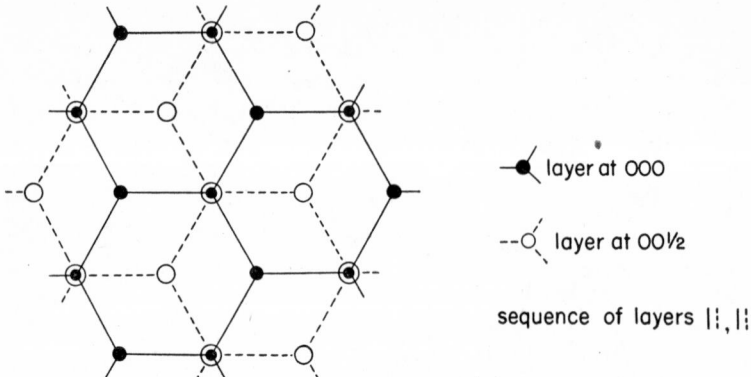

—●⟨ layer at OOO

--○⟨ layer at OO½

sequence of layers |¦, |¦

Fig. 67. Basal projection of the graphite structure.

zero; later work[13] has narrowed the limit still further.   The space group is probably $C_{6v}^4 - P6_3mc$, with two sets of 2-fold special positions occupied; or if the parameter z is assumed exactly zero, it may be described in terms of $D_{6h}^4 - P6_3/mmc$.

Assuming the parameter z to be zero, the structure then consists of regular planar hexagonal networks (Fig. 67) with a carbon-carbon distance of $a/\sqrt{3} = 1.42$ Å.   Alternate carbon atoms, A and B, in the network are crystallographically different (different surroundings), but as there is no evidence of any parameter difference, they may be assumed identical. These infinite hexagonal networks are widely spaced along the c axis, at distances of $c/2 = 3.40$ Å (Fig. 68), and are so arranged that the atoms of any one net lie alternately directly over the atoms of the net beneath and over the centre points of the hexagonal rings, as will be clear from the projection in Fig. 67.

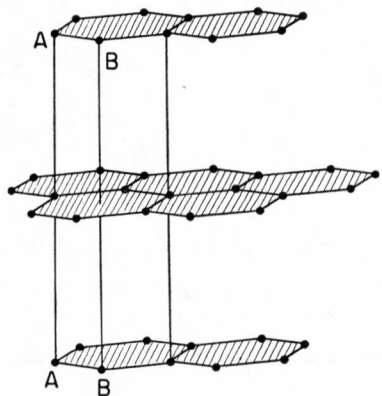

Fig. 68. Graphite structure.

This structure of widely separated infinite sheets immediately explains the outstanding physical properties of graphite.   The

[13] H. Ott, *Ann. Physik*, 1928(5), **85**, 81.

distance of 3.40 Å separating the layers is too great for any covalent linkage, and only weak van der Waals forces can operate. The soft, lubricating properties of graphite are thus due to the ease of relative movement of these sheets, while at the same time considerable hardness and tenacity are to be expected in the sheets themselves. In striking contrast to these properties, the three-dimensional array of covalent linkages in the diamond structure produces the hardest substance known.

The fundamental carbon-carbon distance of 1.42 Å within the graphite networks is considerably less than the diamond distance of 1.54 Å. It is slightly greater than the distance of 1.39 Å found in benzene, but closely similar to the average distance which occurs in the larger aromatic molecules (see Chapter VII). These facts are readily explained in terms of chemical concepts if we assume that in the infinite sheets, where each atom is covalently linked to three neighbours, there is resonance[14] between a large number of structures of the type

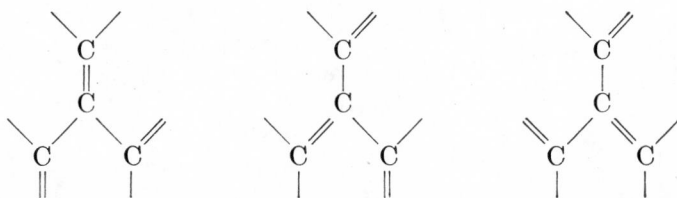

Each carbon-carbon bond will then achieve a 33 per cent double bond character, as against a 50 per cent double bond character in benzene, an average of 45 per cent in naphthalene, and an average of 40 per cent in coronene.

As in the case of diamond, later work has revealed second order complexities in the graphite structure, which probably vary in extent in different specimens. The great difficulty in obtaining good crystals of graphite and the ease with which they can be deformed by displacement across the cleavage plane always tend to produce uncertainties in such work. Certain additional X-ray diffraction lines, however, have been observed, and these have been explained in an interesting way by Lipson and Stokes,[15] who propose a new structure which may be present to the extent

[14] L. Pauling, *The Nature of the Chemical Bond*, Ithaca, N. Y.: Cornell University Press, 1939.

[15] H. Lipson and A. R. Stokes, *Proc. Roy. Soc* (London), 1942, **A181**, 101.

of about 14 per cent in normal crystalline graphite. The new structure, which explains the additional X-ray diffraction lines, postulates a third hexagonal network within the c translation distance, symmetrically related to the other two. The disposition of the carbon-carbon bonds viewed in basal projection would then be as indicated in Fig. 69, where the double lines represent the orientation of the bonds in the additional layer. This structure would have $a = 2.461$ Å, $c = \frac{3}{2} \times 6.709 = 10.064$ Å with atoms at 000; $00\frac{1}{3}$; $\frac{1}{3}\frac{2}{3}0$; $\frac{2}{3}\frac{1}{3}\frac{1}{3}$; $\frac{1}{3}\frac{2}{3}\frac{2}{3}$; $\frac{2}{3}\frac{1}{3}\frac{2}{3}$. An alternative

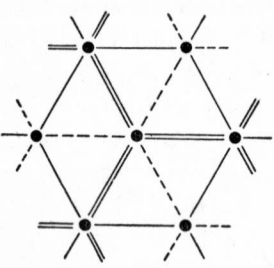

Fig. 69. Basal projection of proposed new graphite structure.

description in terms of a primitive unit would be in the rhombohedral system with $a = 3.642$ Å, $\alpha = 39.5°$, the space group being $D_{3d}^5 - R\bar{3}m$. Two atoms then occupy the special positions xxx; $\bar{x}\bar{x}\bar{x}$; the value of x being $\frac{1}{6}$. Although based on rather slight evidence, it seems likely that this attractive structure may well be present in natural graphite crystals. There is also some evidence that it may arise naturally from the fundamental manner of growth of the crystals.

2. MOLECULAR CRYSTALS

Although considerable progress had been made by 1921 in the determination of the structures of simple inorganic and mineral substances, very few organic crystals had until then been examined by the X-ray method. The most simple substances of this kind which could easily be obtained in crystalline form were known to contain a dozen or more atoms in the molecule; and the unit cells of the crystals would presumably contain several molecules. The forbidding complexity of the problem acted as a deterrent, although it was clear that the greatest application of the X-ray method would probably lie ultimately in this field.

In 1921 W. H. Bragg[16] published the preliminary results of X-ray examination of a number of fairly complex organic crystals, including naphthalene, anthracene, acenaphthene, α- and β-naphthol, and benzoic acid. In this work he introduced the

[16] W. H. Bragg, *Proc. Phys. Soc.* (London). 1921, **34**, 33; 1922, **35**, 167.

important idea that certain units of structure, like the benzene or naphthalene ring, having definite size and form, might be preserved with little or no alteration in passing from one crystalline derivative to another.   Furthermore the size and shape of these units could be deduced from the already-known structures of diamond and graphite and so compared with X-ray measurements of the unit cell dimensions of the organic crystals.   The effect of known chemical substitutions, for example, hydroxyl for hydrogen in the formation of the α- and β-naphthols, might then be expected to reveal themselves in dimensional changes on passing to these new crystals.   This general line of attack, based on the ideas of organic chemistry, seemed much more promising than any attempt to determine the absolute positions in space of 20 or 30 atoms governed by independent parameters.

These ideas may now appear simple and obvious, but it must be remembered that in the great majority of crystal structures determined up to this time the chemical molecule had not been found to possess any individual existence.   The structural arrangements were determined, rather, by the relative sizes of the atoms or ions, together with their mutual electrical charges.

Applied to naphthalene and anthracene, the new method was immediately successful in determining one of the main features in these structures, namely, that the long axes of the molecules are nearly coincident with the c crystal axes.   Fig. 70 shows the unit cells of these two crystals and the relative positions of the

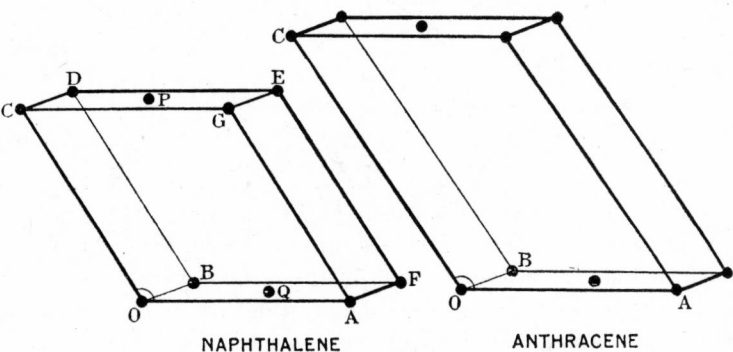

NAPHTHALENE   ANTHRACENE

Fig. 70. Unit cells of naphthalene and anthracene (Bragg).

	OA = a	OB = b	OC = c	AOC = β
Naphthalene	8.24	6.00	8.66 Å	122.9°
Anthracene	8.56	6.04	11.16	124.7°

two molecules contained within them, as deduced by Bragg from ionisation spectrometer measurements. It is obvious that while a, b, and β remain almost the same in the two unit cells, the c axis increases in length by about 2.5 Å in passing from naphthalene to anthracene. Now if the extra ring required by the chemical formula has the dimensions of the rings of carbon atoms found in diamond or graphite, it would account for this addition of about 2.5 Å. Bragg concluded that in both crystals the molecules lie end to end along the c axis and that the structures are similar.

A more convincing demonstration of the correctness of this hypothesis was discovered shortly afterwards.[17] If the molecules lie along the c axis and if the rings are regular, there will be certain periodicities in the distribution of the carbon atoms along this axis. For carbon rings of the dimensions found in diamond or graphite, periodicities of about 1.25 Å and 2.5 Å might be expected (Fig. 71). We should therefore find the X-ray reflections enhanced whenever a set of planes divides the long axis into the fractions $c/1.25$ or $c/2.5$; and hence those planes for which the l index approximates to these values should give strong reflections. On making a statistical survey of the distribution of intensities

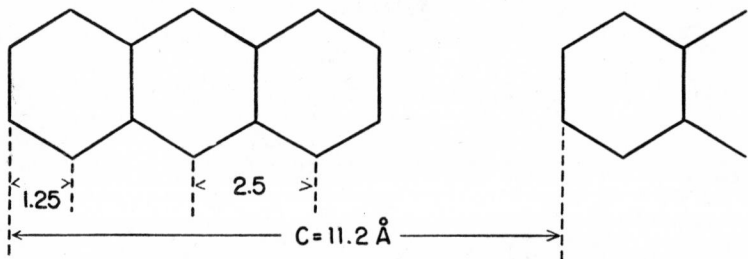

Fig. 71. Expected periodicities in anthracene.

for a large number of planes in naphthalene and anthracene, a very pronounced effect of this kind was discovered. This is shown in Fig. 72, where the sum of the observed intensities for a large number of reflections in each crystal is plotted against the value of the index l, divided by the c axial length. The upper curve refers to naphthalene, where a larger scale factor was employed for the intensities. Distinct maxima occur on both

[17] W. H. Bragg, *Z. Krist.*, 1928, **66**, 22.

curves at approximately 0.4 and 0.8, corresponding to c/l values, or periodicities, of about 2.5 and 1.25 Å, as expected.

The same effect, at least as regards the larger periodicity of 2.5 Å, is shown very clearly on rotation photographs taken about the c axes of these crystals (Fig. 86, p. 194). In anthracene, the fourth layer line, carrying all the reflections for which

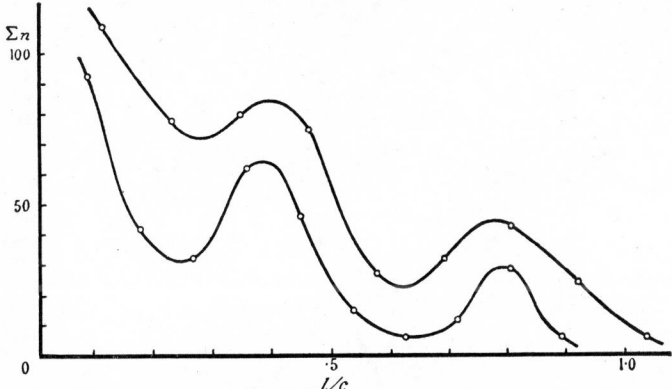

Fig. 72. Statistical distribution of intensities in naphthalene and anthracene.

$l=4$, is very strong; while in naphthalene similar enhancement occurs when $l=3$.

These early investigations have been described at some length because they were fundamental and demonstrated for the first time that in the crystals of organic compounds the molecules, predicted through long and careful study of chemical reactions, behaved like rigid building units and retained their identity. Their over-all size could be estimated from the dimensions of the carbon frameworks in diamond and graphite. It was also apparent that these units must be separated from each other by considerable gaps, such as occur between the carbon networks of the graphite crystal.

While these results were important, and initiated the X-ray study of organic crystals, attempts to extend the methods used for naphthalene and anthracene to other compounds, and so deduce the molecular arrangement and orientation, were disappointing and in many cases misleading. Even in the case of naphthalene and anthracene themselves, although the arrangement along the c axis was established, the sideways disposition of the molecules, deduced from packing considerations and a few isolated intensity observations, proved later to be completely erroneous. In other crystals, changes in cell dimensions occurring on substitution of certain organic groupings for others proved hard to interpret, and structures proposed on this basis have not been confirmed. With a few very notable exceptions, described later in this chapter, most of the work on organic crystals pub-

Fig. 73. Molecules of different shapes packed in identical unit cells (Bunn).

lished before 1932 can be regarded as quite unreliable in so far as it attempts any detailed description of atomic arrangement.

The difficulty which occurs in deducing information concerning the shapes of molecules from unit cell measurements is graphically illustrated in Fig. 73 (from Bunn[18]), which shows how molecules of very different shapes may be packed in identical unit cells. This diagram refers to the simple case of one molecule per cell, which is rare. When it is realised that there are frequently from four to sixteen or more organic molecules in the unit cell, interrelated by various symmetry operations, the full complexity of the problem and the difficulty of making any reliable deductions will be obvious.

[18] C. W. Bunn, *Chemical Crystallography*, Oxford: Clarendon Press, 1945.

164

Further progress in the X-ray analysis of organic crystals had to be based on adequate surveys of the intensities of reflections from many different zones in the crystals, and not confined to a few isolated measurements. The introduction of absolute measurements with the proper evaluation of the structure factors was an important step, together with the further development of photographic methods for recording the complex spectra on moving films. These advances, together with the evolution of the Fourier method for refining the structural parameters, gradually led to the reliable determination of complex organic structures, some of which are described in the following chapters.

3. LONG CHAIN COMPOUNDS

Some of the earliest and most direct evidence concerning atomic distributions in organic crystals was obtained from the long chain compounds (n-hydrocarbons, alcohols, ketones, acids, and esters). In one sense these structures are extremely simple and attractive, but later investigation has shown that they present a number of intricate problems, many of which still remain unsolved. It is always difficult to isolate good single crystals from such material, and consequently the amount of diffraction data available is limited.

Early studies[19] on crystalline waxes of this class revealed the presence of certain long spacings which indicated that the zig-zag chains of the chemical formulas were actually extended in the crystal; and when homologous series of such compounds were examined[20] a regular increase of about 1.25 Å per additional carbon atom was found to occur in these long spacings, showing very clearly that the molecules must be oriented lengthwise between the planes, although not necessarily perpendicular to them.

When the ordinary powder method is applied to such material, a number of strong and fairly constant "side-spacings" of between 3.6 and 5 Å are obtained, together with the very faint

[19] W. Friedrick, *Physik. Z.*, 1913, **14**, 317; M. de Broglie and E. Friedel, *Compt. rend.*, 1923, **176**, 738; S. H. Piper and E. N. Grindley, *Proc. Phys. Soc.* (London), 1923, **35**, 269; **36**, 31; A. Muller, *J. Chem. Soc.*, 1923, **123**, 2043.

[20] G. Shearer, *J. Chem. Soc.*, 1923, **123**, 3151; A. Muller and G. Shearer, *ibid.*, 1923, **123**, 3156.

"long spacings" which alone are characteristic of the individual member of the series. The latter reflections can be recorded more easily if the flaky crystals are pressed as a thin layer on a glass plate and oscillated at grazing incidence to the X-ray beam. The effect of this treatment is to align the basal, or (001) planes, of all the crystals. The full perfection of a single crystal is not obtained in this way, but a large number of orders of the long spacing can be accurately recorded. The first results obtained by this method indicated molecular lengths of the expected order of magnitude, and these were in agreement with earlier results obtained by N. K. Adam[21] by an entirely different physical method involving area and density measurements on unimolecular surface films of the same compounds.

Further X-ray studies revealed a complex situation and showed that the compounds as a class are in general polymorphous, a fact confirmed by thermal data and microscopic investigation. Confining ourselves to the n-hydrocarbons as an example, it is found that when long spacings are plotted against the number of carbon atoms for many specimens, the values group themselves on at least three distinct straight lines, representing types that may be referred to as the A, B, and C modifications. The form adopted depends on temperature, length of chain, and degree of purity, but the A modification appears to be the normal form for most of the longer chain paraffins from C_{23} onwards, provided they have previously been molten.

It is now established that in the A modification the chain axis is perpendicular to the (001) plane in a rectangular cell, and this spacing therefore measures the length of the molecules, including, of course, the gap between the end of one molecule and the next. The B and C modifications are of lower symmetry, and it seems very likely that the chains must be inclined at certain constant angles to the (001) plane, to account for the shorter spacings. However, as no complete single crystal investigations have been made for any of these modifications, precise structural details are unknown. It should be added that the full picture is more complicated than we have outlined, because further reversible changes are known, common to all members of the series, which involve a change of cross-section of the cell, without

[21] N. K. Adam, *Proc. Roy. Soc.* (London), 1922, **A101**, 452; *The Physics and Chemistry of Surfaces*, 2d ed., Oxford, Clarendon Press, 1938.

TABLE IV. LONG SPACING MEASUREMENTS ON n-HYDROCARBONS

	M. pt.	A (normal form)	B	C
C_5H_{12}	$-131.5°$	—	7.36 Å	—
C_6H_{14}	$-\ 94.3°$	—	8.57	—
C_7H_{16}	$-\ 90.5°$	—	10.0	—
C_8H_{18}	$-\ 56.5°$	—	11.0	—
C_9H_{20}	$-\ 53.7°$	—	12.8	—
$C_{10}H_{22}$	$-\ 31°$	—	13.4	—
$C_{11}H_{24}$	$-\ 26°$	15.9 Å	—	—
$C_{12}H_{26}$	$-\ 12°$	—	—	—
$C_{13}H_{28}$	$-\ 6°$	—	—	—
$C_{14}H_{30}$	$5°$	—	—	—
$C_{15}H_{32}$	$10°$	21.0	—	—
$C_{16}H_{34}$	$20°$	—	20.9	—
$C_{17}H_{36}$	$22°$	23.6	—	—
$C_{18}H_{38}$	$28°$	25.4	23.3, 23.0	—
$C_{19}H_{40}$	$32°$	26.3	—	—
$C_{20}H_{42}$	$37°$	27.5	25.5	—
$C_{21}H_{44}$	$40°$	28.8	—	—
$C_{22}H_{46}$	$44°$	—	—	—
$C_{23}H_{48}$	$48°$	31.1	—	—
$C_{24}H_{50}$	$51°$	32.7	—	—
$C_{25}H_{52}$	$54°$	—	—	—
$C_{26}H_{54}$	$56.5°$	35.3, 35.1	32.7	31.11 Å
$C_{27}H_{56}$	$59.0°$	36.5	—	—
$C_{28}H_{58}$	$61.5°$	37	—	33.4
$C_{29}H_{60}$	$63.5°$	38.9, 38.8	—	—
$C_{30}H_{62}$	$65.7°$	40.1, 40.6	—	—
$C_{31}H_{64}$	$67.7°$	41.7, 41.3	—	—
$C_{32}H_{66}$	$69.6°$	42.4	—	37.9
$C_{33}H_{68}$	$71.8°$	—	—	—
$C_{34}H_{70}$	$72.7°$	45.6	—	40.1
$C_{35}H_{72}$	$74.5°$	46.7, 46.3	—	—
$C_{36}H_{74}$	$75.8°$	47.6	—	42.4
$C_{60}H_{122}$	$101°$	78.4	—	—

affecting the molecular tilt or long spacing.[22]

In Table IV a number of long spacing measurements, based mainly on pure synthetic specimens, are collected from the work of various observers, chiefly Muller[23] and Piper.[24]   These data

[22] A. Muller, *Proc. Roy. Soc.* (London), 1932, **A138**, 514.

[23] A. Muller, *Proc. Roy. Soc.* (London), 1930, **A127**, 417; A. Muller and W. B. Saville, *J. Chem. Soc.*, 1925, p. 599.

[24] S. H. Piper, A. C. Chibnall, S. J. Hopkins, A. Pollard, J. A. B. Smith, and E. F. Williams, *Biochem. J.*, 1931, **25**, 2072.

provide a valuable means of identification, but in applying such measurements to natural products care is necessary to ensure the purity of the material. With regard to mixtures, Piper has shown that only one (001) spacing occurs unless the components are separated by at least four carbon atoms. In equimolar binary mixtures differing by more than four atoms distorted crystals of both constituents generally separate. With constituents differing by less than four carbon atoms the observed spacing depends in a somewhat unpredictable way on the composition of the mixture.[24]

The first single crystal of a long chain n-hydrocarbon was obtained from nonicosane, $C_{29}H_{60}$, isolated in a state of high purity from oil of supa.[25] From single crystal photographs and quantitative intensity measurements on certain orders of (001), Muller[26] was able to effect a reasonably complete determination of the structure, which confirmed in detail the various features deduced earlier from powder and pressed layer measurements.

The orthorhombic cell (Table V) was found to contain four molecules of $C_{29}H_{60}$, lying strictly along the c axis, the true length of this axis being double the observed basal spacing. The space group appears to be D_{2h}^{16}, involving a plane of symmetry in the molecule, perpendicular to its length, but this is rather uncertain and the space group may be C_{2v}^{9}, not involving any molecular symmetry. Greatly enhanced reflections from the basal plane were observed in the region of the sixtieth order, and calculations based on the relative intensities of 0, 0, 60 and 0, 0, 62 provided an extremely sensitive measure of the distance s between successive atoms in either row of the chain. Expressed as a fraction of the c axis, the result obtained was $s/c = 0.03286 \pm 0.00002$, or, in absolute measure, $s = 2.542$ Å, to within about 0.5 per cent. This calculation, of course, assumes absolute uniformity along the chain, which may not be true; but it is an interesting demonstration of the extreme sensitivity of very high order reflections in calculations of interatomic distance.

With regard to the lateral disposition of the atoms, the intensity data are not so sensitive. The two rows in the chain appear to be rather widely separated (about 1.4 Å) and the zigzag angle

[25] G. G. Henderson, W. McNab, and J. M. Robertson, *J. Chem. Soc.*, 1926, p. 3077.

[26] A. Muller, *Proc. Roy. Soc.* (London), 1928, **A120**, 437.

is fairly acute (about 84° instead of the expected $109\frac{1}{2}°$). This result, deduced entirely from intensity data, may be due to the scattering effect of the two hydrogen atoms attached to each carbon, or possibly to thermal movements. If we *assume* a tetrahedral angle in the chain, it is significant that an almost normal carbon-carbon distance of 1.55 Å is obtained. In the crystal the chain molecules are separated by large lateral gaps of between 3.6 Å and 3.8 Å, the end to end approach being about 4 Å.

Later single crystal data on other *n*-hydrocarbons have confirmed this general type of structure. Unit cell measurements are collected in Table V, which includes figures for the long chain polymer, polyethylene, $(CH_2)_n$, analyzed by Bunn.[27] Although single crystals are not obtained in this case, the polymer yields powder and fibre diagrams which are capable of detailed and accurate analysis, and a Fourier synthesis of the structural unit has been made. In this case the repeat distance within the carbon chain is given directly by the fibre period on the X-ray photographs. The constancy of the *a* and *b* axial lengths in all these structures is remarkable.

A recent single crystal study[28] of the low temperature form of octadecane, $C_{18}H_{38}$ (not included in Table V), has shown it to be triclinic, with one molecule per unit cell. This structure is probably characteristic of the *B* series (Table IV), and the

TABLE V. CRYSTAL DATA FOR *n*-HYDROCARBONS

	Ref.	Space group	Mol. per unit cell	a	b	c	Repeat distance in chain
				(Å)	(Å)	(Å)	(Å)
$C_{29}H_{60}$	26	D_{2h}^{16}, C_{2v}^{9} (or C_{2h}^{5})	4	7.46	4.98	77.4	2.542
$C_{30}H_{62}$	29	D_{2h}^{16}, C_{2v}^{9} (or C_{2h}^{5})	4	7.46	4.98	81.8	2.53
$C_{35}H_{72}$	30	D_{2h}^{16}, C_{2v}^{9} (or C_{2h}^{5})	4	7.44	4.98	(92.6)	(2.55)
$C_{60}H_{122}$	30	D_{2h}^{16}, C_{2v}^{9} (or C_{2h}^{5})	4	7.45	4.96	(156.7)	2.55
$(CH_2)_n$	27	D_{2h}^{16}, C_{2v}^{9} (or C_{2h}^{5})	4	7.41	4.94	—	2.539

[27] C. W. Bunn, *Trans. Faraday Soc.*, 1939, **35**, 482.

[28] A. Muller and K. Lonsdale, *Acta Cryst.*, 1948, **1**, 129.

[29] R. Kohlkass and K. H. Soremba, *Z. Krist.*, 1938, **100**, 47; *Angew. Chem.*, 1938, **51**, 483.

[30] J. Hengstenberg, *Z. Krist.*, 1928, **67**, 583.

following spacings are recorded at room temperature: $d_{001} = 23.04$, $d_{100} = 4.09$, $d_{010} = 4.60$, $d_{111} = 3.62$ Å.

A great variety of other long chain compounds, including alcohols, ketones, acids, and many esters have been studied. The very complete analyses of potassium caprate, $C_9H_{19}COOK$, and of lauric acid, $C_{11}H_{23}COOH$, by Vand and others[31] are of special interest. In the former structure the carbon chains are inclined in opposite directions and cross each other in the monoclinic cell. Such complete structure determinations are still rare, however, and it is not possible to summarize the results here.

In one very early investigation Shearer[32] studied series of ketones of the general formula $CH_3(CH_2)_mCO(CH_2)_nCH_3$. From a study of the intensity distribution among the orders of the long spacing, in which the Fourier series method was employed, it was found possible to fix the positions of the carbonyl group in the chain with a fair degree of accuracy. This probably represents the first application of the X-ray method to the determination of chemically unknown organic structures, because Shearer, working on an unknown series of these ketones, was able to deduce correctly both the number of carbon atoms in the chains and the positions of the carbonyl groups.[33]

4. SOME COMPLETE STRUCTURE DETERMINATIONS

The early work on organic crystals included a few complete and fundamentally important structure determinations, which are considered briefly in this section. These compounds are in the main chemically unrelated; they should, however, be included here for historical reasons, and also because of the importance of these early results in the subsequent attack on more complex structures. Each of these isolated determinations became possible because of some special simplifying feature in the crystal structure. Consequently, the methods employed are not of universal application, but the quantitative information gained regarding molecular structure is of great importance.

In addition to these purely organic structures, it should be

[31] V. Vand, T. R. Lomer, and A. Lang, *Acta Cryst.*, 1949, **2**, 214; V. Vand, W. M. Morley, and T. R. Lomer, *ibid.*, 1951, **4**, 324.

[32] G. Shearer, *Proc. Roy. Soc.* (London), 1925, **A108**, 655.

[33] R. Robinson, *Nature*, 1925, **116**, 45.

noted that a large amount of accurate work was carried out on metallo-organic compounds, and particularly on substituted ammonium salts[34,35] during this period.   These structures are often of higher symmetry than the average organic compound. They lie somewhat outside our present field, and for a description reference should be made to the *Strukturbericht*[36] or to Wyckoff.[37]

4A. Hexamethylene tetramine and the cyclohexane hexahalides.—This remarkable determination by Dickinson and Raymond[38] in 1923 represents the earliest complete solution of a structure in the whole organic field.   A slightly later determination by Gonell and Mark[39] led to rather less accurate values for the parameters.

The compound, $C_6H_{12}N_4$, is one of the comparatively few organic substances which crystallise in the cubic system.   It can be prepared by the action of aqueous ammonia on aqueous formaldehyde,

$$4 \ NH_3 + 6 \ CH_2O = (CH_2)_6N_4 + 6 \ H_2O,$$

and the formula

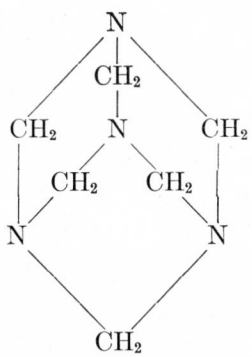

[34] S. B. Hendricks, *Z. Krist.*, 1928, **67**, 106, 119, 465, 472; **68**, 189.

[35] R. W. G. Wyckoff, *Z. Krist.*, 1928, **67**, 91, 550; **68**, 231.

[36] P. P. Ewald and C. Hermann, *Strukturbericht, 1913–1928* and later volumes.

[37] R. W. G. Wyckoff, *The Structure of Crystals,* 2d ed., New York: Chemical Catalog Co., 1931, and later editions.

[38] R. G. Dickinson and A. L. Raymond, *J. Amer. Chem. Soc.*, 1923, **45**, 22.

[39] G. W. Gonell and H. Mark, *Z. physik. Chem.*, 1923, **107**, 181.

although generally accepted was by no means conclusively proved by the chemical evidence until the structure was established by X-ray analysis.

Two molecules of $C_6H_{12}N_4$ occur in a body-centred cubic lattice, with $a = 7.03$ Å. The space group may be $T_d^3 - I\bar{4}3m$, or $T_d^4 - P\bar{4}3n$, but this does not introduce any ambiguity because in both cases the molecules are situated at $000; \frac{1}{2}\frac{1}{2}\frac{1}{2}$; the eight nitrogen atoms at $xxx; x\bar{x}\bar{x}; \bar{x}x\bar{x}; \bar{x}\bar{x}x; \frac{1}{2}+x, \frac{1}{2}+x, \frac{1}{2}+x; \frac{1}{2}+x, \frac{1}{2}-x, \frac{1}{2}-x; \frac{1}{2}-x, \frac{1}{2}+x, \frac{1}{2}-x; \frac{1}{2}-x, \frac{1}{2}-x, \frac{1}{2}+x;$ and the twelve carbon atoms at $x00; 0x0; 00x; \bar{x}00; 0\bar{x}0; 00\bar{x}; \frac{1}{2}+x, \frac{1}{2}, \frac{1}{2}; \frac{1}{2}, \frac{1}{2}+x, \frac{1}{2}; \frac{1}{2}, \frac{1}{2}, \frac{1}{2}+x; \frac{1}{2}-x, \frac{1}{2}, \frac{1}{2}; \frac{1}{2}, \frac{1}{2}-x, \frac{1}{2}; \frac{1}{2}, \frac{1}{2}, \frac{1}{2}-x$. Recent and probably the most accurate values for the two parameters are given as[40] $x_N = 0.120_4$

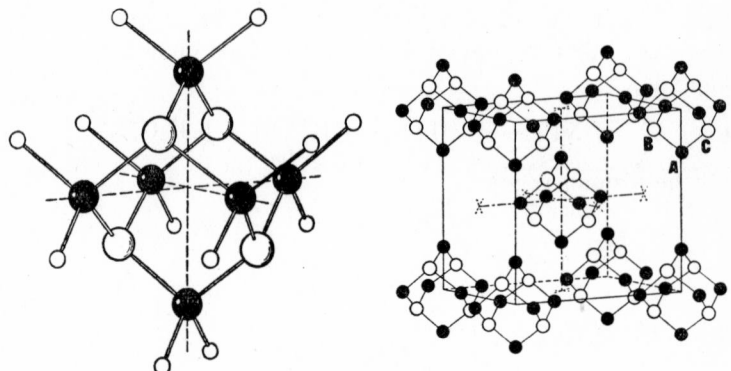

Fig. 74. Single molecule of $C_6H_{12}N_4$ showing hydrogens. (Dickinson and Raymond).

Fig. 75. Arrangement of $C_6H_{12}N_4$ molecules in the crystal.

± 0.0008, $x_C = 0.235_3 \pm 0.0004$, but these figures are identical, within the limits given, with those of the original investigation, which was carried out from a study of the Laue diagrams. Spectrometric studies were made in the later work. The carbon-nitrogen distance within the molecule is 1.44 Å, in reasonable agreement with electron diffraction experiments on the vapour,[41] which give this distance as 1.47 Å. The nearest

[40] R. Brill, H. G. Grimm, C. Hermann, and C. Peters, *Ann. Physik*, 1939(5), **34**, 393; R. W. G. Wyckoff and R. B. Corey, *Z. Krist.*, 1934, **89**, 462.

[41] G. C. Hampson and A. J. Stosick, *J. Amer. Chem. Soc.*, 1938, **60**, 1814.

approach of CH_2 groups of different molecules is 3.72 Å, again in good agreement with later studies on other organic crystals.

The spatial arrangements of the atoms in the molecule and of the molecules in the crystal are shown in Figs. 74 and 75. If we think of the structure in terms of the chemical formula given above, it should be noted that the outer hexagon is "chair" shaped, the three inner CH_2 groups are vertically above the three outer nitrogen atoms, and a trigonal axis of the cube is perpendicular to the paper through the central nitrogen atom. Alternatively, the configuration may be described as consisting of four nitrogen atoms tetrahedrally, and six CH_2 groups octahedrally, grouped about a common centre. The molecular symmetry displayed in the crystal, $Td - \overline{4}3m$, is the maximum possible for this molecule.

Two other organic crystals belonging to the cubic system are β-benzene hexachloride, $C_6H_6Cl_6$, and hexabromide, $C_6H_6Br_6$. These structures were also very thoroughly examined at an early date.[42] The space group in each case is $T_h^6 - Pa3$, with four molecules centred on the points 000; $0\frac{1}{2}\frac{1}{2}$; $\frac{1}{2}0\frac{1}{2}$; $\frac{1}{2}\frac{1}{2}0$. The 24 halogen atoms occupy the 24 general positions in this space group and are thus governed by only three parameters, which have been quite accurately determined. The result leads to regularly staggered arrays containing six bromine atoms, 3.40 Å apart, grouped about the above points. Owing to the great difference in scattering power, no direct X-ray evidence was obtained regarding the carbon positions, but the results are in agreement with a molecule containing a cyclohexane ring of "tetrahedral" carbon atoms, 1.54 Å apart, situated within the halogen ring. Such a postulated structure leads to carbon-bromine distances of 1.94 Å, and carbon-chlorine distances of 1.81 Å, and is almost certainly correct. The halogens of neighbouring molecules are separated by minimum distances varying between 3.7 and 4.0 Å.

Although the carbon positions are not directly determined, each molecule must possess a 3-fold axis and a centre of symmetry, and this, taken in conjunction with the known halogen positions, is sufficient to define the chemical configuration of these isomers, which may be represented as

[42] S. B. Hendricks and C. Bilicke, *J. Amer. Chem. Soc.*, 1926, **48**, 3007; R. G. Dickinson and C. Bilicke, *ibid.*, 1928, **50**, 764.

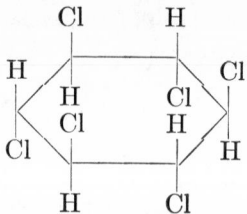

the carbon ring being of the "chair" or Z-form. Again, as in hexamethylene tetramine, the molecular symmetry displayed is the maximum possible.

4B. Urea and thiourea.—Among organic crystals, tetragonal symmetry is more common than cubic, but it is still comparatively rare. Urea is one of the few simple compounds belonging to this system. Early studies[43] did not completely solve the structure, but by 1928 a reasonably accurate determination was completed by Hendricks,[44] and the structure has been further refined by Wyckoff[45] and others.

The tetragonal crystals, with $a = 5.681$, $c = 4.735$ Å, contain two molecules of $CO(NH_2)_2$ per unit cell. The space group being $D_{2d}^3 - P\bar{4}2_1m$, the molecule must possess 4-fold symmetry consisting of two mutually perpendicular planes of symmetry and a 2-fold axis ($C_{2v} - mm$). This reduces the number of parameters governing the positions of the four heavier atoms from twelve in the general case to only four, and so makes the problem manageable. Analytically, the carbon and oxygen atoms are at $0\frac{1}{2}z$; $\frac{1}{2}0\bar{z}$; with $z_C = 0.335$ and $z_O = 0.60$, while the four nitrogen atoms are at x, $\frac{1}{2}-x$, z; $\bar{x}$, $\frac{1}{2}+x$, z; $\frac{1}{2}+x$, x, $\bar{z}$; $\frac{1}{2}-x$, $\bar{x}$, $\bar{z}$; with $x_N = 0.145$ and $z_N = 0.18$. This leads to the structure shown in projection in Fig. 76, the molecular symmetry planes being parallel to the diagonals of the square base. The carbon and oxygen atoms are superimposed in the projection, carbon being above oxygen, and oxygen above carbon in alternate molecules.

The molecular dimensions obtained are C—O = 1.25 Å, C—N = 1.37 Å, with the angle N—C—N = 116°. Between

[43] K. Becker and W. Jancke, *Z. physik. Chem.*, 1921, **99**, 242, 267; H. Mark and K. Weissenberg, *Z. Physik*, 1923, **16**, 1.

[44] S. B. Hendricks, *J. Amer. Chem. Soc.*, 1928, **50**, 2455.

[45] R. W. G. Wyckoff, *Z. Krist.*, 1930, **75**, 529; 1932, **81**, 102; R. W. G. Wyckoff and R. B. Corey, *ibid.*, 1934, **89**, 462; P. Vaughan and J. Donohue, *Acta Cryst.*, 1952, **5**, 530.

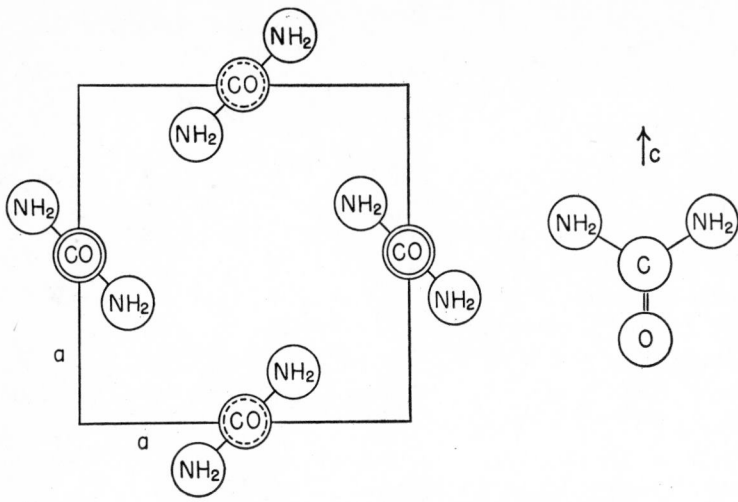

Fig. 76. Basal projection of urea structure.

neighbouring molecules the closest approach, NH $\cdots$ O, is about 3.0 Å. (Later work gives C—O $= 1.26$ Å and C—N $= 1.34$ Å.)

The crystal structure of thiourea, $SC(NH_2)_2$, is slightly more complicated, but it was also established at about the same time.[44,46] In this case there are four molecules in an orthorhombic cell, the space group being D_{2h}^{16}. The molecule has a plane of symmetry passing through the sulphur and carbon atoms, the nitrogen atoms occupying equivalent general positions. There are now seven parameters to determine, as against four in the case of the urea structure. From an accurate refinement of these parameters by the double Fourier series method,[46] the following interatomic distances were obtained: C—N $= 1.35$ Å, C—S $= 1.64$ Å, with intermolecular contacts of $NH_2 \cdots S = 3.45$ Å, and $NH_2 \cdots NH_2 = 3.85$ Å. Although the molecular symmetry displayed in the crystal does not require a coplanar molecule, as in the case of urea, yet the values obtained for the parameters show that the sulphur, carbon, and nitrogen atoms do lie quite accurately in one plane. Moreover, the exact crystallographic equivalence of the nitrogen atoms is in keeping with the formula

[46] L. Demény and I. Nitta, *Bull. Chem. Soc. Japan*, 1928, **3**, 128; R. W. G. Wyckoff and R. B. Corey, *Z. Krist.*, 1932, **81**, 386.

$$S=C \overset{\displaystyle NH_2}{\underset{\displaystyle NH_2}{\Big\langle}} \quad .$$

4C. Hexamethylbenzene and durene.—The structure determinations so far dealt with in this section were rendered possible largely because high crystallographic symmetry reduced the number of independent parameters to manageable proportions. High symmetry, however, is exceptional in the organic field, and the first outstanding determination of a more typical structure was achieved by K. Lonsdale[47] in 1928 in the analysis of hexamethylbenzene. This work established the true geometry of the benzene ring and paved the way for future X-ray studies on the aromatic hydrocarbons.

Hexamethylbenzene crystals are triclinic, and there is only one molecule of $C_6(CH_3)_6$ in the unit cell. The space group may be either C_1-P1 or $C_i-P\bar{1}$, which cannot be distinguished by any direct X-ray evidence, so that the 12 carbon atoms, occupying general positions, are really governed by 36 independent parameters. There are, however, simplifying features of a different nature in this structure. A well-pronounced cleavage plane, (001), gives an exceptionally strong X-ray reflection, and the higher orders of this reflection are found to fall off uniformly in intensity, and in the same manner as the reflections from the basal plane in graphite. It follows with a high degree of certainty that the atoms must all lie on this plane, at least to within very narrow limits. Moreover, an examination of the ($hk0$) reflections reveals a marked similarity in the structure factors for planes differing by successive rotations of $\pi/3$ about this zone axis, which indicates most clearly a hexagonal arrangement of the atoms about this zone axis.

From these considerations a simple model can be constructed, based on a regular planar benzene hexagon lying in the (001) plane. This can be described in terms of three parameters, one defining its orientation in the plane, one the benzene carbon-carbon distance, or ring radius, and one giving the distance between the benzene ring carbon and the methyl group carbon.

[47] K. Lonsdale, *Nature*, 1928, **122**, 810; *Proc. Roy. Soc.* (London), 1929, **A123**, 494; *Trans. Faraday Soc.*, 1929, **25**, 352.

The analysis carried out on this basis showed the benzene ring to be of about the same size as the carbon rings in graphite (1.42 Å) and that the methyl groups lay in the same plane but at a somewhat greater carbon-carbon distance.

A later refinement of this structure by the double Fourier series method[48] gave the electron density projection shown in Fig. 77, from which the interatomic distances can be measured with some accuracy. Structure factor calculations based on models of differing bond lengths, in the range between 1.39 Å

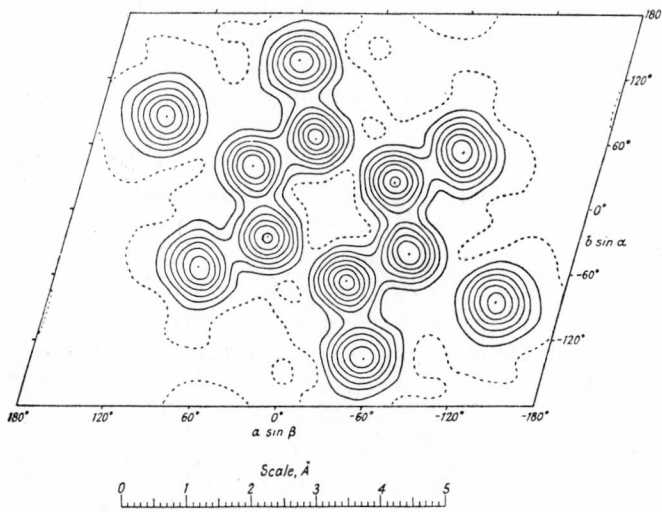

Fig. 77. Hexamethylbenzene. Electron density map for c axis projection. Contours are drawn at unit electron intervals.

and 1.54 Å, yield the same phase constants, and so the results of the Fourier synthesis are independent of the precise nature of the assumptions made regarding the initial model. The final results give a bond length of 1.39 Å within the benzene ring and 1.53 Å between the methyl group carbon and the benzene carbon. This result is in agreement with the expectation that a methyl group bond attached to a fully substituted benzene ring should be of about the same length as in saturated aliphatic hydrocarbons.

It should be noted that the projection shown in Fig. 77 is drawn on a plane perpendicular to the c axis along which the

[48] L. O. Brockway and J. M. Robertson, *J. Chem. Soc.*, 1939, p. 1324.

projection is made, and owing to the triclinic axes this plane does not coincide with the (001) plane in which the molecule lies but is inclined to it at a fairly large angle (about 46°). Most of the distances are thus foreshortened in the projection, but the two clearly resolved and isolated methyl groups, near the obtuse angles of the unit cell, and their adjacent benzene carbon atoms, happen to lie almost in the projection plane, and here the distances can be measured directly.

It may also be noted here that Fig. 77, as well as most of the similar electron density projections that follow, is based on an approximately absolute scale of structure factors. The contour lines are then conveniently drawn at unit electron intervals, that is, each line denotes an increment of electron density amounting to approximately one electron per $Å^2$. The first line, generally ill-defined, is dotted. On this scale the peak density at the centre of each carbon atom is generally in the region of from six to eight electrons per $Å^2$. The precise value, however, is found to vary rather widely in different crystals, and is obviously dependent on the temperature factor, or degree of thermal movement which the atom undergoes. This peak value is not to be confused with the number of electrons associated with each scattering centre. This latter quantity can only be obtained by integrating the density function over the area of the atom. When this is done a value close to the atomic number for the atom concerned is generally obtained.

The mutual arrangement of the hexamethylbenzene molecules in a single (001) layer of the crystal is shown by the skeleton drawing in Fig. 78. The closest approach of the methyl group carbons in adjoining molecules is about 3.9 to 4.0 Å. No direct X-ray evidence is available regarding the hydrogen atoms, but these separations are undoubtedly governed by repulsions between the methyl group hydrogens. Assuming a C—H distance of about 1.1 Å and rotation of the methyl groups, the H · · · H distance will be about 2.0 Å. In the crystal successive layers of the kind shown in Fig. 78 are translated with respect to each other so that the methyl groups do not lie directly over one another. This allows a closer packing of the molecules than would otherwise be possible while maintaining the hydrogen-hydrogen intermolecular separations at about 2.0 Å. The actual distance between successive layers is given by the (001) spacing of 3.66 Å.

178

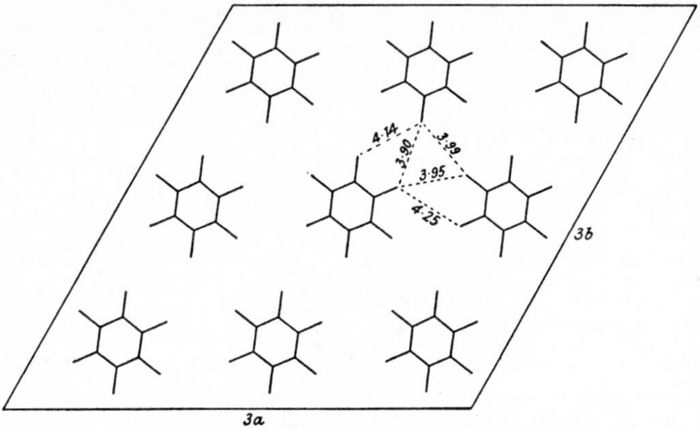

Fig. 78. A single layer of molecules in the (001) plane.

These packing considerations probably account for the fact that full hexagonal symmetry is not utilized in building up the crystal structure. The detailed analysis also indicates that the molecular planes are tilted very slightly out of the (001) crystal plane, by just over 1°. This slight deviation from regularity can be explained in the same way, as it permits a somewhat greater hydrogen separation. It is interesting to note that recent work[49] has revealed a higher temperature modification of hexamethylbenzene, which conforms to the orthorhombic system and in which a closer approximation to hexagonal symmetry is attained by the crystal structure.

Other hexa-substituted benzene derivatives, including hexachlorobenzene,[50,51] hexabromobenzene,[50] and hexaminobenzene,[52] were examined at an early date, but they did not reveal the same simplicity of crystal structure as hexamethylbenzene. The later analysis of hexachlorobenzene[51] is detailed and interesting, including what is perhaps the first published electron density map of an organic crystal by the double Fourier series method, but the precise orientation of the molecule in the crystal was not ascertained.

Although the structure of durene, or *sym.* tetramethyl-

[49] T. Watanabe, Y. Saito, and H. Chihara, *Scientific Papers from Osaka University*, 1949, no. 2, 9–14.

[50] H. Mark, *Ber.*, 1924, **57**, 1826; W. G. Plummer, *Phil. Mag.*, 1925, **50**, 1214.

[51] K. Lonsdale, *Proc. Roy. Soc.* (London), 1931, **A133**, 536.

[52] I. E. Knaggs, *Proc. Roy. Soc.* (London), 1931, **A131**, 612.

benzene, $C_6H_2(CH_3)_4$, was determined a little later,[53] it is convenient to describe it briefly here, for comparison with hexamethylbenzene. The crystals are monoclinic, space group $C_{2h}^5 - P2_1/a$, with two molecules in the unit cell. It follows that the molecules must each possess a centre of symmetry, and that these two molecular centres are situated at 000; $\frac{1}{2}\frac{1}{2}0$. The dimensions of the unit cell, however, provide no clue regarding the orientation of the molecules in the crystal, and further analysis has to depend entirely upon a study of the X-ray intensities. Here a further difficulty arises, because a survey of these intensities reveals no particularly outstanding feature, such as might be expected if a planar molecule coincided even approximately with any of the simple crystallographic planes.

The method adopted in this case and for many later analyses of organic compounds consisted in the first place of setting up a molecular model based on the known chemical structure. The detailed dimensions do not matter greatly. It is generally sufficient to adopt some mean value for the carbon-carbon distances, close to that of graphite (1.42 Å), or of diamond (1.54 Å) if aliphatic groups are involved. The atomic co-ordinates can then be referred to three mutually perpendicular axes drawn through some fixed point in the molecule, such as a centre of symmetry if one is present. The orientation of the molecule in the crystal may be specified by the angles that these molecular axes make with the crystallographic axes, and the crystal co-ordinates corresponding to any given orientation of the model are then easily evaluated. This process effectively reduces the problem to three parameters for any given model, if the molecular centres are fixed by space group considerations. But in the non-centrosymmetric case, or if the molecules occupy general positions, a further three parameters specifying the position of the molecule are usually involved.

In such work the use of absolute intensity measurements is of considerable help, because the calculation of a few structure factors is then usually sufficient to narrow down the possible orientation or position of the proposed model considerably. The general procedure is to determine the orientation roughly from the low index reflections and refine by a careful study of the higher

[53] J. M. Robertson, *Proc. Roy. Soc.* (London), 1933, **A141**, 594; **A142**, 659.

index reflections, which are more sensitive to small movements.

As soon as some sensible measure of agreement is obtained, further refinement and actual measurement of the bond lengths and valency angles can proceed by the application of the Fourier series method. In many cases, as in the durene structure, there will be some projection available on which a certain number of atoms are resolved, and the double Fourier series method can be employed, at least in the first stages of the work.

0 1 2 3 4 5 Å

Fig. 79. Durene. Electron density map for *b* axis projection, showing arrangement of molecules in the crystal.

These general methods were employed in the analysis of the durene structure. The electron density projection showing greatest resolution is the one made along the symmetry axis *b*, and this is reproduced in Fig. 79. It shows the mutual arrangement of a group of molecules in the crystal, and the large open spaces around each molecule are immediately evident. In spite of the more complex arrangement, the methyl group contacts between adjoining molecules are almost exactly the same as in hexamethylbenzene. The three closest approaches are 3.9, 4.1, and 4.2 Å, the two longer contacts being across the cleavage plane, which coincides with the large horizontal gap in Fig. 79.

The durene molecules are planar, but their planes are inclined at about 49° to the projection plane, (010), used in Fig. 79. Successive molecules in each row are oppositely inclined to this

181

plane, as depicted by the end view given in Fig. 80; but in the *b* axis projection these oppositely inclined molecules are identical, as required by the glide plane *a*.   The complete orientation of the molecule in the crystal can be worked out by studying projections obtained by the Fourier method along the different crystallographic axes.   These other projections, however, do not resolve the separate atoms.

The steep inclination of the molecule to the principal projection plane prohibits any very accurate measurement of the bond lengths, but the mean carbon-carbon distance in the benzene ring is 1.39 Å as before.   The distance of the methyl group carbon from the benzene carbon was at first obtained as 1.47 Å, but later calculations[48] show that the structure factors are insensitive to variations between 1.47 and 1.54 Å in this distance. These figures may be taken as limits.   The true value is likely to be over 1.50 Å, recent accurate electron diffraction studies[54] on toluene vapour giving a methyl group bond distance of 1.51 Å for the monosubstituted derivative.

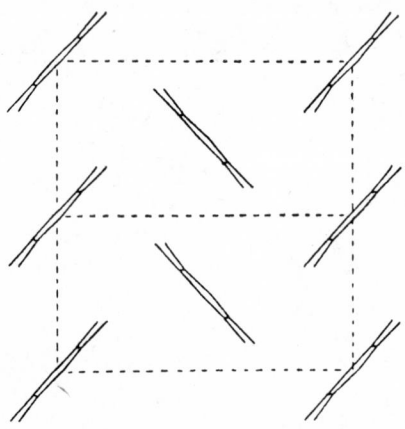

Fig. 80. Durene. Projection on (001).

A further feature observed on the electron density projection of Fig. 79 is a slight mutual displacement of the methyl groups away from each other towards the unsubstituted positions of the benzene ring.   The displacement is about 3° from the symmetrical position, and is probably a real effect.

[54] F. A. Keidel and S. H. Bauer, private communication, 1951.

VIII

Bond Length Measurements in Condensed Ring Hydrocarbons and Some Related Structures

Even when the chemical structure of a compound is well established and the relative positions of all the atoms are known, detailed X-ray analysis can still provide much new and fundamental information. Examples have been given in the previous chapter. In general, a rigorous determination of the molecular geometry can only be achieved by X-ray methods. This includes the task of measuring the bond lengths and angles quantitatively, as well as that of attempting a complete survey of the electron distribution. A fairly large number of organic compounds has now been examined with sufficient accuracy to provide this information. In the present chapter attention is confined chiefly to the condensed six-membered ring aromatic hydrocarbons, because here the survey of the principal types is reasonably complete. The comparatively simple atomic arrangements in these molecules have also attracted a considerable amount of careful theoretical study, and so it is possible in many cases to present rather detailed comparisons between theory and experiment in this field. Crystallographic data, for the compounds mentioned and a few others, are collected in Table VI (p. 216).

1. THE NAPHTHALENE-ANTHRACENE SERIES

The early X-ray work on naphthalene and anthracene has been described in the previous chapter. Although the alignment of the molecules in the c axial direction was established and the fundamental periodicities had been observed, the full solution of these structures was not obtained until later. The analysis of hexamethylbenzene first demonstrated the strictly planar arrangement of the atoms in the aromatic ring, while a study of the optical and magnetic anisotropies of naphthalene crystals by Bhagavantam[1] showed that the rings must lie closer to the *bc* than to the *ac* plane in this crystal. Optical and magnetic measurements of this kind later proved to be of great importance as an aid to the determination of many organic crystal structures, by giving a clue to, and sometimes a very precise indication of, the molecular orientation in the crystal.[2] The orientation indicated

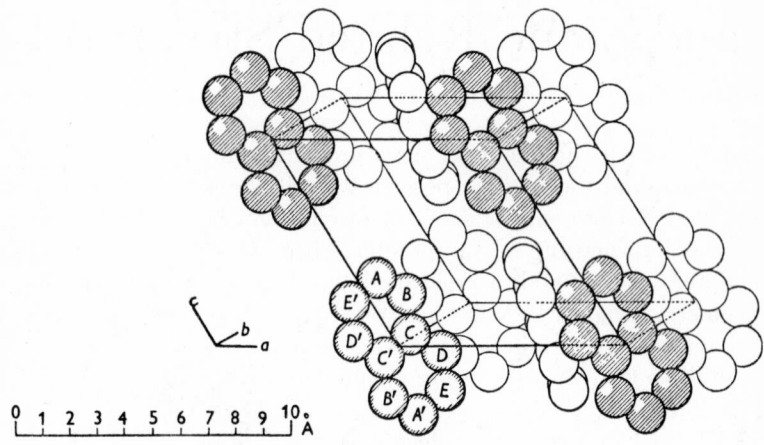

Fig. 81. Naphthalene crystal structure

for naphthalene and anthracene was confirmed by a number of absolute X-ray intensity measurements made by W. H. Bragg and by a further X-ray study by Banerjee.[3]

The first accurate determination of these structures, based on a large number of absolute intensity measurements, was completed by the double Fourier series method shortly afterwards,[4]

[1] S. Bhagavantam, *Proc. Roy. Soc.* (London), 1929, **A124**, 545.

[2] K. Lonsdale and K. S. Krishnan, *Proc. Roy. Soc.* (London), 1936, **A156**, 597; L. Pauling, *J. Chem. Phys.*, 1936, **4**, 673; W. A. Wooster. *Crystal Physics*, Cambridge: Cambridge University Press, 1938.

[3] K. Banerjee and J. M. Robertson, *Nature*, 1930, **125**, 456; K. Banerjee, *Indian J. Phys.*, 1930, **4**, 557.

[4] J. M. Robertson, *Proc. Roy. Soc.* (London), 1933, **A140**, 79; 1933, **A142**, 674.

184

and the results established the strictly planar form of the molecules to within narrow limits and gave their orientation in the crystal to within about 1°.   The arrangement in the unit cell is shown for naphthalene in Fig. 81.   The long axes of the molecules are not exactly coincident with the crystal c-axes, as at first supposed, the deviation of naphthalene being about 14° and of anthracene about 9°.   In both cases the molecular planes are

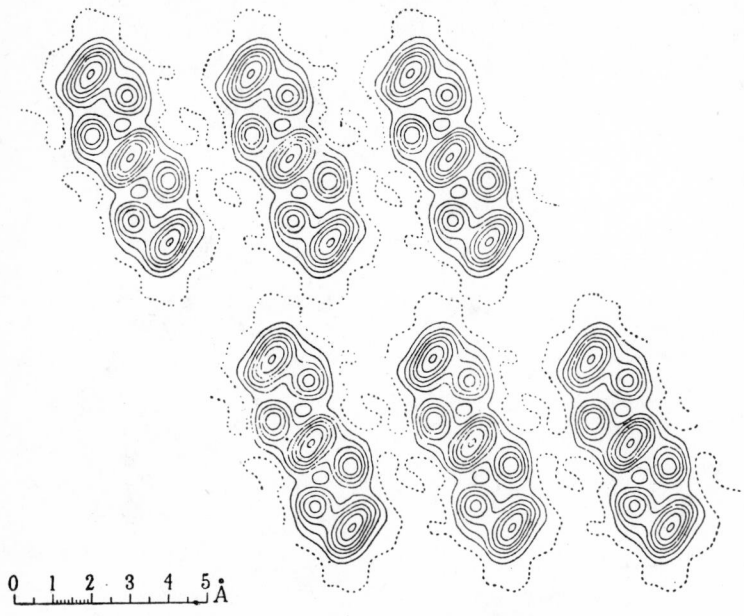

0 1 2 3 4 5 Å

Fig. 82. Naphthalene. Electron density projected on (010).

steeply inclined, at an angle of about 64°, to the (010) or *ac* plane.   As a consequence, the electron density projection on this plane, which is illustrated for naphthalene in Fig. 82, does not succeed in resolving all the atoms, although the over-all dimensions of the rings can be accurately measured.   Other atoms can be resolved in other projections, but in no case is a complete picture obtained.   For both naphthalene and anthracene an average carbon-carbon distance of 1.41 Å was found by combining the results from the projections made along the three principal crystal axes in each case.

Although this investigation was quantitative and detailed, it

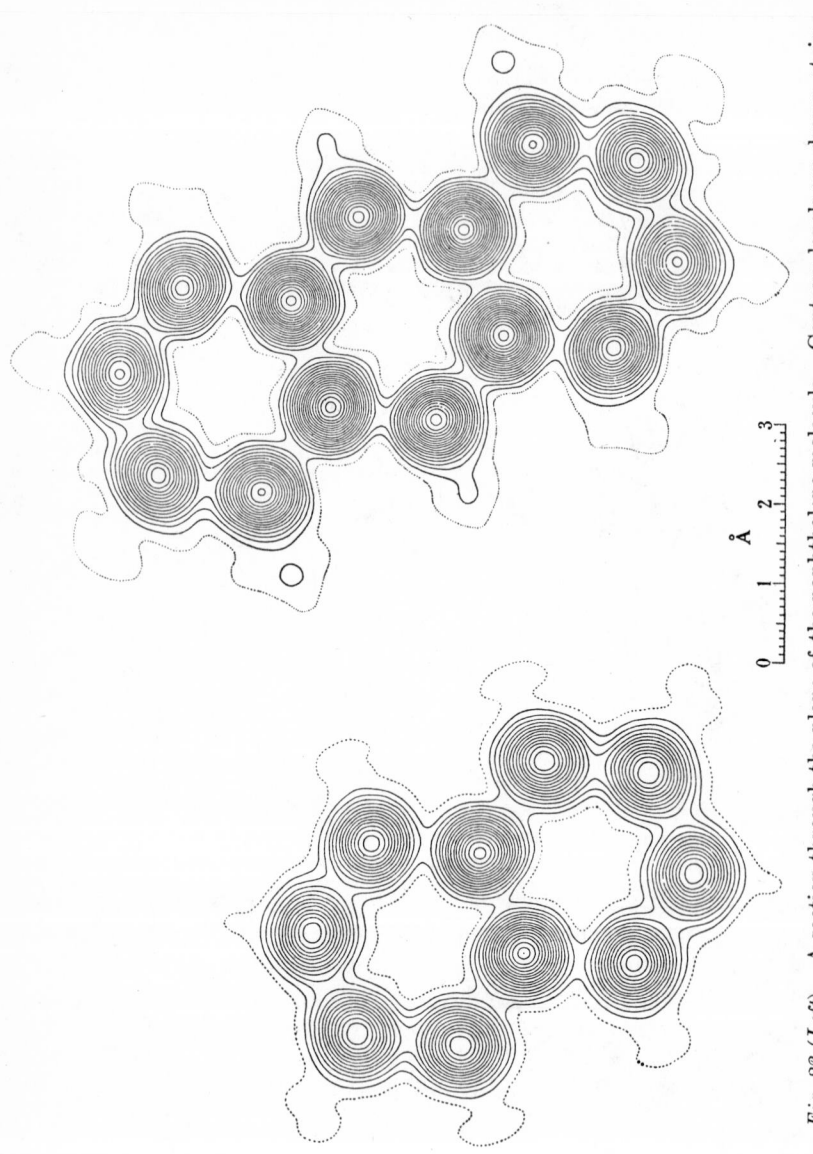

Fig. 83 (Left). A section through the plane of the naphthalene molecule. Contour levels are drawn at intervals of ½ electron per Å³, the half electron line being dotted. *Fig. 84 (Right).* Section through the plane of the anthracene molecule. Scale is as in Fig. 83.

still does not represent the X-ray method at its full power, and the results fail to answer the more interesting questions concerning these molecular structures, such as whether the bond lengths are all equal or whether they vary slightly in different parts of the molecule, as theory predicts. To answer these questions a more comprehensive investigation has been undertaken, for both naphthalene[5] and anthracene.[6] In this work a complete survey was made of all the X-ray reflections given by copper radiation from these two crystals, totalling 691 for anthracene and 644 for naphthalene. By using the atomic co-ordinates of the earlier two-dimensional analysis as a first approximation, reliable phase constants could be calculated for most of the structure factors. The electron density distribution was then evaluated by the triple Fourier series method at 54,000 points over the asymmetric crystal unit (half a chemical molecule) in each structure.

The vast amount of information contained in these surveys, derived from the addition of about 70,000,000 terms in the two series employed, has not yet been fully explored. A number of sections have been made, but the most interesting are those evaluated through the mean molecular planes, and reproduced in Figs. 83 and 84. In both cases the molecules are found to be coplanar to within 0.01 Å, and the valency angles are all 120° to within 2°. As the molecular planes do not coincide with any simple crystallographic plane, the evaluation of the electron density function in this manner is troublesome, but it has the advantage of giving an extremely direct picture of the molecular structure, on which the bond lengths and angles can be measured directly. The resolution is high, and it will be noted that the hydrogen electron clouds are clearly visible, although distorted, at the half electron per $Å^3$ level. In anthracene, which has a somewhat lower temperature factor, the hydrogen resolution is better.

In the crystal these molecules display only a centre of symmetry. Consequently independent measurements can be made on bonds which are chemically equivalent. We expect these bonds to be of the same length, and the measurements show this

[5] S. C. Abrahams, J. M. Robertson, and J. G. White, *Acta Cryst.*, 1949, **2**, 233, 238.

[6] A. McL. Mathieson, J. M. Robertson, and V. C. Sinclair, *Acta Cryst.*, 1950, **3**, 245, 251.

to be the case to within 0.01 Å.   Although there may be some very small real differences in these bond lengths, due to their different crystallographic environments, the deviations observed certainly lie within the probable limits of error, and the figures for equivalent bonds may be averaged.

The Fourier series employed to derive the maps shown in Figs. 83 and 84 were very complete, and the terminating F values generally small.   Corrections for finite series effects (see Chapter V, Section 9A) are therefore expected to be small, and calculations, which have recently been completed by Cruickshank and his coworkers on the naphthalene and anthracene data, show this to be the case.

The final values for the bond lengths in these two molecules are given below, as directly measured on the electron density maps and after applying Cruickshank's corrections for the finite series effect:

Naphthalene	Bond	Direct measurement	Corrected for finite series
A	A	1.359 Å	1.365 Å
B	B	1.420	1.425
C	C	1.395	1.393
D	D	1.395	1.404

Anthracene	Bond	Direct measurement	Corrected for finite series
A	A	1.364 Å	1.370 Å
B	B	1.419	1.423
C	C	1.391	1.396
D	D	1.440	1.436
E	E	1.390	1.408

The mean finite series correction per co-ordinate is 0.006 Å for both naphthalene and anthracene, with the largest effect on the end bonds in both cases.   Somewhat larger average corrections might have been anticipated, but the graphical interpolation methods employed for deriving the peak positions from the electron density maps are self-correcting in so far as they tend to smooth certain minor irregularities due to series termination effects.   The final corrected values for the bond lengths given in the last columns are probably accurate to within 0.010 Å.

It becomes a matter of considerable interest to compare these observations with theoretical predictions.   Soon after the development of the concept of resonance of molecules among several valence bond structures, by the work of Slater, Hückel, and Pauling, the structures of the aromatic molecules were discussed in some detail.   It became clear that in the condensed ring aromatic hydrocarbons the carbon-carbon distances might be expected to vary over a small range in different parts of the molecule.[7]   In naphthalene there are three non-excited, or Kekulé structures:

and in anthracene four:

To a first approximation the percentage double bond character in any bond may be assessed by a simple averaging of these stable resonance structures, with the following results:

The relation between double bond character, defined in this way, and interatomic distance can be expressed by means of the empirical curve[8] shown in Fig. 85.   This is calibrated by four fixed points.   At one end is the pure single bond of 1.54 Å, as in diamond and aliphatic compounds; at the other the pure double bond of 1.34 Å, as in ethylene.   The two intermediate points represent benzene, 1.39 Å,[9] with 50 per cent double bond character assessed from the two Kekulé structures, and graphite, 1.42 Å, with 33 per cent double bond character (p. 159).   A

[7] L. Pauling, *Proc. Nat. Acad. Sci. U. S.*, 1932, **18**, 293; L. Pauling, L. O. Brockway, and J. Y. Beach, *J. Amer. Chem. Soc.*, 1935, **57**, 2705.

[8] L. Pauling and L. O. Brockway, *J. Amer. Chem. Soc.*, 1937, **59**, 1223.

[9] L. Pauling and L. O. Brockway, *J. Chem. Phys.*, 1934, **2**, 867.

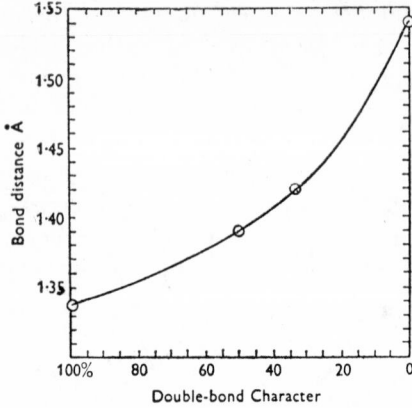

Fig. 85. Relation between interatomic distance and double bond character (after Pauling and Brockway).

smooth curve can be drawn through these points and used to predict the carbon-carbon bond lengths in various types of molecule.

Applying this method to naphthalene and anthracene, we see that it immediately accounts for the main features of the X-ray measurements. In both molecules the shortest distances are observed at the chemically reactive 1,2-bonds, and these have the highest double bond character. The distances predicted from the curve are as follows:

Calc. 1.42 | 1.42   Calc. 1.44 | 1.44

while the observed distances, averaged over chemically equivalent bonds, are:

Obs. 1.39 | 1.40   Obs. 1.44 | 1.41

The over-all agreements are astonishingly good, and in fact much better than might be expected from this over-simplified and very empirical method of approach.

A more accurate treatment would take account of the excited structures as well. All the distinct and independent pairing schemes should be considered, and their relative weights determined on the basis of their coefficients in the wave function for the whole system.[10] For naphthalene there are 42 independent

[10] L. Pauling and G. W. Wheland, J. Chem. Phys., 1933, 1, 362.

(uncrossed) bond diagrams, or canonical structures, and the wave function in terms of these structures has been calculated by Sherman.[11] When the double bond character is assessed in this way, a better description of the structure in its normal state should be obtained.

A different method of calculation has been introduced by Penney,[12] and further modified and extensively applied by Vroelant and Daudel.[13] In this method a bond "order" is defined in terms of energy, and is determined by the mean value of the angle between the spin vectors of adjacent atomic orbitals. In Vroelant and Daudel's treatment this is referred to as the method of "spin states."

Finally, the method of molecular orbitals,[14] which has recently undergone extensive development, can be applied to the estimation of bond lengths. In this treatment the orbital of each π-electron extends over the whole molecule, and from the contributions of these electrons a bond order can be calculated for each link. In this case again the bond order is measured in terms of the energy that may be associated with the bond.

The application of these various methods to the naphthalene and anthracene bond length data has recently been very fully discussed,[15] together with possible further refinements of the theoretical treatment. It appears that the results obtained by the molecular orbital method agree most closely with the experimental measurements, although when complete sets of structures are used the other methods also give very good agreements.

The results of the molecular orbital calculations for naphthalene and anthracene are given on the following page.

---

[11] J. Sherman, *J. Chem. Phys.*, 1934, **2**, 488.

[12] W. G. Penney, *Proc. Roy. Soc.* (London), 1937, **A158**, 306.

[13] C. Vroelant and R. Daudel, *Bull. soc. chim. France*, 1949, **16**, 36, 217; *Compt. rend.*, 1949, **228**, 399; N. P. Buu-Hoi, P. and R. Daudel, and C. Vroelant, *Bull. soc. chim. France*, 1949, **16**, 211.

[14] E. Hückel, *Z. Elektrochem.*, 1937, **43**, 752; J. E. Lennard-Jones and C. A. Coulson, *Trans. Faraday Soc.*, 1939, **35**, 811; C. A. Coulson, *Proc. Roy. Soc.* (London), 1939, *A*169, 413; C. A. Coulson, *Valence*, Oxford, 1952.

[15] C. A. Coulson, R. Daudel, and J. M. Robertson, *Proc. Roy. Soc.* (London), 1951, **A207**, 306; for a theoretical study of the electron distribution in benzene, see N. H. March, *Acta Cryst.*, 1952, **5**, 187.

Naphthalene, $C_{10}H_8$	Bond	X-ray meas. (corrected value)	M.O. calcs.	Δ
	A	1.365 Å	1.384 Å	0.019 Å
	B	1.425	1.416	0.009
	C	1.393	1.424	0.031
	D	1.404	1.406	0.002

Root mean square Δ, over whole molecule, = 0.016 Å

Anthracene, $C_{14}H_{10}$	Bond	X-ray meas. (corrected value)	M.O. calcs.	Δ
	A	1.370 Å	1.382 Å	0.012 Å
	B	1.423	1.420	0.003
	C	1.396	1.406	0.010
	D	1.436	1.430	0.006
	E	1.408	1.410	0.002

Root mean square Δ, over whole molecule, = 0.008 Å

The agreement between theory and experiment is extremely close, especially for the anthracene molecule, where the average discrepancies are of the same order as the expected experimental errors. Larger discrepancies occur in naphthalene, where theory gives too great a value for the short bond A and for the central bond C. There is the possibility of a larger experimental error in the measurement of the bond C, which has only one occurrence in the molecule. The bond distance here thus depends on only one measurement of length. Allowing for this, however, the divergence between theory and experiment in the case of naphthalene is greater than might be expected and is hard to explain. It suggests that in naphthalene the symmetrical Kekulé structure (Erlenmeyer)

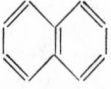

should be given a greater weight than the resonance calculations indicate. The structure is actually the most significant in these calculations, but the results suggest that for some unknown reason its weight should be still further increased.

Perhaps the most important single result in the detailed X-ray investigation of naphthalene and anthracene is the fact that the molecules are shown to be coplanar to within about 0.01 Å, and that the angles are all extremely close to 120°. These facts are fundamental to the whole wave mechanical treatment as applied to aromatic compounds.

Details of the electron distribution in these molecules have been studied, and it will be seen from Figs. 83 and 84 that distinct bridge values in the density exist along the various bonds, rising to a value of about 2 electrons per $Å^3$ on the short 1,2-bonds. There is, however, another effect which was first observed in the original two-dimensional analysis of anthracene[4] and has since been found to occur in many other crystal structures. The peak value of the density at the various atoms is found to diminish progressively as we pass away from the centre of the molecule. In anthracene it falls from 9.0 to about 7.5 electrons per $Å^3$, and in naphthalene from 8.0 to 6.5 electrons per $Å^3$ on the outermost atoms. Part of this variation in peak height has been shown to be due to the finite series effect, but even after making corrections (based on models with identical atoms, but different for naphthalene and anthracene) the variation still persists. It is accompanied by a distortion of the electron clouds of these outer atoms, which suggests that the effect may be a thermal one, due to some vibration of the molecule as a whole about a fixed centre. These effects tend to obscure the more chemically interesting aspects of the charge distribution, and further investigation at lower temperatures is required before reliable deductions can be made.

The crystal structures of the higher linear benzologues of the naphthalene-anthracene series, naphthacene (or tetracene)$C_{18}H_{12}$, pentacene $C_{22}H_{14}$, and hexacene $C_{26}H_{16}$

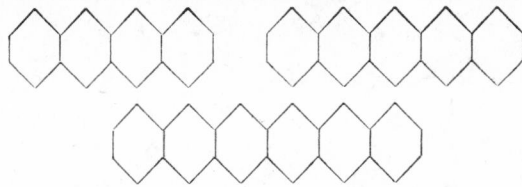

have been examined.[16] They are triclinic, with two molecules

[16] V. C. Sinclair, Ph.D. thesis, Glasgow, 1949; R. B. Campbell, private communication. For chemical properties and preparation, see E. Clar, *Ber.*, 1939, **72**, 1817; 1942, **75**, 1283; *Aromatische Kohlenwasserstoffe* Berlin: Springer-Verlag, 1941.

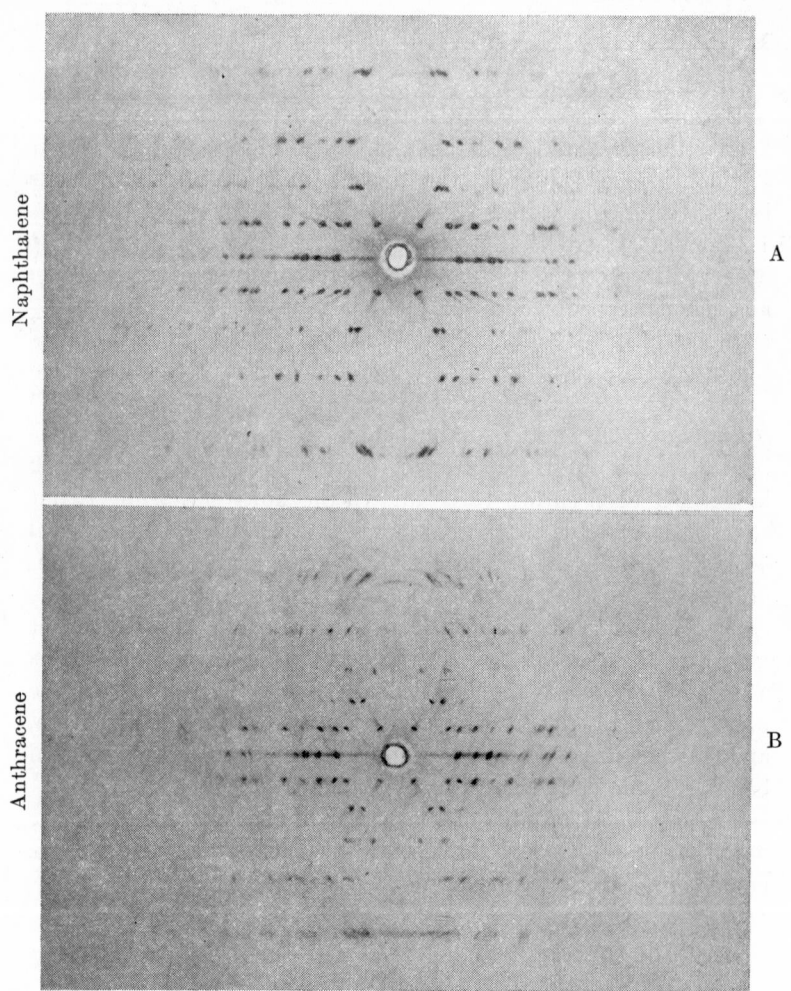

Naphthalene

Anthracene

A

B

Fig. 86. c-Axis rotation photographs

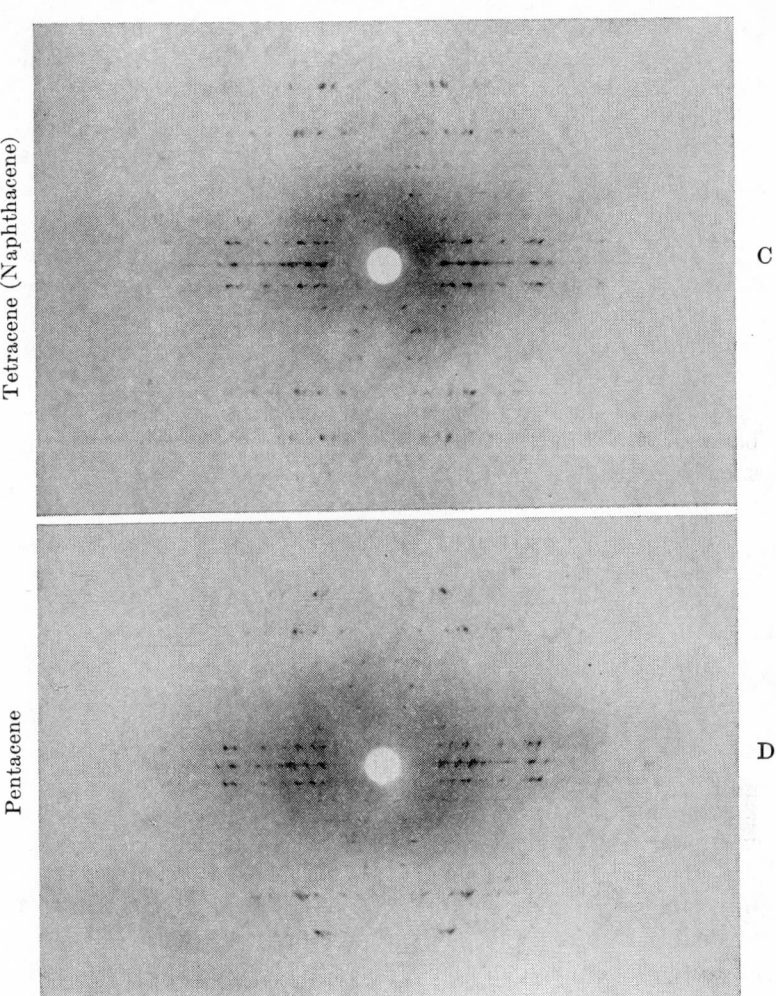

in the naphthalene-anthracene series.

per unit cell.   In spite of this decrease in crystallographic symmetry, the structural types resemble the naphthalene-anthracene series very closely.   In particular, the molecules are found to be accurately aligned in pairs along one of the principal axes.   This is demonstrated conclusively by rotation photographs taken about this axis (c), which are reproduced in Fig. 86 and compared with the corresponding naphthalene and anthracene photographs. (A good photograph for hexacene is not available, but the reflections which are observed indicate a similar structure type.)

We note that in naphthalene the third and fourth layer lines are enhanced; in anthracene, the fourth and fifth; in tetracene, the fifth and sixth; and in pentacene, the sixth and seventh. This enhancement occurs in the same region of the photograph in each case, and indicates a fundamental periodicity of about 2.45 Å in all these structures, corresponding to the width of the benzene ring.   Further analysis of the reflections shows that the smaller periodicity of 1.22 Å is also present (see Figs. 71, 72, pp. 162, 163).

Although the c axis alignment must be fairly accurate, the two molecules in these triclinic cells occupy general positions, and this fact complicates the analysis.   The main outlines of the structures have been established, and electron density projections obtained, but it is not yet possible to make any detailed carbon-carbon bond length measurements.

2. CHRYSENE AND DIBENZANTHRACENE

None of the other condensed ring hydrocarbons has been examined in the same degree of detail as naphthalene and anthracene, but many measurements have been recorded (Table VI). In a number of cases approximate structures have been refined by double Fourier series methods, and from the electron density projections obtained in this way individual bond length estimates can sometimes be made.   These vary considerably in accuracy in the different structures, and in the remainder of this chapter the more significant determinations are reviewed.

The crystal structure of chrysene,[17] $C_{18}H_{12}$,

----

[17] J. Iball, *Proc. Roy. Soc.* (London), 1934, **A146**, 140.

differs from that of anthracene, there being four molecules in a unit cell based on a centred lattice ($C_{2h}^6 - I2/c$ or $C_s^4 - Ic$). One fairly well-defined projection has been recorded, from which it is clear that the molecule consists of almost regular planar hexagons, with a mean carbon-carbon distance of 1.41 Å, but it has not been possible to estimate individual bond lengths with any certainty.

The structure of 1:2:5:6-dibenzanthracene, $C_{22}H_{14}$,

has received more detailed study.[18] Two crystalline modifications of this compound are known, a two-molecule monoclinic form of low symmetry ($C_2^2 - P2_1$) and a four-molecule orthorhombic form ($D_{2h}^{15} - Pcab$). Detailed analyses of both these varieties have been completed, and the difference is found to be due to rather minor variations in the manner of packing of the molecules in the unit cells. The alternative arrangements correspond to only slight energy differences and would not be expected to involve any significant change in molecular structure or in bond lengths.

In the orthorhombic modification the molecules display centres of symmetry and are inclined at only about 30° to the (100) plane. This gives an excellent opportunity to obtain high resolution of the atoms by double Fourier series methods, and the electron density map, as a projection on this plane, is shown in Fig. 87. Unfortunately, the benefits of the nearly planar ar-

[18] J. Iball and J. M. Robertson, *Nature*, 1933, **132**, 750; K. S. Krishnan and K. Banerjee, *Z. Krist.*, 1935, **91**, 170, 173; J. Iball, *Nature*, 1936, **137**, 361; J. M. Robertson and J. G. White, *J. Chem. Soc.*, 1947, p. 1001, and unpublished data.

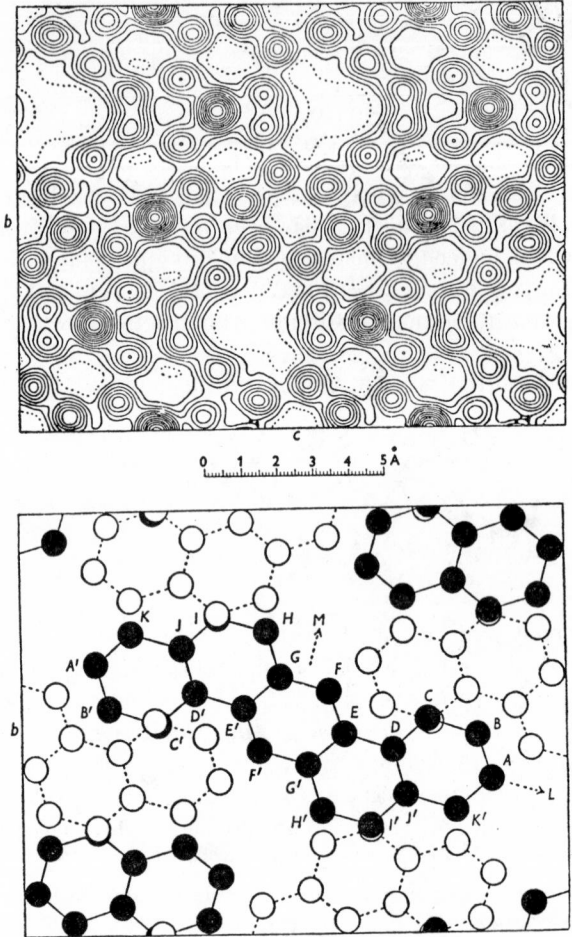

Fig. 87. Dibenzanthracene (orthorhombic). Structure
is projected on (100).

rangement are partly lost because other symmetry-related molecules at different levels in the crystal overlap in this projection
of the structure, causing certain groups of atoms to approach
too closely for separate resolution.   Nevertheless, a very complete analysis can be carried out, and the results are consistent
with a strictly planar molecular structure.   Nine of the eleven
crystallographically independent atoms can be resolved.

   In the monoclinic modification the principal electron density map has a rather similar appearance, but in this case no centre of symmetry is required by the space group and all 22 atoms are crystallographically independent. Of these 18 can be separately resolved. The results of the measurements show that although the centre of symmetry is not demanded by the space group, it is actually present in the molecule to a high degree of approximation. The bond lengths that can be independently measured in these two crystal structures agree to within the expected limits of accuracy, which vary from ± 0.02 to ± 0.04 Å.

The results of the measurements, based mainly on the monoclinic structure, are shown below. Simple assessment of double bond character from the 12 possible Kekulé structures does not give a very good account of these results, although it indicates that the longest bonds should be B, H, K, M, and N, and the shortest should be J. More detailed calculations of bond length have been made by the spin states method of Vroelant and Daudel,[13] and these results are shown in the second column. The general measure of agreement is encouraging, and certainly as good as can be expected in view of the rather unsatisfactory experimental measurements which are only good enough to detect the more extreme variations.

Dibenzanthracene, $C_{22}H_{14}$

	X-ray measurement	S.S. calcs. V. and D.
A	1.39 Å	1.41 Å
B	1.40	1.44
C	1.40	1.41
D	1.39	1.39
E	1.41	1.40
F	1.40	1.39
G	1.40	1.42
H	1.40	1.42
J	1.38	1.38
K	1.44	1.42
L	1.38	1.41
M	1.44	1.44
N	1.45	1.44

It is well known that the central 9 and 10 positions are chemically reactive, but the positions at bond J are also attacked on oxidation, with the formation of a 3:4-quinone as a minor product. Any direct correlation of chemical properties with bond length measurements is, however, a very difficult matter.

3. PYRENE, TRIPHENYLENE AND PERYLENE

The pyrene crystal structure belongs to the familiar space group $C_{2h}^5 - P2_1/a$, and in this respect is similar to naphthalene, anthracene, and many of the other hydrocarbons in this series. There are, however, four pyrene molecules in the unit cell, and so no inherent molecular symmetry is required in building up the crystal structure. This complicates the X-ray analysis, but a very complete determination has been made by two-dimensional methods.[19] The results are again consistent with planar mole-

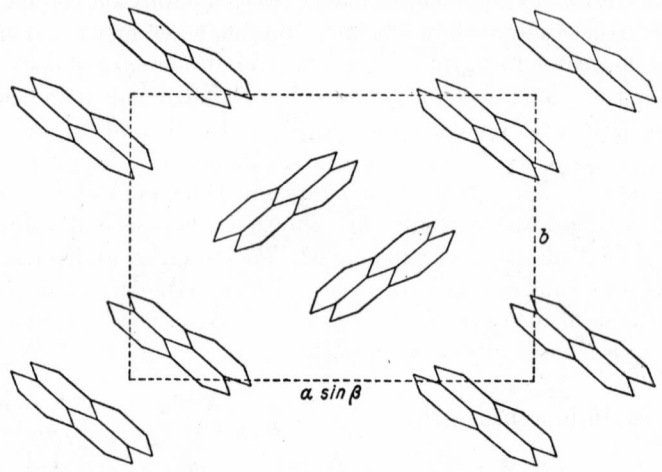

Fig. 88. The pyrene crystal structure viewed in projection along the c axis.

cules and approximately regular hexagonal carbon rings. The arrangement of the molecules as viewed along the c axis is shown in Fig. 88, where they are grouped in pairs about the symmetry centres of the crystal.

This projection, however, gives a very foreshortened, end-on view of the molecules. The principal evidence regarding bond lengths is derived from the final electron density projection made along the b axis which is shown in Fig. 89. In this view of the structure the molecular planes are inclined to the projection plane at about 40° and many of the atoms can be resolved, although certain positions are unfortunately obscured by the inter-leaving effect of other molecules at different levels in the unit

[19] J. M. Robertson and J. G. White, *J. Chem. Soc.*, 1947, p. 358.

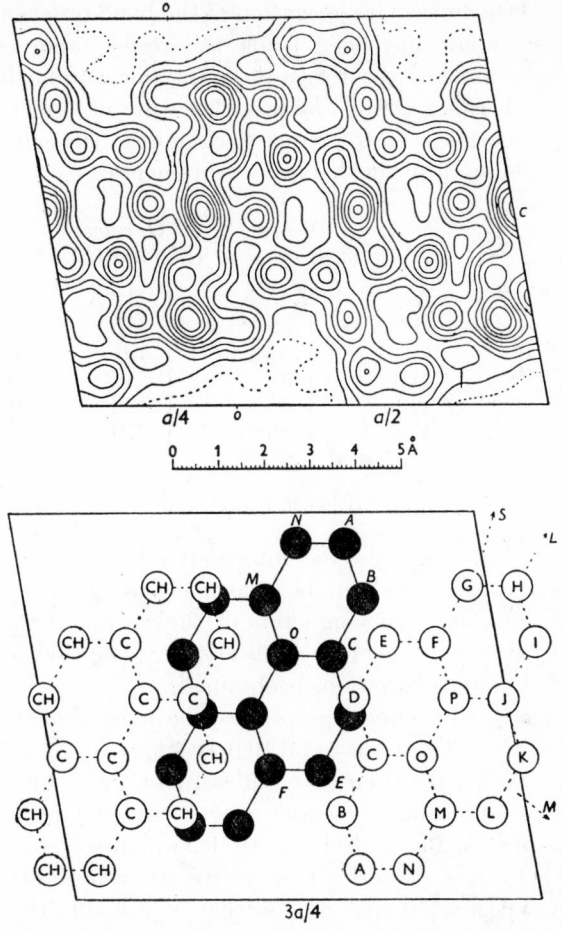

Fig. 89. Pyrene. Electron density projection on (010).

cell.   But careful measurement of this electron density map, and of the one corresponding to Fig. 88, shows the molecule to possess the expected symmetry about the two axes L and M (Fig. 89).   Assuming this symmetry to be exact, and allowing for the inclination of the molecular plane in the projection, it is possible to estimate all the interatomic distances.   These results cannot be very accurate, and may contain errors as great as 0.03 or 0.04 Å, but even so, the conclusion that the bonds do vary in length in different parts of the molecule seems definite.

A careful theoretical investigation of the bond orders and bond lengths in pyrene has been made by Moffitt and Coulson.[20] Their results are compared with the X-ray measurements below, and although there is not any large measure of agreement, the theoretical work is seen to predict variations in the same general direction as found by the X-ray measurements.

Pyrene, $C_{16}H_{10}$		X-ray measurement	M.O. calcs. (M. and C.)	Δ
	A	1.45 Å	1.412 Å	0.038 Å
	B	1.39	1.415	0.025
	C	1.42	1.401	0.019
	D	1.39	1.388	0.002
	E	1.45	1.420	0.030
	F	1.39	1.370	0.020

Root mean square Δ, over whole molecule, $= 0.023$ A

These results may also be interpreted in terms of the six Kekulé structures that can be drawn on the pyrene formula. When the double bond characters of the various links are assessed in this way, a general qualitative explanation of the observed bond length variations is obtained.

With regard to chemical properties, the most reactive centres are generally the 3 and 8 or 3 and 10 positions, these usually being the ones attacked when two substituents are introduced.[21] However, when osmium tetroxide reacts with pyrene, a complex is obtained which is hydrolysed to 1:2-dihydroxy-1:2-dihydropyrene.[22] In this case the reactive positions are situated, as might be expected, at the ends of what is probably the shortest bond (F) in the molecule.

A rather more interesting structure which has recently been examined in considerable detail[23] is that of the hydrocarbon triphenylene, $C_{18}H_{12}$. A consideration of the nine available Kekulé structures shows that very long bonds (11 per cent double bond character) may be expected in the A positions. The crystal

[20] W. E. Moffitt and C. A. Coulson, *Proc. Phys. Soc.* (London), 1948, **60**, 309.

[21] H. Vollmann, H. Becker, M. Corell, H. Struck and G. Langbein, *Ann.*, 1937, **531**, 1.

[22] J. W. Cook and R. Schoental, private communication.

[23] A. Klug, *Acta Cryst.*, 1950, **3**, 165, 176.

structure is rather complicated, the space group being $P2_12_12_1$ with four apparently planar molecules occupying general positions. The structure was elucidated by an elegant application of the Fourier transform method and refined by means of electron density projections along the axial directions. The accuracy of the individual bond length measurements is not expected to be very high, being estimated at ± 0.03 Å, but on averaging over chemically equivalent bonds the agreements obtained with molecular orbital calculations are extremely good, as shown below.

Triphenylene, $C_{18}H_{12}$		X-ray measurements	M.O. calcs.[23]	Δ
	A	1.47 Å	1.436 Å	0.034 Å
	B	1.435	1.409	0.026
	C	1.39	1.400	0.010
	D	1.375	1.384	0.009
	E	1.385	1.394	0.009

Root mean square Δ, over whole molecule, $= 0.018$ Å

One very puzzling feature of this structure concerns the intermolecular approach distances, where a number of lateral contacts varying between 2.6 and 3.2 Å are reported. It has been suggested that these may arise through dipole interactions, but it seems extremely difficult to account for such distances on this basis alone. A further critical examination of the structure should probably be made to see whether alternative explanations are possible.

The hydrocarbon perylene, $C_{20}H_{12}$, is also of particular interest from the point of view of bond length study. When we attempt to formulate the stable valence bond structures for this molecule, we find that in no case can a double bond be drawn between the two naphthalene nuclei. The nine Kekulé structures available are in fact merely the three Kekulé structures for each naphthalene nucleus placed together in every possible combination. The double bond character and expected bond lengths are thus easily assessed. To a first approximation they should resemble those calculated for naphthalene, with "single" bonds connecting the nuclei. Detailed calculations are not yet available, but they would certainly modify this simple picture.

X-ray examination[24] shows the crystal structure to be similar to that of pyrene, with the same four-molecule complexity and absence of crystallographic molecular symmetry. The principal electron density projection on (010) is somewhat similar in appearance to that obtained for pyrene (Fig. 89), but the perylene molecules are inclined at a steeper angle of about 55° to this plane. Making the same assumptions about molecular symmetry, the bond lengths can be evaluated, with the results shown below.

Perylene, $C_{20}H_{12}$		*X-ray measurements*	*Distances assessed from Kekulé structures alone*
	A	1.38 Å	1.37 Å
	B	1.38	1.42
	C	1.45	1.42
	D	1.45	1.42
	E	1.38	1.37
	F	1.45	1.42
	G	1.50	1.54

These results show variations in the expected directions and in particular the presence of a very long bond at G is confirmed. It might be expected that this molecular structure could depart somewhat from the strictly coplanar form, to provide better clearances between the CH groups at the DE positions, but the agreements obtained between calculated and observed intensities for all zones of reflections show that any such deviation must be very slight.

4. CORONENE AND BENZPERYLENE

The large and highly symmetrical condensed ring hydrocarbon coronene, $C_{24}H_{12}$, possesses a beautifully simple crystal structure which is very suitable for detailed analysis.[25] While the expected hexagonal symmetry is not displayed in the crystal, there are only two molecules in the monoclinic cell and they each have

[24] D. M. Donaldson, J. M. Robertson, and J. G. White, *Proc. Roy. Soc.* (London) (to be published).

[25] J. M. Robertson and J. G. White, *Nature*, 1944, **154**, 605; *J. Chem. Soc.*, 1945, p. 607.

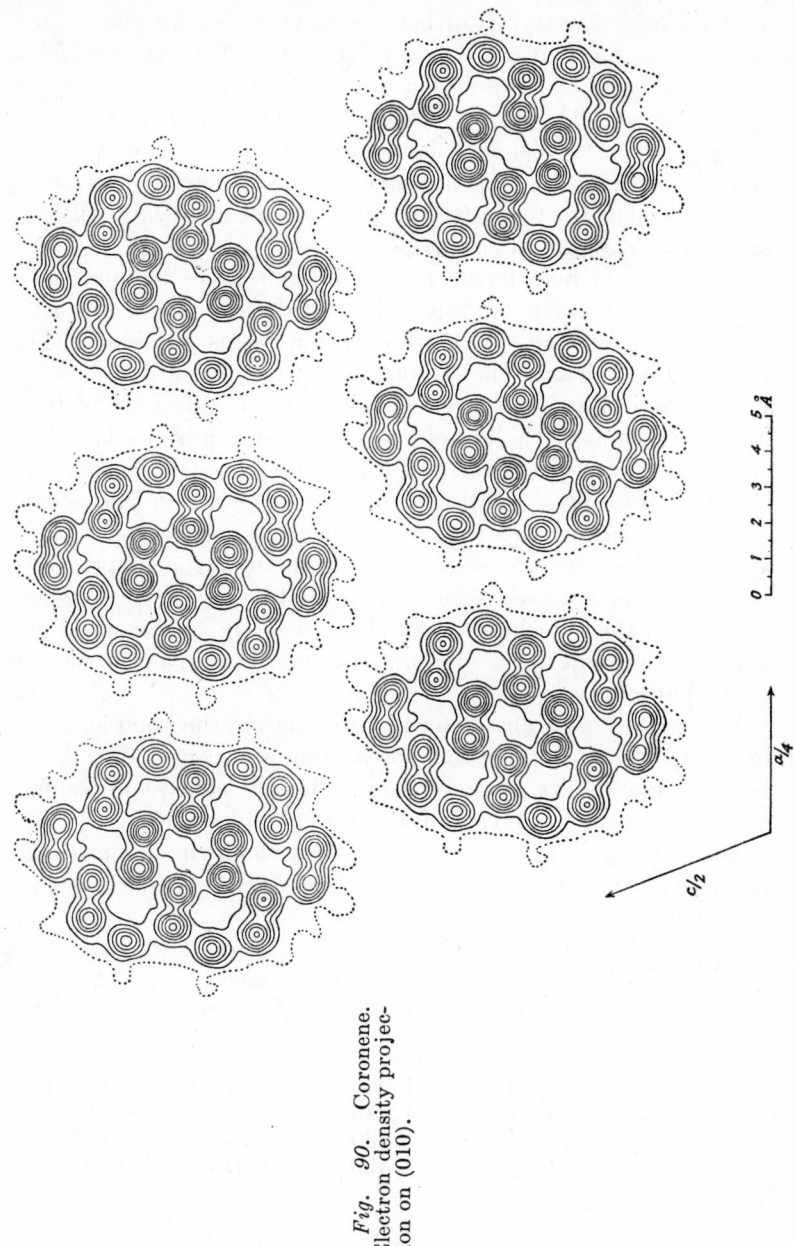

Fig. 90. Coronene. Electron density projection on (010).

a centre of symmetry.   Further, the large molecular planes are found to be inclined to the symmetry plane (010) at only 44°, thus permitting good resolution in a two-dimensional projection on this plane.   The length of the b axis, 4.695 Å, gives the distance between two parallel molecules, and, allowing for the inclination of 44°, the perpendicular distance between successive molecular planes is found to be 3.40 Å, almost identical with the interplanar distance in graphite.

After several refinements by the Fourier series method, the electron density map as projected on (010) is shown in Fig. 90. It will be noted that, unlike the previous cases, the individual molecules as well as all the carbon atoms are now clearly resolved. A large gap of empty space, containing some indication of the hydrogen atoms at the one-electron level, surrounds each molecule.   Alternate molecules in each row are oppositely inclined to this projection plane, and so the molecular planes in successive stacks of molecules are almost perpendicular to each other.   This appears to be a feature of the crystal structure type adopted by molecules which are disc-like in shape and large in area as compared with their thickness.   An almost exactly similar mode of packing is present in the ovalene structure, $C_{32}H_{14}$, and in many of the phthalocyanines.

When allowance is made for the orientation, the bond lengths in this molecule can be estimated with an accuracy of 0.02 Å or better.   The results, averaged over chemically equivalent bonds, are shown below.   A careful theoretical investigation of this molecule has also been made by Coulson,[26] employing the method of molecular orbitals, and the results of these calculations are also given.

Coronene, $C_{24}H_{12}$		*X-ray* measurements	*M.O.* calcs. (*Coulson*)	Δ
	P	1.385 Å	1.372 Å	0.013 Å
	Q	1.415	1.411	0.004
	R	1.430	1.411	0.019
	S	1.430	1.415	0.015

Root mean square Δ, over whole molecule, = 0.015 Å

[26] C. A. Coulson, *Nature*, 1944, **154**, 797; W. E. Moffitt and C. A. Coulson, *Proc. Phys. Soc.* (London), 1948, **60**, 309.

The measure of agreement achieved with the theoretical work in this case is very encouraging. Although the bond length variations are small, they are distinct enough to be easily observed on the two-dimensional electron density maps. This is evident from the enlargement of a portion of the molecule shown in Fig. 91. Any structure based on completely regular hexagons would require the atoms shown to lie strictly on a straight line. The deviations from such a line are quite obvious.

Fig. 91. Enlargement of a portion of the coronene molecule, showing distortion.

The molecule of 1:12-benzperylene, $C_{22}H_{12}$, differs by only

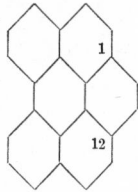

two carbon atoms from that of coronene, and the crystal structures might be expected to resemble each other. This proves not to be the case, and the four-molecule unit cell (Table VI, p. 217) resembles the pyrene and perylene structures more closely. Detailed X-ray investigation by White[27] has succeeded in determining many of the bond lengths, but the more interesting ones, which are expected to occur on the "open" side of the molecule, are unfortunately obscured in projection by the overlapping effect of adjoining molecules. Three-dimensional analysis would be required to determine these values. In that part of the molecule which is resolved, however, the bond length variations are found to resemble those of coronene quite closely, and are in good agreement with those predicted by theory.

[27] J. G. White, *J. Chem. Soc.*, 1948, p. 1398.

5. OVALENE (OCTABENZONAPHTHALENE)

The still larger condensed ring hydrocarbon ovalene, $C_{32}H_{14}$, has recently been synthesized by Clar,[28] and a detailed X-ray analysis of this structure[29] has given results which are again in striking agreement with theoretical predictions based on molecular orbital treatment.

The crystal structure resembles that of coronene rather closely, the unit cell containing two centrosymmetric molecules inclined at $\pm 43°$ to (010). The electron density projection on this plane is shown in Fig. 92. A comparison of this diagram with Fig. 91 will show the manner in which the additional rings are accommodated by the crystal structure, while maintaining the same wide and very uniform van der Waals gap around each molecule.

This molecule provides greater scope for measurement of bond length variations than coronene, because now there are twelve chemically distinct types of bond instead of the four present in coronene. When the measurements are averaged over these bonds, the results shown below are obtained. In the second column the molecular orbital calculations of A. J. Buzeman[30] are compared.

Ovalene, $C_{32}H_{14}$		X-ray measurements	M.O. calcs. (Buzeman)	Δ
	A	1.404 Å	1.410 Å	0.006 Å
	B	1.441	1.427	0.014
	C	1.345	1.380	0.035
	D	1.433	1.425	0.008
	E	1.403	1.418	0.015
	F	1.428	1.421	0.007
	G	1.426	1.424	0.002
	H	1.435	1.428	0.007
	J	1.416	1.422	0.006
	K	1.461	1.430	0.031
	L	1.442	1.425	0.017
	M	1.383	1.387	0.004

Root mean square Δ, over whole molecule, $=0.015$ Å

[28] E. Clar, *Nature*, 1948, **161**, 238.

[29] D. M. Donaldson and J. M. Robertson, *Nature*, 1949, **164**, 1002; J. M. Robertson, *Proc. Roy. Soc.* (London), 1951, **A207**, 101; D. M. Donaldson and J. M. Robertson, *Proc. Roy. Soc.* (London), in press.

[30] A. J. Buzeman, *Proc. Phys. Soc.* (London), 1950, **63**, 827.

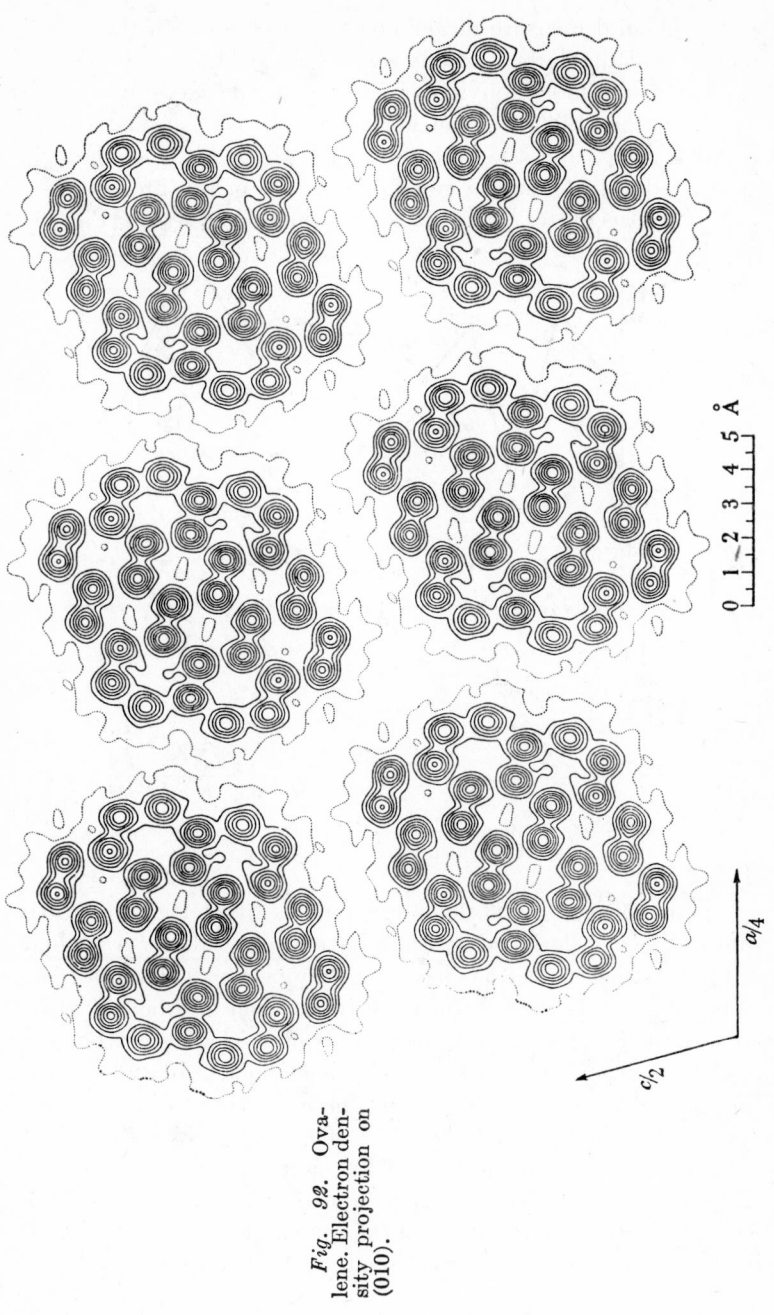

Fig. 92. Ovalene. Electron density. Electron density projection on (010).

The general measure of agreement found between the experimental and theoretical values is quite as good as in the case of coronene, but more convincing in view of the larger variety of bond lengths studied. The longest bond (K) exceeds the calculated value, and the shortest (C) is distinctly less than the calculated value, but the general sequence of the predictions is extremely accurate. This is shown more clearly in Fig. 93,

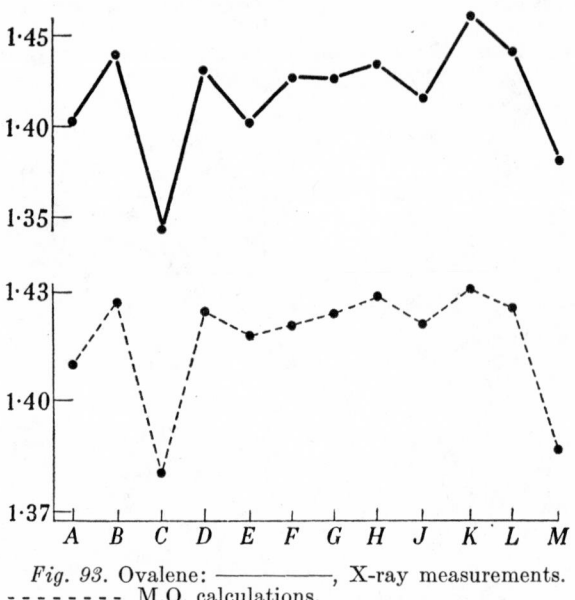

Fig. 93. Ovalene: ————, X-ray measurements.
- - - - - - - -, M.O. calculations.

where the measured and theoretical values are plotted on slightly different scales, designed to make the extreme values correspond.

Finite series corrections have recently been calculated for this structure. As is the case of naphthalene and anthracene, these corrections are relatively small, the average change per bond in ovalene being only 0.010 Å. The long bonds, K and L, are, however, considerably reduced by these corrections and brought closer to the theoretical values. The effect over the whole molecule is to reduce the root mean square discrepancy between the observed and the theoretical values from 0.015 Å to 0.012 Å. The agreements are therefore of the same order as those achieved in the case of the three-dimensional analysis of naphthalene and anthracene.

It is worthy of note that both in the cases of coronene and ovalene a simple assessment of double bond character in the various links by linear superposition of the possible Kekulé structures is capable of giving quite an accurate account of the bond length variations. The number of Kekulé structures available for hydrocarbons of this type, based on three-tier strips without indented ends, is given by the general formula

$$(n + 1)(n + 2)^2(n + 3)/12$$

where n is the number of rings in the basal tier.[31] For coronene, $n = 2$ and the number of structures is 20, while for ovalene, $n = 3$ and the possible Kekulé structures[32] number 50.

It is true, of course, as pointed out by Daudel and Pullman,[33] that for multi-ring hydrocarbons the possible excited structures far outnumber the stable forms. Nevertheless, in the case of very symmetrical hydrocarbons it is possible to make a good estimate of the bond lengths by this simple method.

In this connection, chemical evidence suggests that structures with benzenoid rings are more important than structures with quinonoid rings, a generalisation known as the Fries rule. The usual effect of weighting the contributory structures in this manner is to make some qualitative improvement in the bond length predictions. This is in general agreement with quantum-mechanical calculations, but it has been pointed out by Moffitt and Coulson[20] that the Fries rule can only have a relative validity. An improved version of the rule which would appear to have more general application is that structures which show the greatest number of double bonds in "exposed" positions, such as the bonds C and M in ovalene or the bond P in coronene, have the greatest weight. This modification of the Fries rule, proposed by Moffitt and Coulson, is supported by nearly all the X-ray measurements we have quoted, which generally reveal abnormally short bonds in these positions.

6. NON-PLANAR CONDENSED RING HYDROCARBONS AND SOME RELATED STRUCTURES

Both chemical and X-ray evidence agree in assigning planar structures to the condensed ring hydrocarbons so far considered.

---

[31] M. Gordon and W. H. T. Davison, *J. Chem. Phys.*, 1952, **20**, 428.

[32] J. M. Robertson, *Proc. Roy. Soc.* (London), 1951, **A207**, 101.

[33] R. Daudel and B. Pullman, *J. phys. radium*, 1946, **7**, 105.

This simplifying feature has greatly facilitated the X-ray studies and has also permitted a fairly accurate theoretical treatment of those molecules to be carried through in many cases.   However, several hydrocarbons of this series have now been synthesised for which coplanar molecular structures are either impossible or very unlikely on steric grounds.   Examples are 3:4-benzphenanthrene, $C_{18}H_{12}$; 3:4:5:6-dibenzphenanthrene, $C_{22}H_{14}$; and tetrabenznaphthalene, $C_{26}H_{16}$.

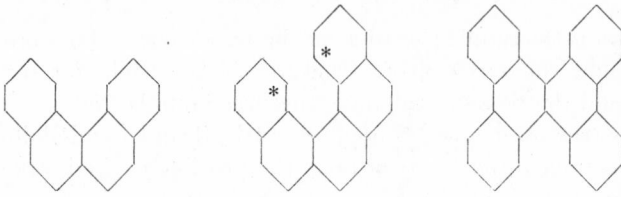

If these structures were planar, as drawn above, it is clear that the closest approach between certain non-bonded atoms would be much smaller than the usual van der Waals gap of about 3.4 Å.   This "intramolecular overcrowding" has been the subject of much careful chemical investigation,[34] and in the case of the first two compounds mentioned above definite evidence of non-planar forms has been obtained by the resolution of certain derivatives of these hydrocarbons into optical isomers.   A review of the chemical properties has been given by J. W. Cook.[35]

Preliminary X-ray data have been obtained from crystals of these hydrocarbons,[36] but only in the case of 3:4:5:6-dibenzphenanthrene has a detailed analysis been effected.[37]   The crystals exist in two polymorphic modifications, and the one studied most fully has a complicated structure.   The space group is $C_{2h}^{6} - A2/a$ with twelve molecules of the hydrocarbon in the unit cell.   Now the number of asymmetric units required to complete this symmetry is eight.   The analysis shows that there are in fact eight molecules in the general positions, together with four

[34] M. S. Newman and W. R. Wheatley, *J. Amer. Chem. Soc.*, 1948, 70, 1913; F. Bell and D. H. Waring, *J. Chem. Soc.*, 1949, p. 2689.

[35] J. W. Cook, *Ann. Repts. on Progress Chem.* (Chem. Soc. London), 1942, 39, 155, 173.

[36] E. Harnik, F. H. Herbstein, and G. M. J. Schmidt, *Nature*, 1951, 168, 160.

[37] A. O. McIntosh, J. M. Robertson, and V. Vand, *Nature*, 1952, 169, 322.

other molecules in special positions on the 2-fold symmetry axes. This occurrence of crystallographically distinct molecules in a crystal structure is somewhat unusual, but it is also found in the case of biphenylene[38] and dibiphenylene ethylene.[39]

The results of the analysis, after refinement by a large number of difference syntheses, show that the molecule of dibenzphenanthrene has assumed a non-planar form mainly by lateral distortion, accompanied by very little change in the valency angles. The distance between the non-bonded carbon atoms, which are starred in the formula, is about 3.0 Å, representing a considerable compression below the usual van der Waals distance. The fact that this is achieved with but very little change in valency angles possibly explains the retention of aromatic character in spite of

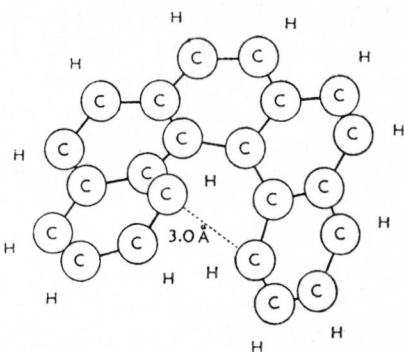

Fig. 94. Molecule of 3:4:5:6-dibenzphenanthrene showing distortion from planar form.

the severe distortion. Unfortunately the individual carbon-carbon distances cannot be estimated with much accuracy, but they do not appear to differ greatly from the usual values. The general shape of the molecule is indicated in Fig. 94.

Departures from a planar form may occur in certain five-membered ring systems, such as fluorene, $C_{13}H_{10}$,[40] and, of course, when several separate benzene rings are present in the molecule, as in the various phenyl benzenes, such departures are common. These structures, however, lie beyond the scope of this chapter, which is intended to deal with the six-membered condensed ring hydrocarbons. But one particularly interesting distortion of a benzene ring in another type of hydrocarbon may be mentioned. This occurs in the compound known as di-*p*-xylylene, obtained by pyrolysis of *p*-xylene, and which appears to have the formula

[38] J. Waser and C. S. Lu, *J. Amer. Chem. Soc.*, 1944, **66**, 2035.

[39] C. P. Fenimore, *Acta Cryst.*, 1948, **1**, 295.

[40] J. Iball, *Z. Krist.*, 1936, **94**, 397.

$$CH_2\text{——}CH_2$$

$$CH_2\text{——}CH_2$$

The crystals are tetragonal, and a detailed X-ray investigation[41] shows that the aliphatic carbon-carbon bonds are of about the usual length (1.54 Å). The benzene rings themselves, however, are not quite flat, the two substituted carbon atoms in each ring being displaced from the plane of the other four by about 0.13 Å. The non-bonded approach distance between the unsubstituted atoms on the two rings is about 3.09 Å, and between the substituted atoms, 2.83 Å.

These results appear to confirm what has been found in the case of dibenzphenanthrene: that even under conditions involving the most severe distortion, the usual aromatic van der Waals contact distance of 3.4 Å, in directions normal to the rings, cannot be reduced appreciably below about 3.0 Å. Structures involving smaller non-bonded distances in these directions are likely to be incapable of stable existence.

On the other hand, lateral contacts may sometimes be reduced further, as in dibiphenylene-ethylene,[39] where a distance of 2.5 Å is reported between the starred atoms, with a structure that appears to be at least approximately coplanar. This is certainly

an unexpected and unusually close contact, and involves a pair of hydrogen atoms in an impossibly close approach unless they are displaced from the ring planes. It should be remarked, however, that the minimum approach distance in general will depend on the positions of the approaching atoms with respect to the orientation of the covalent bonds of the other atoms. In

[41] C. J. Brown and A. C. Farthing, *Nature*, 1949, **164**, 915.

directions close to the already existing bond directions, the minimum approach distances are certainly less; and in particular, as has been pointed out by Pauling,[42] atoms which are bonded to the same atom can approach each other much more closely than the sum of the van der Waals radii, without undue strain. Thus in naphthalene (and in hydrocarbons in general) the juxtaposition of the atoms 1 and 8 is not a disturbing factor in the

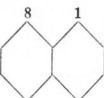

structure. Intermolecular distances, as distinct from these intramolecular distances, are discussed in the next chapter.

7. COVALENT RADII

In this chapter attention has been confined to the relatively simple aromatic hydrocarbon structures, for which it is possible to discuss carbon-carbon bond lengths in some detail. It is not possible within the compass of this book to review the many other organic and simple molecular crystal structures involving atoms additional to carbon which have now been determined with sufficient accuracy to permit a reliable evaluation of bond lengths. Some of these crystal structures are mentioned in other contexts in the following chapters, but for convenience of reference it is desirable to list here the principal data which have been obtained for bond lengths between the atoms which occur most frequently in organic compounds.

These bond lengths are conveniently summarised by listing the *covalent radii* for the atoms concerned, in the particular valency configuration (single, double, or triple bond) which may apply. Such tables of covalent radii were derived by Pauling and Huggins[43] from a study of crystal data and from the results

[42] L. Pauling, *The Nature of the Chemical Bond*, Ithaca, N. Y.: Cornell University Press, 1939.

[43] M. L. Huggins, *Phys. Rev.*, 1926, **28**, 1086; L. Pauling and M. L. Huggins, *Z. Krist.*, 1934, **87**, 205; see also L. Pauling, *Proc. Nat. Acad. Sci. U. S.*, 1932, **18**, 293; N. V. Sidgwick and E. J. Bowen, *Ann. Repts. on Progress Chem.* (Chem. Soc. London), 1931, **28**, 384. For a recent and very complete review of bond lengths obtained by electron diffraction in the gas phase, see P. W. Allen and L. E. Sutton, *Acta Cryst.*, 1950, **3**, 46.

TABLE VI. CRYSTAL DATA FOR AROMATIC HYDROCARBONS

Compound	For-mula	M. pt.	Space group	Mol. per cell	a / α	b / β	c / γ	Ref.
Benzene	C_6H_6	5.5°	Pbca	4	7.44	9.65	6.81	44
Naphthalene	$C_{10}H_8$	80°	$P2_1/a$	2	8.24 / 90°	6.00 / 122.9°	8.66 / 90°	5
Anthracene	$C_{14}H_{10}$	218°	$P2_1/a$	2	8.56 / 90°	6.04 / 124.7°	11.16 / 90°	6
Naphthacene (tetracene)	$C_{18}H_{12}$	341°	$P\bar{1}$	2	7.98 / 98.0°	6.14 / 112.4°	13.57 / 92.5°	16
Pentacene	$C_{22}H_{14}$	high	$P\bar{1}$	2	7.90 / 101.3°	6.06 / 111.8°	15.95 / 94.4°	16
Hexacene	$C_{26}H_{16}$	high	$P\bar{1}$	2	7.96 / 97.8°	6.16 / 110°	∼18.1 / 95.3°	16
Fluorene	$C_{13}H_{10}$	116°	Pna	4	8.47	5.70	18.87	45
Phenanthrene	$C_{14}H_{10}$	100°	$P2_1/a$	4	8.66 / 90°	6.11 / 98°	19.24 / 90°	46
Pyrene	$C_{16}H_{10}$	150°	$P2_1/a$	4	13.60 / 90°	9.24 / 100.2°	8.37 / 90°	19
Triphenylene	$C_{18}H_{12}$	199°	$P2_12_12_1$	4	13.20	16.84	5.28	23
Chrysene	$C_{18}H_{12}$	254°	$I2/c$	4	8.34 / 90°	6.18 / 115.8°	25.0 / 90°	17
1:2-Benzanthracene	$C_{18}H_{12}$	159°	$P2_1/m$ or $P2_1$	4	7.91 / 90°	6.43 / 99°	23.96 / 90°	47
3:4-Benzphenanthrene	$C_{18}H_{12}$	168°	$P2_12_12_1$	4	14.69	14.19	5.75	36
Perylene	$C_{20}H_{12}$	274°	$P2_1/a$	4	11.35 / 90°	10.87 / 100.8°	10.31 / 90°	24
1:2-Benzpyrene (A)	$C_{20}H_{12}$	177°	$P2_1/c$	4	4.52 / 90°	20.32 / 97.4°	13.47 / 90°	48
1:2-Benzpyrene (B)	$C_{20}H_{12}$	177°	$P2_12_12_1$	4	7.59	7.69	22.38	48
1:2:5:6-Dibenzanthra-cene (A)	$C_{22}H_{14}$	267°	$P2_1$	2	6.59 / 90°	7.84 / 103.5°	14.17 / 90°	18

[44] E. G. Cox, *Nature*, 1928, **122**, 401; *Proc. Roy. Soc.* (London), 1932, **A135**, 491.
[45] J. Iball, *Z. Krist.*, 1936, **94**, 397.
[43] J. D. Bernal and D. Crowfoot, *J. Chem. Soc.*, 1935, p. 93.
[47] J. Iball, *Z. Krist.*, 1938, **99**, 230.
[48] J. Iball, *Z. Krist.*, 1936, **94**, 7.

TABLE VI. (*Continued*)

Compound	Formula	M. pt.	Space group	Mol. per cell	a α	b β	c γ	Ref.
1:2:5:6-Dibenzanthracene (B)	$C_{22}H_{14}$	267°	*Pcab*	4	8.22	11.39	15.14	18
Picene	$C_{22}H_{14}$	365°	*Aba*	4	8.21	6.16	28.8	46
3:4:5:6-Dibenzphenanthrene (A)	$C_{22}H_{14}$	178°	$P2_1/a$	4	17.46	14.24	5.83	36
3:4:5:6-Dibenzphenanthrene (B)	$C_{22}H_{14}$	178°	$A2/a$	12	26.17 90°	8.94 105.1°	19.57 90°	37
1:12-Benzperylene	$C_{22}H_{12}$	273°	$P2_1/a$	4	11.72 90°	11.88 98.5°	9.89 90°	27
Tetrabenznaphthalene	$C_{26}H_{16}$	218°	$P2_1/a$	4	20.45 90°	7.74 119.5°	12.18 90°	36
Coronene	$C_{24}H_{12}$	435°	$P2_1/a$	2	16.10 90°	4.70 110.8°	10.15 90°	25
Ovalene	$C_{32}H_{14}$	473°	$P2_1/a$	2	19.47 90°	4.70 105°	10.12 90°	29
Di-*p*-xylylene	$C_{16}H_{16}$	285°	$P4/mnm$	2	7.82	7.82	9.33	41
Dibiphenylene-ethylene	$C_{26}H_{12}$	190°	*Pcan*	12	17.22	36.9	8.23	39

of gas diffraction experiments. These radii are chosen so that
their sum gives the average interatomic distance observed, at
normal temperatures, between the specified atoms, in their usual
covalency state. For tetrahedral and octahedral co-ordination,
and for purely ionic crystals, other sets of radii should be
employed.[42]

The covalent radii given in Table VII have been widely tested
on measurements from many compounds, and they are usually
found to give fairly accurate predictions of the observed bond
lengths. For bonds between like atoms, deviations from the
predicted values generally indicate some measure of bond multiplicity of the kind already discussed for carbon-carbon bonds.
But for a bond between unlike atoms, the ionic character of the
bond also becomes important and causes a further deviation from
the predicted values. Schomaker and Stevenson[49] have shown

[49] V. Schomaker and D. P. Stevenson, *J. Amer. Chem. Soc.*, 1941, **63**,
37; J. M. Robertson, *J. Chem. Soc.*, 1945, p. 249.

TABLE VII. COVALENT RADII AND ELECTRONEGATIVITIES

Element	Single bond radius, Å	Pauling electronega-tivity	Double bond radius, Å	Triple bond radius, Å
H	0.37	2.1	—	—
Li	1.34	1.0	—	—
B	0.88	2.0	0.76	0.68
C	0.771	2.5	0.665	0.602
N	0.74	3.0	0.60	0.547
O	0.74	3.5	0.55	0.50
F	0.72	4.0	0.54	—
Na	1.54	0.9	—	—
Si	1.17	1.8	1.07	1.00
P	1.10	2.1	1.00	0.93
S	1.04	2.5	0.94	0.87
Cl	0.99	3.0	0.89	—
K	1.96	0.8	—	—
Ge	1.22	1.7	1.12	—
As	1.21	2.0	1.11	—
Se	1.17	2.4	1.07	—
Br	1.14	2.8	1.04	—
Rb	2.11	0.8	—	—
Sn	1.40	1.7	1.30	—
Sb	1.41	1.8	1.31	—
Te	1.37	2.1	1.27	—
I	1.33	2.4	1.23	—
Cs	2.25	0.7	—	—

that this deviation is generally proportional to the absolute value of the Pauling electronegativity difference for the atoms concerned.   They deduce the empirical relation

$$r_{AB} = r_A + r_B - \beta |x_A - x_B|$$

where r is the single bond covalent radius (Table VII, column 2), x is the Pauling electronegativity value (Table VII, column 3), and the constant β has a value of 0.09.   This relation has been tested fairly extensively, and although it tends to give low values in some cases, there is a considerable measure of general agreement.   When double and triple bonds occur, the radius values listed should be used without correction.

IX

Molecular Arrangement and Hydrogen-bonded Structures

1. THE GROUPING OF MOLECULES IN CRYSTALS

Of the new knowledge made available by X-ray crystallographic studies some of the most interesting and important concerns the grouping and mutual arrangement of molecules in crystals. When a crystal structure is fully known, the co-ordinates of the atoms provide a complete picture of the structure in three dimensions and enable us to state all the intermolecular contacts with the same precision as the bond lengths within the molecule itself. In general, this precise information cannot be obtained by any other method. Although limited to the solid state, the conclusions are of the greatest importance in accounting for the physical properties of substances and explaining their general behaviour. In the liquid state, the factor of mobility enters, but the average contact distances between organic molecules must remain very nearly the same, because no great change in density occurs.

A survey of organic crystal structures reveals the presence of two very distinct types of interaction between molecules. In the first of these, typified by the hydrocarbons, ordinary packing considerations dominate. The atoms of adjoining molecules approach one another to within certain minimum and fairly well-defined limits, known as the van der Waals, or non-bonded approach distances. These distances are comparatively large and correspond to the weak residual attractions between molecules; the substances are generally volatile and of low melting point.

An entirely different picture of molecular interaction is presented when we study structures which contain certain strongly electronegative groups together with replaceable hydrogen atoms, such as carboxyl, hydroxyl, or amino groups.   In such structures a more compact type of molecular grouping often occurs, and the distances between the reactive centres (oxygen, nitrogen, or sometimes halogen) may be about 1 Å less than the closest approaches between the carbon atoms of hydrocarbon molecules.   The crystals of such substances are generally more brittle and have higher melting points than those of the first class.   The hydrogen atom obviously plays a key role in these types of molecular association, which have long been known to chemistry and are generally referred to as *hydrogen-bonded* or *hydrogen-bridged* structures.   Such bonds are of the greatest importance in chemistry, and particularly in biology.   Their small energy of formation (5 to 6 kcal. per mole) means that they can readily be formed or broken in processes taking place at body temperature, and yet they can have a profound effect in modifying the properties of substances.

In this section some general illustrations of these two contrasting types of organic crystal structure are given.   A discussion

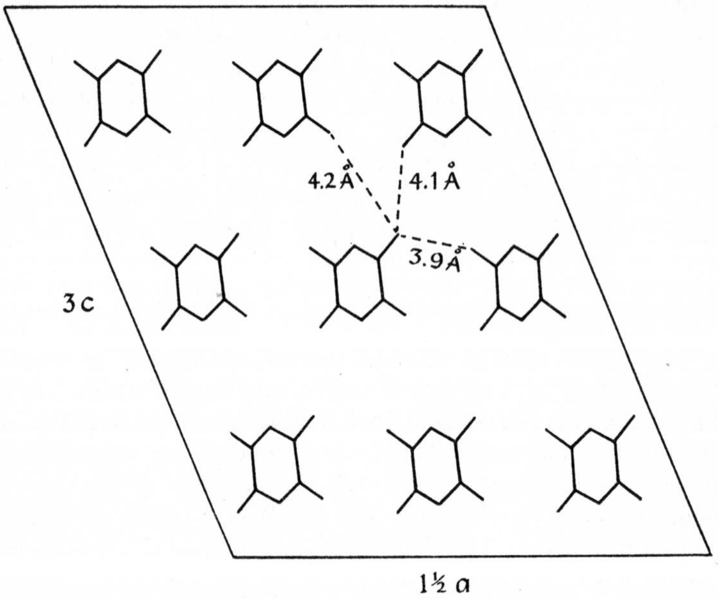

Fig. 95. Grouping of durene molecules in the crystal, seen in projection along the *b* axis. (For an electron density map, see Fig. 79.)

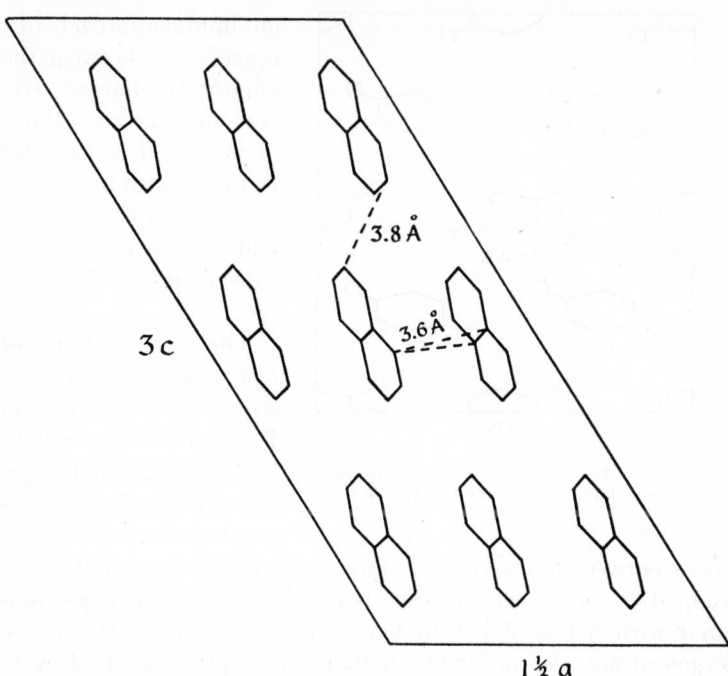

Fig. 96. Grouping of naphthalene molecules in the crystal, seen in projection along the *b* axis.

of the distances involved, and of the different types of intermolecular linkage, is given in the later sections.

Some of the hydrocarbon structures already described are typical of the first class, where only van der Waals attractive forces operate. Examples are given in Figs. 78 (p. 179), 95, and 96. These projections reveal the enormous amount of comparatively empty space that exists between the carbon frameworks. The hydrogen atoms are not shown in these drawings, and they do, of course, help to fill this space. But the non-bonded distances between the hydrogen atoms themselves are still large, generally varying from 2.0 to 2.5 Å or more. It appears likely that the lateral approach distances between the molecules are largely governed by the contacts between these marginal hydrogen atoms.

Intermolecular approach distances less influenced by hydrogen contacts are to be found in the benzoquinone crystal.[1] Here the

[1] J. M. Robertson, *Proc. Roy. Soc.* (London), 1935, **A150**, 106.

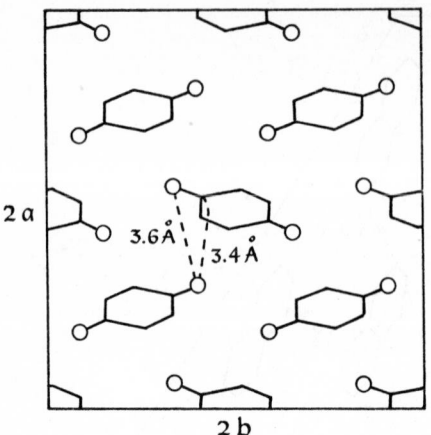

2 a

3.6 Å
3.4 Å

2 b

Fig. 97. Arrangement of benzoquinone molecules in the crystal, viewed along the *c* axis.

minimum approach distances occur between the oppositely charged oxygen and carbon atoms, as indicated in the projection shown in Fig. 97, but these distances are still large (3.4 Å) and the substance is quite volatile.

The second type, of *hydrogen-bonded crystal structures*, is illustrated in Fig. 98 for oxalic acid dihydrate[2] and in Fig. 99 for the alcohol pentaerythritol.[3] The distances between the oxygen atoms in these structures have now dropped to values which range between 2.5 and 2.8 Å, as compared with 3.4 to 3.6 Å in benzoquinone, for example, where oxygen atoms also occur but without the replaceable hydrogen.

It is well known that the physical properties of such associated substances are considerably modified. In solution they give a variety of abnormal data, leading to high molecular weight values. The carboxylic acids generally exist as dimers, and in the case of formic and acetic acid the double molecules persist even in the vapour at the boiling point. Alcohols, phenols, and compounds involving nitrogen, and occasionally halogen, with replaceable hydrogen, may display a more variable association, which depends in degree on the concentration of the solution. In the crystalline state the physical properties also differ markedly from those of non-hydrogen-bonded structures. Thus, formic acid melts at 8°, but the hydrocarbon propane, of similar molecular weight, is a gas at normal temperatures, the boiling point being −42° and melting point −190°. Pentaerythritol melts at

[2] W. H. Zachariasen, *Z. Krist.*, 1934, **89**, 442; J. M. Robertson and I. Woodward, *J. Chem. Soc.*, 1936, p. 1817; J. D. Dunitz and J. M. Robertson, *ibid.*, 1947, p. 142.

[3] F. J. Llewellyn, E. G. Cox, and T. H. Goodwin, *J. Chem. Soc.*, 1937, p. 883.

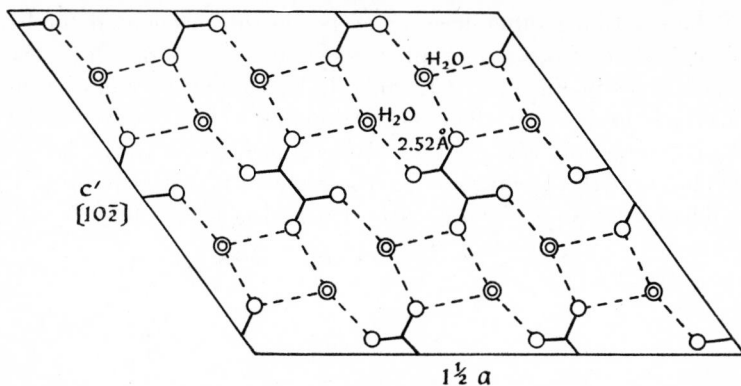

Fig. 98. Structure of oxalic acid dihydrate. Hydrogen bonds are indicated by dotted lines.

253°, while the hydrocarbon decane melts at about $-30°$.   Naphthalene, of similar molecular weight, melts at 80°, but the compactness of the aromatic molecule permits a relatively close packing.   Benzoquinone has the fairly high melting point of 115°, but hydroquinone, a hydrogen-bonded structure, melts at 171°.   Resorcinol, which is discussed below, is a peculiarly open structure, and melts at 110° although strongly hydrogen bonded.

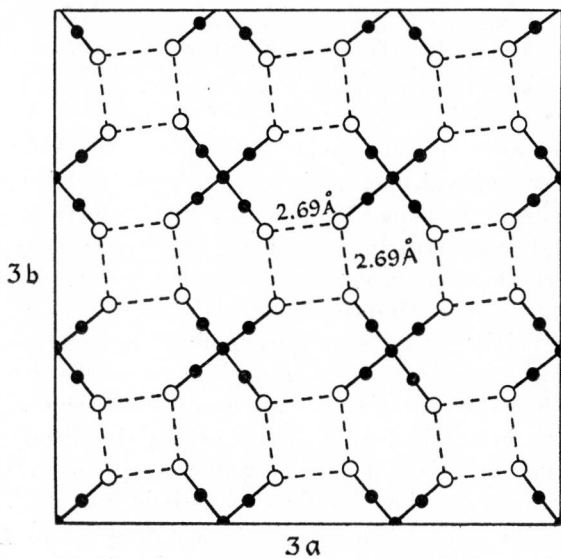

Fig. 99. Structure of pentaerythritol.

These striking differences in properties are accounted for by the existence of the strong intermolecular linkages shown by the dotted lines in Figs. 98–101.   The linkages may be regarded as intermediate between the covalent linkage and the weak van der Waals attraction, although, of course, much closer to the latter. Evidence of the greater compactness and firmness of hydrogen-bonded structures as compared with hydrocarbons may be ob-

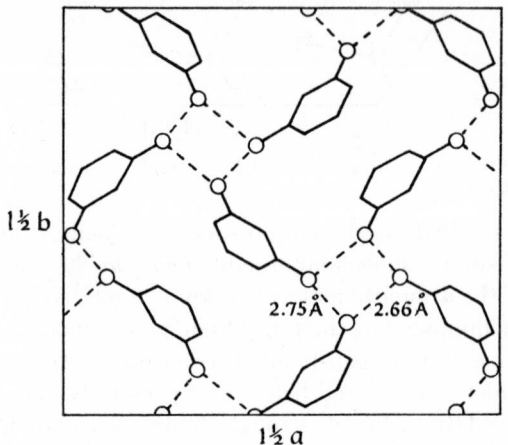

Fig. 100. Structure of α-resorcinol crystal. (The hydrogen bonds do not form closed circuits, as appear in this projection, but extend in infinite helices throughout the crystal.)

tained in a rather direct manner by comparing the X-ray diffraction patterns of typical crystals.   It is found that the reflections from small-spacing planes are generally stronger in the case of the hydrogen-bonded structures.   This is due to some reduction in the average thermal movement of the atoms because of the closer packing and stronger intermolecular linkages.

A study of the crystal structures involved, however, shows that the hydrogen bond is something much more specific than merely a rather stronger type of attraction between molecules.   It is effective only in certain definite directions, and this directive power is sometimes capable of maintaining an unusually open structure, where ordinary packing considerations would indicate the possibility of alternative structures of higher density.   A good example is found in the open structure of the ice crystal, where each oxygen is surrounded tetrahedrally by four other oxygen atoms at the hydrogen-bonded distances of 2.76 Å. On melting, this structure collapses to form water, which is of considerably higher density.

A still better example of such a transition, where the precise positions of all the atoms (except hydrogen) in both forms has been determined by the X-ray method, occurs in the case of re-

sorcinol.[4] The ordinary or α-form of resorcinol has the unusually low density of 1.27, the hydrogen bonds maintaining the very open structure which is shown in projection in Fig. 100. On heating, a somewhat slow transition commences at about 74°, to give the polymorphic β-resorcinol, which has the much higher density of 1.33. The structure of the β-modification is shown in Fig. 101, from which it is clear that with the weakening of the hydrogen bonds at the higher temperature, the open structure has collapsed. When the molecules have regrouped themselves in this

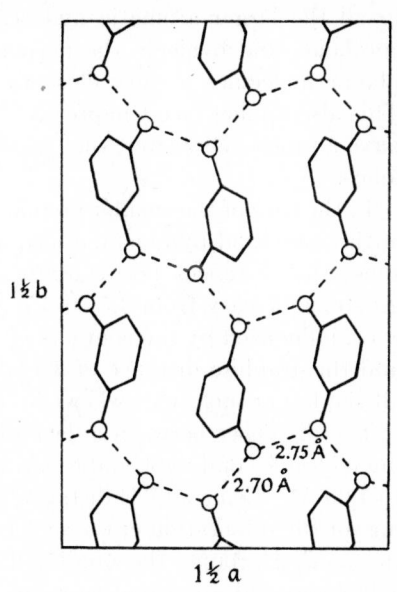

Fig. 101. The collapsed form of the resorcinol structure, stable at a higher temperature (β-resorcinal).

denser and more parallel packing array, the hydrogen bonds form again, in infinite chains instead of the infinite helices of the α-form.

These illustrations of some typical van der Waals and hydrogen-bonded molecular groupings in organic crystals have been confined to hydrocarbons and to oxygen-hydrogen-oxygen linkages. Other molecular groupings and different kinds of hydrogen bond are described in later sections.

[4] J. M. Robertson, *Proc. Roy. Soc.* (London), 1936, **A157**, 79; J. M. Robertson and A. R. Ubbelohde, *ibid.*, 1938, **A167**, 122.

2. VAN DER WAALS, OR NON-BONDED, RADII

The way in which the physical properties of diamond and graphite depend upon the atomic arrangement has already been pointed out. The interplanar distance of 3.4 Å in graphite represents a contact distance between the domains of the π-electrons which is fundamental not only for graphite but for all truly aromatic structures. In the previous chapter it has been shown that in all the larger aromatic hydrocarbons, such as benzperylene, coronene, and ovalene, the perpendicular distance between the planar molecules is very close to the graphite value of 3.4 Å. This also applies (see Chapter X) to other large molecules which have aromatic character, such as the phthalocyanines and tropolones.

In the case of the smaller aromatic molecules, like naphthalene, anthracene, and pyrene, and also in durene and hexamethylbenzene, the distances between the molecular planes are a little greater and vary from 3.5 to 3.7 Å. These distances are, however, influenced by the contacts of the marginal hydrogen atoms, and the graphite distance of 3.4 Å seems to apply in the case of all the larger molecules where direct overlap occurs.

The distances between adjacent methyl groups on neighbouring molecules of durene and hexamethylbenzene vary from 3.95 Å to 4.25 Å. Using a C–H distance of 1.09 Å as in methane, allowing for the orientation of the molecules, and assuming rotation of the methyl groups, the non-bonded H $\cdots$ H distances can be calculated and are found to range upwards from a minimum of about 2.0 Å. The minimum of 2.0 Å also applies to the hydrogen contacts in solid methane,[5] assuming molecular rotation. Hydrogen atom contacts in aromatic molecules have been studied carefully by Mack,[6] who finds a somewhat larger separation of about 2.6 Å to apply in many cases. The contact distance appears to vary in different compounds over the range 2.0 to 2.7 Å, but these conclusions are rather indefinite because the precise positions of the hydrogen atoms cannot generally be determined by X-ray analysis.

For atoms other than carbon, the usual minimum approach

[5] H. H. Mooy, *Nature*, 1931, **127**, 707; *Proc. Koninkl. Nederland. Akad. Wetenschap.*, 1931, **34**, 550, 660.

[6] E. Mack, Jr., *J. Amer. Chem. Soc.*, 1932, **54**, 2141; *J. Phys. Chem.*, 1937, **41**, 221.

distances can be summarised in a useful form by listing what are known as the van der Waals, or non-bonded, radii for different atoms.   These may be defined as one-half of the equilibrium internuclear distance between two similar atoms when the residual van der Waals attractive forces are just balanced by the repulsive forces.   In a careful discussion of these distances, Pauling[7] has pointed out that the van der Waals radius must be considerably greater than the covalent radius, and should approximate to the ionic radius.   In the line of approach between two chlorine molecules, for example,

$$:\overset{..}{\underset{..}{Cl}}:\overset{..}{\underset{..}{Cl}}:\qquad :\overset{..}{\underset{..}{Cl}}:\overset{..}{\underset{..}{Cl}}:$$

it is clear that there are two pairs of valency electrons, instead of the single pair required for the covalent bond; and electronically the molecular surroundings are similar to those which apply to

separate chlorine ions, $:\overset{..}{\underset{..}{Cl}}:^-$.

These conclusions are substantiated by the study of a number of inorganic and organic crystal structures, and the van der Waals

TABLE VIII. VAN DER WAALS RADII

H, 1.0–1.4 Å

N, 1.5 Å	O, 1.40 Å	F, 1.35 Å
P, 1.9	S, 1.85	Cl, 1.80
As, 2.0	Se, 2.00	Br, 1.95
Sb, 2.2	Te, 2.20	I, 2.15

Methyl group, CH_3, or methylene group, CH_2, 2.0 Å. Half thickness of aromatic molecule, 1.7 Å.

radii given in Table VIII are taken largely from Pauling's data. These figures are based for the most part on the known ionic radii of the atoms concerned.   The values are not intended to be exact, as the effective radius will depend on the particular attractive forces present in any given structure; they are usually reliable, however, to within about 0.1 Å.   The hydrogen radius is probably more variable than the others.   For the half thickness of the

[7] L. Pauling, *The Nature of the Chemical Bond*, Ithaca, N. Y.: Cornell University Press, 1939, sec. 24.

aromatic molecule we have listed the graphite value, which appears to apply generally when hydrogen contacts are not operative.

One important reservation, already mentioned in the previous chapter, must be made with regard to these radii. The effective value depends to a considerable extent on the position of the approaching atom with respect to the direction of the covalent bonds already formed. In directions close to these bonds the effective radius is reduced, owing to the localisation of the electrons in the bond path. According to Pauling, the reduction is about 0.5 Å in directions lying within about 35° to the bond direction. This effect explains why atoms which are bonded to the same atom can approach each other to well within the normal van der Waals distance without undue strain.

3. THE HYDROGEN BOND

3A. Chemical aspects.—Some time before the development of modern valency theory, the evidence of organic chemistry pointed strongly to the existence of a kind of intermolecular or inter-radical linkage involving the hydrogen atom. In a recent review of the chemical aspects, Hunter[8] attributes the first reference to Oddo and Puxeddu,[9] who ascribed the abnormal properties of o-hydroxyazo compounds to a divided hydrogen valency. Later Moore and Winmill[10] postulated a hydrogen bond in formulating the structure of trimethylammonium hydroxide as

$$(CH_3)_3N—H—OH.$$

In this way they explained the weakly basic properties of such substances as compared with the more powerfully basic quaternary ammonium hydroxides.

Other early formulations involving a hydrogen bond were made by Pfeiffer[11] and by Morgan and Reilly[12] to explain the reduced combining power of hydroxyl in certain o-hydroxy-quinones, enolised β-diketones, and similar compounds:

[8] L. Hunter, W. C. Price, and A. R. Martin, *Report of Symposium on the Hydrogen Bond*, Roy. Inst. Chem. Publication, 1950, no. 1.

[9] G. Oddo and E. Puxeddu, *Gazz. chim. ital.*, 1906, **36**, ii, 1.

[10] T. S. Moore and T. F. Winmill, *J. Chem. Soc.*, 1912, **101**, 1635.

[11] P. Pfeiffer, *Ann.*, 1913, **398**, 137.

[12] G. T. Morgan and J. Reilly, *J. Chem. Soc.*, 1913, **103**, 1494.

The essential role of the hydrogen atom in forming linkages between molecules and the very widespread occurrence of this phenomenon were, however, most clearly presented by Latimer and Rodebush[13] and by Huggins[14] in their discussion of the abnormal physical properties of associated liquids, especially the high values of their dielectric constants. These authors pointed out that in water, for example, the tendencies to add and to give up hydrogen are almost equally balanced, as is shown by the formation of such ions as H_3O^+ and OH^-; thus a pair of electrons on the H_2O molecule might exert sufficient attraction on the hydrogen of another molecule to bind the two together. In terms of the Lewis valency theory this was represented:

In such a situation the hydrogen was expected to be loosely bound and capable of considerable displacement by an electric field. The high value of the dielectric constant could then be explained. It was also noted that the process need not be limited to the formation of double or triple molecules, but that larger aggregates might be formed; and that in the liquid such aggregates would continually break and reform under the influence of thermal agitation. In the case of other liquids, such as acetic acid and formic acid, however, definite dimers would be formed, and the properties of such liquids would be different.

This type of co-ordination involving a hydrogen atom was also

---

[13] W. M. Latimer and W. H. Rodebush, *J. Amer. Chem. Soc.*, 1920, **42**, 1419.

[14] M. L. Huggins, p. 1431 of Ref. 13, also undergraduate thesis, University of California, 1919; *Phys. Rev.*, 1921, **18**, 333; 1922, **19**, 346.

utilized by Sidgwick[15] and his co-workers to explain the peculiar properties of certain ortho-substituted phenols and other compounds of the type studied by Pfeiffer and by Morgan and Reilly. The association which would normally be expected for such compounds is decreased or eliminated because the hydrogen-bonding capacity is utilized intramolecularly.

3B. Mechanism of hydrogen bond formation.—These early chemical formulations of the hydrogen bond provided a fairly adequate account of the behaviour of the substances involved; there could be no doubt that the fundamental idea of the hydrogen atom strongly attracting *two* other nearest atoms was correct. However, the true physical nature of the linkage remained rather obscure. It soon became clear that the valency picture which attributed two electron pairs to the hydrogen atom was quite inadmissible. In particular, it was shown by Pauling[16] that in hydrogen there is really only one orbital (1s) available for bond formation; the outer orbitals are so much less stable that strong bonds could not be formed. A structure such as that drawn for the water molecule (see above) would require so much energy for its formation that it would be quite unstable.

One of the difficulties attending investigations of the hydrogen bond is that reliable determinations of the position of the hydrogen atom are difficult to make. A symmetrical bond with the hydrogen at the centre was formerly thought to be likely, but, with a few exceptions which are noted later, a large amount of evidence is now against this view. For example, Pauling[17] has shown that the observed entropy of ice requires the existence of discrete water molecules, with the protons unsymmetrically placed between the oxygen atoms. Spectroscopic evidence is also definite. The normal O—H distance in the gas molecule (H_2O or alcohols) is 0.96 Å, whereas the distance between the hydrogen-bonded oxygen atoms in the ice crystal is 2.76 Å. A symmetrical bridge would thus require the O—H distance to be increased from 0.96 Å to 1.38 Å. Infra-red spectroscopy shows that the frequency associated with the stretching vibration of the

[15] N. V. Sidgwick, W. S. Spurrell, and T. E. Davis, *J. Chem. Soc.*, 1915, **107**, 1202; N. V. Sidgwick and W. M. Aldous, *ibid.*, 1921, **119**, 1001.
[16] L. Pauling, *Proc. Nat. Acad. Sci. U. S.*, 1928, **14**, 359; *J. Amer. Chem. Soc.*, 1931, **53**, 1367.
[17] L. Pauling, *J. Amer. Chem. Soc.*, 1935, **57**, 2680.

O—H bond, in the case of the gas molecule or in dilute solution, is indeed reduced in the liquid or solid state, but not far enough. The sharp frequency characteristic of the free O—H group is replaced by a broad band, which has a maximum on the low frequency side. Analysis shows that the change is not sufficient to allow an increase in the distance of from 0.96 to 1.38 Å, but only to about 0.99 Å in the case of water.[18]   In the case of acetic acid[19] the O—H bond length may increase from about 0.97 Å in the

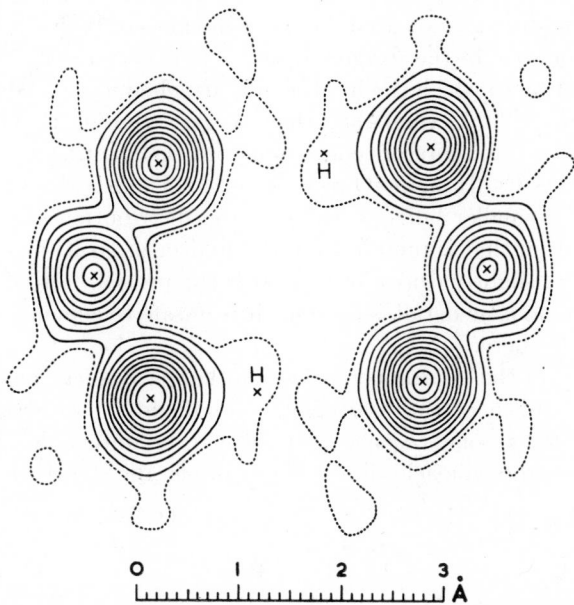

Fig. 102. Section through the central plane of the carboxyl groups of two succinic acid molecules linked by hydrogen bonds. The dotted contour line represents a density of ½ electron per Å³, and the other lines are drawn at intervals of 1 electron per Å³. The section passes through the centres of the oxygen atoms.

monomer to about 1.07 Å in the dimer, but even this is much less than what is required to effect a symmetrical bridge.

Although the scattering power of hydrogen for X-rays is extremely low, recent careful analyses of certain crystal structures

[18] P. C. Cross, J. Burnham, and P. A. Leighton, *J. Amer. Chem. Soc.*, 1937, **59**, 1134.
[19] R. C. Herman and R. Hofstadter, *J. Chem. Phys.*, 1938, **6**, 534.

provide rather definite evidence regarding the hydrogen positions.[20] Fig. 102 shows a section of the electron density distribution evaluated in the central plane of the carboxyl groups of two succinic acid molecules[21] which are linked by hydrogen bonds of length 2.66 Å. In this diagram evidence regarding the hydrogen positions is certainly far from precise. There is, however, a very distinct bulge in the one-electron contour level around one of the oxygen atoms, and the outer contours of the oxygen atoms are also distended. The evidence is consistent with placing the hydrogen atom as indicated, i.e., at a distance of between 1.0 and 1.1 Å from one of the oxygen atoms. It is noteworthy that the oxygen atom to which the hydrogen is attached has a longer bond to carbon (1.29 Å) than the other oxygen atom (1.24 Å). The observations are therefore consistent with the identification of the groups as —O—H and =O respectively. There is also some evidence of a smaller bulge in the electron distribution around the oxygen atom that is remote from the hydrogen, as if the electron cloud were slightly drawn out towards the proton. However, as considerable false detail is present, it is unsafe to come to definite conclusions here.

In view of these various results, another early suggestion[22] that the mechanism of the hydrogen bond might be formulated in terms of the resonance concept is seen to be untenable. In the case of a carboxylic acid dimer, for example, the structures

fail to have the identical nuclear configuration necessary for resonance. In addition to the unsymmetrical bridge, the C—O bond distances are known to be non-equivalent, as shown by Fig. 102 and by a number of other investigations.[23]

[20] See, for example, Figs. 83 and 84, where some of the hydrogen atoms are clearly defined in the naphthalene and anthracene molecules.

[21] J. D. Morrison and J. M. Robertson, *J. Chem. Soc.*, 1949, p. 980; and further unpublished three-dimensional work.

[22] N. V. Sidgwick, *Ann. Repts. on Progress Chem.* (Chem. Soc. London), 1934, **31**, 37.

[23] J. Karle and L. O. Brockway, *J. Amer. Chem. Soc.*, 1944, **66**, 574.

It is now agreed that the hydrogen bond is essentially ionic in nature and that an adequate explanation of its properties can be given in terms of electrostatic attractions between the positively charged hydrogen and the negative charge on the adjoining oxygen, nitrogen, or halogen atom.[24]   At first sight it seems difficult to explain the unique and universal character of the hydrogen bond on this basis when in various compounds, such as benzoquinone or acetone, equally large net charges may be present on the oxygen and carbon atoms without effecting any molecular association.   The explanation of this lies in the peculiar character of the hydrogen atom, which possesses no inner electrons.   The positive hydrogen ion, being a vanishingly small cation, is able to make a close approach to a neighbouring anion without steric interference and thus form a relatively powerful link.   Once this link is formed between two electronegative atoms, steric factors will in general prevent any other atoms from coming sufficiently close to be permanently bound.   The apparent divalency of hydrogen in these circumstances is thus explained.

It is also clear that this mechanism accounts for the fact that hydrogen bonds are only formed between the strongly electronegative atoms, fluorine, oxygen, nitrogen, and chlorine, and that the most powerful bonds will be expected when the atoms are most strongly electronegative.   Examples of the various types of hydrogen bond are given in the next section, and it will be found that these expectations are generally fulfilled.   The other mechanisms for the hydrogen bond which have been suggested afford no explanation of these important facts; they fail to explain, for example, why hydrogen bonds are not formed between carbon and oxygen.

3C. Symmetrical and non-symmetrical bonds: the isotope effect.—It has been mentioned above that the hydrogen bond is in general not symmetrical, the hydrogen being bound more closely to one of the two electronegative atoms involved than to the other.   It does not follow, however, that the hydrogen remains permanently attached to one of these atoms; especially for the shorter hydrogen bonds, tautomerism of the type

$$X—H \cdots X \rightleftarrows X \cdots H—X$$

[24] L. Pauling, *The Nature of the Chemical Bond*. Ithaca, N. Y.: Cornell University Press, 1939, chap. ix.

will be expected to occur.   In the longer bonds the potential function of the hydrogen atom between the two other atoms will have two distinct minima (not necessarily equivalent).   As the bond gets shorter, the intervening barrier will decrease, but no conclusive evidence for a symmetrical bond in organic crystals has yet been obtained.[25]   In a few cases, for example, in potassium hydrogen bisphenylacetate[26] and in potassium hydrogen di-p-hydroxybenzoate hydrate,[27] a symmetrical bond appears to be demanded by considerations of crystallographic symmetry. However, in view of the bond lengths involved in these cases (2.55 and 2.61 Å) it is likely that the crystals have random structures and that the bond symmetry is merely statistical.

Only in the case of the inorganic crystal, potassium hydrogen fluoride, does there appear to be really good evidence for a completely symmetrical bond, in the ion [F—H—F]$^-$.   Westrum and Pitzer[28] have shown, from a careful study of the thermodynamic, spectroscopic, and dielectric data, that the proton in this case appears to lie in a single potential minimum half-way between the two fluorine atoms.   The observed hydrogen bond length of 2.26 Å[29] is the shortest known.   An earlier infra-red study of Ketelaar,[30] who found two intense peaks at 1222 and 1450 cm.$^{-1}$, had indicated a double minimum potential curve even in this case, with the proton oscillating between the two equilibrium positions, but it now seems likely that these two peaks should be ascribed to longitudinal and transverse vibrations of the proton, rather than to a double minimum in the potential curve.[31]

By analogy with the case of the hydrogen fluoride ion and making use of Pauling's relation[32] between atomic radius and

[25] For a review of the physical aspects, see M. Davies, *Ann. Repts. on Progress Chem.* (Chem. Soc. London), 1946, **43**, 5.

[26] J. C. Speakman, *J. Chem. Soc.*, 1949, p. 3357.

[27] J. M. Skinner and J. C. Speakman, *J. Chem. Soc.*, 1951, p. 185.

[28] K. S. Pitzer and E. F. Westrum, *J. Chem. Phys.*, 1947, **15**, 526; E. F. Westrum and K. S. Pitzer, *J. Amer. Chem. Soc.*, 1949, **71**, 1940.

[29] R. M. Bozorth, *J. Amer. Chem. Soc.*, 1923, **45**, 2128; L. Helmholz and M. Rogers, *ibid.*, 1939, **61**, 2590.

[30] J. A. A. Ketelaar, *J. Chem. Phys.*, 1941, **9**, 775; *Rec. trav. chim.*, 1941, **60**, 523.

[31] D. Polder, *Nature*, 1947, **160**, 870.

[32] L. Pauling, *J. Amer. Chem. Soc.*, 1947, **69**, 542.

bond type, Donohue[33] has calculated the probable distance for a symmetrical O—H—O bond to be about 2.30 Å. This is distinctly less than any of the observed bonds between oxygen atoms, the shortest recorded being 2.42 Å in nickel dimethylglyoxime[34] (an interesting intramolecular bond that may prove to be symmetrical) and 2.50 Å in urea oxalate.[35] It therefore seems likely that with the possible exception of nickel dimethylglyoxime the hydrogen bonds so far discovered between oxygen atoms are unsymmetrical. It may be inferred, however, that in the case of the shorter hydrogen bonds the potential barrier will be small and that the proton will be capable of executing relatively large vibrations along a broad, trough-like potential function[36] between the two electronegative atoms.

In connection with the problem of proton transfer in the hydrogen bond, some interesting results are obtained when deuterium is substituted for hydrogen.[37] The normal effect of such a substitution, where no hydrogen bonds are involved, will be to permit a slightly closer packing of adjoining molecules owing to the smaller space requirement of deuterium. The electronic situation is, of course, unchanged, but there is an important vibrational difference, in that deuterium has a smaller zero point energy than hydrogen, in the ratio $1 : \sqrt{2}$. Intermolecular contacts such as C—H $\cdots$ H—C, for example, should contract slightly, because of the smaller amplitude of vibration of the heavier deuterium atom. In lithium hydride,[38] where the structure is ionic (Li$^+$ H$^-$), the result of the substitution is also a contraction, the edge of the cubic lattice being found to diminish from 4.080 Å to 4.060 Å on substitution of deuterium for hydrogen.

[33] J. Donohue, *J. Phys. Chem.*, 1952, **56**, 502.

[34] L. E. Godycki, R. E. Rundle, R. C. Voter, and C. V. Banks, *J. Chem. Phys.*, 1951, **19**, 1205.

[35] A. F. Schuch, L. Merritt, and J. H. Sturdivant (to be published).

[36] M. L. Huggins, *J. Chem. Phys.*, 1935, **3**, 473; *J. Org. Chem.*, 1937, **1**, 407; *J. Amer. Chem. Soc.*, 1936, **58**, 694; *J. Phys. Chem.*, 1936, **40**, 723; A. R. Ubbelohde and I. Woodward, *Proc. Roy. Soc.* (London), 1946, **A185**, 448.

[37] A. R. Ubbelohde, *Trans. Faraday Soc.*, 1936, **32**, 525; J. M. Robertson and A. R. Ubbelohde, *Nature*, 1937, **139**, 504; *Proc. Roy. Soc.* (London), 1939, **A170**, 222, 241.

[38] E. Zintl and A. Harder, *Z. physik. Chem.*, 1935, **B28**, 478.

Now in the case of hydrogen-bonded structures it is found that this normal effect of a small lattice contraction is reversed, and instead an expansion occurs in the direction of the hydrogen bond. The effect is more marked in the case of the shorter and stronger hydrogen bonds. Oxalic acid dihydrate, for example, has one very short hydrogen bond of 2.52 Å. On substituting deuterium the lattice expands by about 0.6 per cent, mainly in the direction of this hydrogen bond. If the whole of this expansion does occur within the hydrogen bond itself, as seems probable, the effect is fairly large and amounts to an increase of about 0.04 Å in the O · · · O distance. No other structure examined has shown such a large effect as this, but smaller expansions have been observed in succinic and benzoic acids, α-resorcinol, and sodium bicarbonate. In phthalic acid, β-resorcinol, and ice, the effect is almost zero. It should be noted, however, that no effect means that the contraction which would normally be expected has been compensated in some way.

The cause of this isotope effect has not yet been fully elucidated, but it appears to be connected with the tautomeric transfer of the proton or deuteron between its two possible positions in the path of the hydrogen bond. (The term *resonance* or *protonic resonance* has been used to describe this motion, but such a term is probably best reserved for use in its usual quantum-mechanical and electronic context.) At first it was thought that the expansion effect provided evidence that the nuclear motion stabilized the structure. Because if this were so, the higher frequency and smaller mass of the proton as compared with the deuteron would lead to a stronger and closer bonding. This is precisely what is observed. However, this factor, although possibly of some significance, does not appear to be great enough to explain the effect observed in the case of oxalic acid dihydrate. Stabilization due to the motion of a particle with mass as great as a proton can be shown to be extremely small.

In a recent study, Nordman and Lipscomb[39] explain the isotope effect from a consideration of the motion of the proton and deuteron in a one-dimensional potential field between the two end atoms. The mean amplitudes of vibration, averaged over a Boltzmann distribution of vibrational states, are calculated. It is then found that the mean amplitude is greater for D than for H

[39] C. E. Nordman and W. N. Lipscomb, *J. Chem. Phys.*, 1951, **19**, 1422.

in a broad, boxlike potential function such as that expected in
the case of the shorter hydrogen bonds. But when the potential
function is parabolic, as will apply in the case of the hydrides,
for example, the reverse is true, the amplitude being greater for
H than for D. Details of these calculations cannot be given
here, but they appear to afford a satisfactory explanation of the
effect.

4. TYPES OF HYDROGEN BOND

A fairly large number of different types of hydrogen bond have
now been observed in organic crystals, and the results of various
X-ray measurements are collected in this section. The bond
lengths observed are to some extent characteristic of the different
electronegative groups involved, those between adjoining carboxyl groups, for example, being considerably shorter than those
connecting hydroxyl groups. However, a certain range of distances is often found to occur in any one kind of hydrogen bond,
as tables IX to XII (pp. 240, 243, 245, 246) show.

Determinations of the energy values[40] of hydrogen bonds are
less numerous and less exact, but such data as are available show
that the energy decreases rapidly with increasing length for any
one type of bond. Thus, for the carboxylic acid dimers, when
the length is about 2.65 Å, bond energy values of between 7 and
8 kcal. per mole have been found. For the alcohols and phenols
the length is about 2.70 Å and the energy about 6 kcal.; while in
ice, when the bond length is 2.76 Å, the energy value is probably
about 4.5 kcal.[40] Bonds between nitrogen atoms are generally
relatively weak, with energy values of between 1 and 3 kcal.,
although stronger bonds occur between nitrogen and oxygen.

A study of the large number of organic structures involving
intermolecular hydrogen bonds which have now been determined
reveals some very important general principles regarding conditions of formation. In the first place, certain steric conditions
have to be satisfied. When hydroxyl groups, water molecules,
or —NH_3^+ ions are involved, an approximately tetrahedral dis-

[40] M. Davies, *Ann. Repts. on Progress Chem.* (Chem. Soc. London),
1946, **43**, 5; L. Pauling, *The Nature of the Chemical Bond*, Ithaca, N. Y.:
Cornell University Press, 1939. For the O—H · · · O bond in ice a
higher energy value of 6.4 ±0.5 kcal. is calculated by A. W. Searcy,
J. Chem. Phys., 1949, **17**, 210. From the energy of dimerization of steam,
calculated from the second virial coefficient, values of from 4.5 to 5 kcal.
are obtained (J. S. Rowlinson, *Trans. Faraday Soc.*, 1949, **45**, 974).

tribution of the bonds about the electronegative atom or ion is required. The careful analysis of Donohue[41] shows that the angles R—O $\cdots$ O, R—N $\cdots$ O, R—N $\cdots$ N, or R—N $\cdots$ Cl, generally lie between 90° and 120°; if more extreme values are involved, the bond is long and weak.

In the case of the bonds between carboxyl groups, coplanar structures are usually required. The C—O distances and R—C—O angles are, however, generally unequal, which suggests that the structure

$$
\begin{array}{c}
\text{O} \cdots \text{H-O} \\
\text{R—C} \qquad \qquad \text{C—R} \\
\text{O-H} \cdots \text{O}
\end{array}
$$

makes the most important contribution to the normal molecular state. In the dicarboxylic acid dihydrates there appears to be greater equality of the C—O distances, indicating a larger contribution from the structure

$$
\text{R—C}
\begin{array}{c}
\text{O}^- \\
\text{O}^+\text{—H}
\end{array}
$$

or possibly the formation of COO^- and H_3O^+ ions.

In the amides and imides, a large number of recent accurate measurements[42] show that the three bonds around the nitrogen atom are also coplanar. In the group

$$
\begin{array}{c}
\text{O} \quad \text{H} \\
\text{—C—N—}
\end{array}
$$

the C—N distance is much shorter than the normal single bond length of 1.47 Å, owing to a considerable contribution from the structure

$$
\begin{array}{c}
\text{O}^- \quad \text{H} \\
\text{—C=N}^+\text{—.}
\end{array}
$$

[41] J. Donohue, *J. Phys. Chem.*, 1952, **56**, 502.
[42] R. B. Corey, *Advances in Protein Chemistry*, New York: Academic Press, 1948, **4**, 385.

This is of great importance in formulating possible structures for the proteins, the approximately planar arrangement of the atoms in the peptide grouping

placing severe restrictions on the possible positions of any atom to which it can be linked by hydrogen bonds. These principles have been utilized by Pauling, Corey, and Branson[43] in formulating probable helical configurations for the polypeptide chain.

Another generalisation derived from a study of these various crystal structures is what may be termed the principle of maximum hydrogen bonding. All the available hydrogen atoms, attached to the electronegative groups, are generally employed in hydrogen bond formation. Some of the bonds formed may be weaker than others, but the molecular packing is generally capable of adjustment in such a way as to permit the fulfilment of this condition. Sometimes the resulting structure may not be the most compact that might be devised, but this condition and the steric requirements are nevertheless generally obeyed. They are observed, for example, in the case of ice and resorcinol, even at the expense of maintaining unusually open structures of low density. Considerations of this kind are of great value as an aid to the determination of crystal structures. If, for example, a postulated structure is such as to involve only a fraction of the available hydrogen atoms in bond formation, then it can usually be dismissed as improbable.

4A. Oxygen · · · oxygen.—X-ray measurements on hydrogen bonds between oxygen atoms in organic crystals are collected in Table IX. Carpenter and Donohue[44] have suggested a classification based on three main types, depending on the situation of the donor oxygen atom: (1) those involving a carboxyl group

[43] L. Pauling and R. B. Corey, *J. Amer. Chem. Soc.*, 1950, **72,** 5349; L. Pauling, R. B. Corey, and H. R. Branson, *Proc. Nat. Acad. Sci. U. S.*, 1951, **37,** 205.

[44] G. B. Carpenter and J. Donohue, *J. Amer. Chem. Soc.*, 1950, **72,** 2315.

TABLE IX. O—H · · · O BONDS

Compound	O—H · · · O (Å)	Type (donor-acceptor)	Ref.
Oxalic acid dihydrate	2.52	carboxyl-water (or oxonium ion)	2
	2.84, 2.87	water-carboxyl	
Acetylene dicarboxylic acid dihydrate	2.56	carboxyl-water (or oxonium ion)	45
	2.82, 2.89	water-carboxyl	
Diacetylene dicarboxylic acid dihydrate	2.55	carboxyl-water (or oxonium ion)	46
	2.83, 2.90	water-carboxyl	
N-Acetylglycine	2.56	carboxyl-carbonyl	44
Urea oxalate	2.50	carboxyl-carbonyl	35
KH-bisphenylacetate	~2.55	carboxyl-carboxyl	23
KH-di-p-hydroxy-benzoate hydrate	~2.61	carboxyl-carboxyl	27
	2.69	hydroxyl-water	
Succinic acid	2.64	carboxyl-carboxyl	47
Adipic acid	2.68	carboxyl-carboxyl	48
Glutaric acid	2.69	carboxyl-carboxyl	49
Sebacic acid	2.68	carboxyl-carboxyl	50
Tartaric acid	2.74, 2.78 2.87, 2.87	hydroxyl-carboxyl (?)	51
L-Threonine	2.66	hydroxyl-carboxyl	52
DL-Serine	2.67	hydroxyl-carboxyl	53
Hydroxy-L-proline	2.80	hydroxyl-carboxyl	54
Cytidine	2.74	hydroxyl-carboxyl	55
	2.83	hydroxyl-hydroxyl	
α-Resorcinol	2.66, 2.75	hydroxyl-hydroxyl	56
β-Resorcinol	2.70, 2.75	hydroxyl-hydroxyl	57
Quinol·SO₂	2.75, 2.76	hydroxyl-hydroxyl	58
Pentaerythritol	2.69	hydroxyl-hydroxyl	3
Hydrogen peroxide	2.78, 2.78	hydroxyl-hydroxyl	59
Urea·H₂O₂	2.63	hydroxyl-hydroxyl	60
4:6-Dimethyl-2-hydroxy-pyrimidine dihydrate	2.75, 2.93	hydroxyl-water	61
	2.73	water-water	
Ice	2.76	water-water	

[45] J. D. Dunitz and J. M. Robertson, *J. Chem. Soc.*, 1947, p. 148.
[46] J. D. Dunitz and J. M. Robertson, *J. Chem. Soc.*, 1947, p. 1145.
[47] H. J. Verweel and C. H. MacGillavry, *Z. Krist.*, 1939, **102**, 60; J. D. Morrison and J. M. Robertson, *J. Chem. Soc.*, 1949, p. 980.
[48] C. H. MacGillavry, *Rec. trav. chim.*, 1941, **60**, 605; J. D. Morrison and J. M. Robertson, *J. Chem. Soc.*, 1949, p. 987.
[49] J. D. Morrison and J. M. Robertson, *J. Chem. Soc.*, 1949, p. 1001.
[50] J. D. Morrison and J. M. Robertson, *J. Chem. Soc.*, 1949, p. 993.

with its acidic hydrogen atom, where the length varies from about 2.50 Å in oxalic acid and urea oxalate to about 2.65 Å in the carboxylic acid dimers; (2) those involving hydroxyl groups as in the alcohols and phenols, where the lengths vary from about 2.7 to 2.8 Å; (3) those involving water molecules, as in the ice crystal and certain hydrates, where the length varies from 2.76 to about 2.9 Å. There are, however, occasional uncertainties in this classification, as direct evidence is not always available to distinguish between the situation of the donor and acceptor oxygen atom. For example, in oxalic acid dihydrate there are three bonds or close approaches associated with each water molecule, one with one of the carboxyl oxygen atoms and two with the other. As the positions of the hydrogen atoms cannot be observed, the classification can really only be based on the observed distances. The assignments adopted in Table IX usually follow the careful analysis of Donohue.[41]

4B. Nitrogen · · · oxygen.—The hydrogen bonds between nitrogen and oxygen are of great importance because they govern the crystal structure of the amino acids and peptides, and presumably play a fundamental rôle in protein structure. Extremely accurate X-ray studies, refined by elaborate three-dimensional Fourier methods, have recently been carried out on the amino acids, alanine, serine, threonine, acetylglycine, and some related compounds. Space does not permit any detailed description of these rather complicated but very beautiful crystal structures. The most simple amino acid, glycine, has not yet

[51] F. Stern and C. A. Beevers, *Acta Cryst.*, 1950, **3**, 341.

[52] D. P. Shoemaker, J. Donohue, V. Schomaker, and R. B. Corey, *J. Amer. Chem. Soc.*, 1950, **72**, 2328.

[53] D. P. Shoemaker, R. E. Barieau, J. Donohue, and C. S. Lu, *Acta Cryst.*, 1953, **6**, 241.

[54] J. Donohue and K. N. Trueblood, *Acta Cryst.*, 1952, **5**, 414, 419.

[55] S. Furberg, *Acta Cryst.*, 1950, **3**, 325.

[56] J. M. Robertson, *Proc. Roy. Soc.* (London), 1936, **A157**, 79.

[57] J. M. Robertson and A. R. Ubbelohde, *Proc. Roy. Soc.* (London), 1938, **A167**, 122, 136.

[58] D. E. Palin and H. M. Powell, *J. Chem. Soc.*, 1947, p. 208.

[59] S. C. Abrahams, R. L. Collin, and W. N. Lipscomb, *Acta Cryst.*, 1950, **4**, 15.

[60] C. S. Lu, E. W. Hughes, and P. A. Giguere, *J. Amer. Chem. Soc.*, 1941, **63**, 1507.

[61] G. J. Pitt, *Acta Cryst.*, 1948, **1**, 168.

been measured with quite the same accuracy as the others, but the arrangement of the nearly planar molecules may be illustrated by the following scheme:

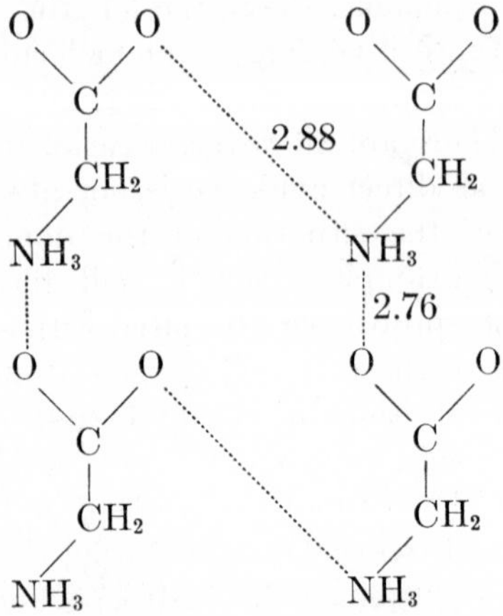

A second layer of molecules is related by a centre of symmetry to the one shown, and is rather closely bound to it, the minimum approach distances of 2.93 and 3.05 Å probably representing very weak hydrogen bonds due to the third hydrogen of —NH$_3^+$ sharing its bond-forming capacity with the two nearest oxygens of the adjacent layer.

This glycine arrangement is somewhat exceptional. Other amino acids also have the "zwitter-ion" structure NH$_3^+$CHR·COO$^-$, but they usually exhibit three strong intermolecular nitrogen-oxygen bonds (Table X), and this results in a very compact molecular grouping. The three bonds are disposed tetrahedrally about the —NH$_3^+$ group, and the atoms of the carboxyl group are usually engaged unequally by different —NH$_3^+$ groups:

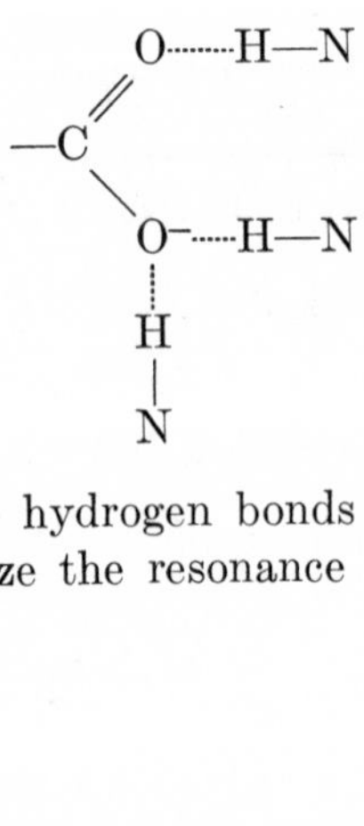

The formation of two hydrogen bonds by one of the oxygens should tend to stabilize the resonance structure shown above,

242

and confirmation of this is obtained from unequal C—O distances and C—C—O angles in the carboxyl group of DL-alanine.[62]

The importance of the amides and imides in connection with protein structure has already been mentioned. The essentially planar arrangement of the atoms in the peptide group limits the possibilities of hydrogen bond formation. Examples of such bonds from the >NH group are given by N-acetylglycine (3.03 Å), β-glycylglycine (3.07 Å), and diketopiperazine (2.88 Å).

TABLE X. N—H · · · O BONDS

Compound	N—H · · · O	Ref.
Glycine	2.76, 2.88 Å (and 2.93, 3.05 Å?)	63
DL-Alanine	2.80, 2.84, .288	62
DL-Serine	2.79, 2.81, 2.87	53
L-Threonine	2.80, 2.90, 3.10	52
N-Acetylglycine	3.03	44
Hydroxy-L-proline	2.69, 3.17	54
β-Glycylglycine	2.68, 2.80, 2.81, 3.07	64
Diketopiperazine	2.85	65
Acetamide	2.83, 2.99	66
Urea	2.99, 3.00	67
Urea oxalate	2.89, 2.97, 3.00, 3.14	35
Urea·H_2O_2	2.94, 3.04	60
Adenine·HCl·$\frac{1}{2}H_2O$	2.87 (N · · · H_2O)	68
4:6-Dimethyl-2-hydroxy-pyrimidine	2.78, 2.89 (O—H · · · N)	31
Cytidine	2.87 (O—H · · · N)	55
Isatin	2.93	69
p-Nitroaniline	3.07, 3.11	70

[52] H. A. Levy and R. B. Corey, *J. Amer. Chem. Soc.*, 1941, **63**, 2095; J. Donohue, *ibid.*, 1950, **72**, 949.

[53] G. Albrecht and R. B. Corey, *J. Amer. Chem. Soc.*, 1939, **61**, 1087.

[64] E. W. Hughes and W. J. Moore, *J. Amer. Chem. Soc.*, 1949, **71**, 2618; for the analysis of cysteylglycine sodium iodide, in which a short NH · · · O bond of 2.55 Å is reported, see H. B. Dyer, *Acta Cryst.*, 1951, **4**, 42.

[65] R. B. Corey, *J. Amer. Chem. Soc.*, 1938, **60**, 1598.

[66] F. Senti and D. Harker, *J. Amer. Chem. Soc.*, 1940, **62**, 2008.

[37] R. W. G. Wyckoff and R. B. Corey, *Z. Krist.*, 1934, **89**, 462; P. Vaughan and J. Donohue, *Acta Cryst.*, 1952, **5**, 530.

[38] J. M. Broomhead, *Acta Cryst.*, 1948, **1**, 324.

[69] G. H. Goldschmidt and F. J. Llewellyn, *Acta Cryst.*, 1950, **3**, 294.

[70] S. C. Abrahams and J. M. Robertson, *Acta Cryst.*, 1948, **1**, 252.

The crystal structures of urea and acetamide also confirm the expected planar arrangement of the atoms concerned. The hydrogen bonds observed in these structures and in some related compounds are collected in Table X.

4C. Nitrogen · · · nitrogen.—Hydrogen bonds between nitrogen atoms are relatively weak, and vary in length from about 3.0 to 3.3 Å (Table XI). They do not play such a significant part in determining crystal structures as the types of hydrogen bond already described, and in fact the principle of maximum hydrogen bonding is no longer of general application where only nitrogen atoms are concerned. For example, the first solid in the series of polymethylene diamines is hexamethylene diamine, $C_6H_{16}N_2$, and it has a melting point of only 42°, indicating very weak intermolecular attractions. Detailed X-ray analysis[71] shows that with four available hydrogen atoms, only two hydrogen bridges, of length 3.21 Å, occur between the terminal nitrogen atoms of successive chains in the crystal.

Similarly in melamine[72]

only four of the six hydrogen atoms are used in hydrogen bond formation between the amino groups and the ring nitrogen atoms of neighbouring molecules.

In crystalline ammonia[73] each nitrogen atom appears to have six nearest neighbours at 3.38 Å, distances so great that the contacts can only be due to extremely weak hydrogen bond formation. On the basis of six bonds per nitrogen atom, a bond energy value of 1.3 kcal./mole is obtained. However, Lambert and Strong[74] have shown, from compressibility measurements and

[71] W. P. Binnie and J. M. Robertson, *Acta Cryst.*, 1950, **3**, 424.

[72] E. W. Hughes, *J. Amer. Chem. Soc.*, 1941, **63**, 1737.

[73] H. Mark and E. Pohland, *Z. Krist.*, 1925, **61**, 532; J. de Smedt, *Bull. Acad. Roy de Belg.*, 1925, **10**, 655.

[74] J. D. Lambert and E. D. T. Strong, *Proc. Roy. Soc.* (London), 1950, **A200**, 566.

evaluation of the second virial coefficient, that dimerization occurs in ammonia vapour, as well as in certain primary and secondary amines. If hydrogen bonds are responsible, then the energy per bond is found to vary from 3.8 kcal. in ammonia to 2.5 kcal. in the secondary amines. These results would be consistent with the observations on crystalline ammonia if it could be assumed that each nitrogen made only two hydrogen bonds, but this seems difficult to reconcile with the crystal structure.

The better-established N—H · · · N bond length measurements in organic crystals are collected in Table XI.

TABLE XI. N—H · · · N BONDS

Compound	N—H · · · N	Ref.
Hexamethylene diamine	3.21 Å	[71]
Melamine	3.00, 3.02, 3.05, 3.10	[72]
Hydrazine	3.19, 3.30	[75]
Ammonium azide	2.94, 2.99	[76]
Dicyandiamide	2.94, 3.02, 3.04, 3.16	[77]
Adenine · HCl · $\frac{1}{2}$H$_2$O	2.99	[68]
2-Amino-4:6-dichloro-pyrimidine	3.21, 3.37	[78]
4-Amino-2:6-dichloro-pyrimidine	3.09, 3.28	[79]

4D. Nitrogen · · · chlorine.—Hydrogen bonds between nitrogen and chlorine are of rather frequent occurrence and have been measured with some accuracy in the chlorides of a number of organic bases. The bond lengths given in Table XII, as in most of the other tables in this section, are taken from the critical survey made by Donohue,[32] who has recalculated and corrected many of the values from the earlier original data. The contacts listed may be ascribed to hydrogen bond formation, because the angle R—N · · · Cl in each case is in the region of the tetrahedral value, as expected for the hydrogen positions on the —NH$_3^+$ ion. Donohue points out that in addition to these hydrogen bonds there is, in many of the structures, a fourth short N · · · Cl contact with the R—N · · · Cl angle near 180°. This cannot be due to hydrogen bond formation, and probably arises

[75] R. L. Collin and W. N. Lipscomb, *Acta Cryst.*, 1951, **4**, 10.
[76] L. K. Frevel, *Z. Krist.*, 1936, **94**, 197.
[77] E. W. Hughes, *J. Amer. Chem. Soc.*, 1940, **62**, 1258.
[78] C. J. B. Clews and W. Cochran, *Acta Cryst.*, 1948, **1**, 4.
[79] C. J. B. Clews and W. Cochran, *Acta Cryst.*, 1949, **2**, 46.

merely from the packing of the large chloride ions, which generally dominate the structure. No accurate energy values are available, but the strength of the N—H $\cdots$ Cl bond appears to be quite small.

TABLE XII. N—H $\cdots$ Cl BONDS

Compound	N—H $\cdots$ Cl	Ref.
Hexamethylene-diamine·2HCl	3.15, 3.25, 3.25 Å	80
	3.01, 3.07, 3.33	
Adenine·HCl·$\frac{1}{2}$H$_2$O	3.11, 3.21	58
Hydrazine·2HCl	3.10	81
Hydroxylamine·HCl	3.16, 3.23, 3.26	82
Geranylamine·HCl	3.17, 3.22, 3.22	83
m-Tolidine·2HCl	3.10, 3.22, 3.26	84

5. MOLECULAR COMPOUNDS: CLATHRATES

Although the hydrogen bond is perhaps the most universal and certainly the most studied form of molecular association, there are other cases where this particular mechanism either does not apply at all, or is only a minor factor. The molecular compounds, usually highly coloured, which are formed between the aromatic hydrocarbons and polynitro-compounds, are one example.[85] Addition compounds formed between polycyclic hydrocarbons and certain halogens or antimony pentachloride are probably of a similar nature. A large number of loose addition complexes between unsaturated compounds and copper, silver, and mercury salts are also known.[86] Finally, an entirely new class of molecular compounds, known as the clathrates, have recently been intensively studied by Powell.[87] In these, two types of molecule may be firmly united without the operation of any

[80] W. P. Binnie and J. M. Robertson, *Acta Cryst.*, 1949, **2**, 180; values listed recalculated by J. Donohue.

[81] J. Donohue and W. N. Lipscomb, *J. Chem. Phys.*, 1947, **15**, 115.

[82] B. Jerslev, *Acta Cryst.*, 1948, **1**, 21.

[83] G. A. Jeffrey, *Proc. Roy. Soc.* (London), 1945, **A183**, 388.

[84] F. Fowweather and A. Hargreaves, *Acta Cryst.*, 1950, **3**, 81.

[85] See, for example, J. W. Cook, *Ann. Repts. on Progress Chem.* (Chem. Soc. London), 1942, **39**, 167.

[86] S. Winstein and H. J. Lucas, *J. Amer. Chem. Soc.*, 1938, **60**, 836.

[87] For a general review, see H. M. Powell, *Research*, 1948, **1**, 353.

attractive forces between them at all, by a process of molecular imprisonment or enclosure of one species by the other.

Complete X-ray analysis of these various complexes, where they exist in the solid state, would probably quickly settle many points regarding their structure which have been the subject of much speculation in the past. Unfortunately it is usually difficult to attain the necessary precision on account of the large numbers of atoms involved, but several structures which throw considerable light on the problem have now been studied.

The p-iodoaniline-s-trinitrobenzene complex has been analysed very completely,[88] by taking advantage of the phase-determining power of the iodine atom and then carrying out refinements by the triple Fourier series method. As a result, all the atomic positions have been determined with an accuracy of about 0.1 Å. The structure is a very complex one, the monoclinic unit cell containing four molecules of each of the components. It consists essentially of alternate planar molecules of s-trinitrobenzene and p-iodoaniline stacked together in columns. The most striking feature is that there is no particularly close approach between these component molecules, and certainly nothing in the nature of a covalent bond connecting them. The carbon separations between adjacent benzene rings vary from 3.5 to 3.7 Å. There is some evidence of hydrogen bond formation between the amino groups of p-iodoaniline and the oxygen atoms of neighbouring trinitrobenzene molecules, where distances of 3.1 and 3.2 Å are observed. Such hydrogen bonds provide a mechanism of interaction between the molecules, but they cannot be considered an essential feature for this type of molecular compound, because dimethylaniline, without available hydrogen, forms similar complexes. One contact which may be of significance occurs between one of the nitro-groups of trinitrobenzene and the carbon ring of the iodoaniline molecule, where an N $\cdots$ C distance of 3.25 Å and an O $\cdots$ C distance of 3.35 Å are found. These groups become adjacent owing to the inclination of the molecular planes. In this respect the structure bears some resemblance to the p-nitroaniline self-complex that is mentioned below.

The molecular compounds formed between picryl chloride and hexamethylbenzene, and between picryl iodide and hexamethyl-

---

[88] H. M. Powell, G. Huse, and P. W. Cooke, *J. Chem. Soc.*, 1943, p. 153.

benzene, have also been studied,[89] and these are of interest because no hydrogen bonding in the usual sense is possible with such components. The crystal structures are extremely complex, and to some extent disordered, but the main features are clear. They consist essentially of alternate layers of the two components, with a separation of about 3.5 Å, which indicates only a relatively weak form of attraction between the molecules.

The crystal structure of p-nitroaniline[90] is a comparatively simple one that has considerable bearing on the problem of molecular compound formation, and the atomic positions in this case are known with much higher accuracy than in the structures discussed above. In the first place, as might be expected, there is a simple end-to-end linking of the molecules by two rather weak hydrogen bonds extending from the amino nitrogen of one molecule to the oxygen atoms of the nitro-groups of two adjacent molecules. The important feature of the structure, however, is the unusually close lateral approach of two molecules, where the nitro-group of one overlies part of the carbon ring of the other, with O · · · C distances of 2.66, 2.99, and 3.03 Å. The line of closest O · · · C approach (2.66 Å) lies almost normal to the plane of the benzene ring, so there is clearly no question of any hydrogen bonding being involved here. These distances are far below the normal van der Waals approach of about 3.4 Å, and indicate considerable attraction between the molecules. In the crystal the nitroaniline molecules are apparently engaged in the formation of a self-complex, similar in type to the molecular compounds that are so often found between nitro-aromatic compounds and substituted aniline derivatives.

[89] H. M. Powell and G. Huse, *J. Chem. Soc.*, 1943, p. 435.

[90] S. C. Abrahams and J. M. Robertson, *Acta Cryst.*, 1948, 1, 252; P. J. A. McKeown, A. R. Ubbelohde, and I. Woodward, *ibid.*, 1951, 4, 391; S. C. Abrahams, *J. Amer. Chem. Soc.*, 1952, 74, 2692.

The actual mechanism of formation of such molecular compounds has been much discussed, particularly by Bennett,[91] Weiss,[92] Dewar,[93] Brackman,[94] and recently Mulliken,[95] the latter having given a general quantum-mechanical treatment of the problem. The evidence appears to favour the view that the formation of such molecular compounds is due to the combination of an electron donor molecule, or base, with an electron acceptor molecule, or Lewis acid.[96] In the case of certain simple molecules, e.g. in the combination of BF_3 (acid) and NH_3 (base), stable compounds can be formed by means of a dative or co-ordinate link, H_3N^+—B^-F_3, the acid in this sense accepting the pair of electrons made available by the base, with the production of a stable bond.[97] In the case of the molecular compounds with which we are concerned, particularly those involving aromatic molecules, there is no question of the formation of any unique or stable covalent bond between the components. However, the mechanism may be of this general type, the attraction being of a Lewis acid-base nature between the donating π electrons of a benzene ring and the accepting orbitals of a nitro-group. The strength of the interaction will vary, depending on the aromatic substituents present, from complete electron transfer to slight overlap of the orbitals, with resonance between the no-bond structure and the charge transfer, or ionic, structure.

On this theory we need not expect any very close interatomic approaches, one of the ions or partial ions being a complete aromatic ring, of flat cylindrical shape. The distances observed, which vary from 2.7 to 3.5 Å, appear to be consistent with the idea. It has also been pointed out[98] that theoretical evidence indicates that in these structures the component aromatic rings,

[91] G. M. Bennett and G. H. Willis, *J. Chem. Soc.*, 1929, p. 256; G. M. Bennett and R. L. Wain, *ibid.*, 1936, p. 1108.

[92] J. Weiss, *J. Chem. Soc.*, 1942, p. 245.

[93] M. J. S. Dewar, *Nature*, 1945, **156**, 784; *J. Chem. Soc.*, 1946, pp. 406, 777.

[94] W. Brackman, *Rec. trav. chim.*, 1949, **68**, 147.

[95] R. S. Mulliken, *J. Amer. Chem. Soc.*, 1950, **72**. 600; 1952, **74**, 811.

[96] G. N. Lewis, *J. Franklin Inst.*, 1938, **226**, 293; W. F. Luder and S. Zuffanti, *The Electronic Theory of Acids and Bases*, New York: John Wiley & Sons, 1946.

[97] The B—N separation in ammonia-boron trifluoride and related compounds has been measured by J. L. Hoard, S. Gellar, and W. M. Cashin, *Acta Cryst.*, 1951, **4**, 396; the distance, 1.60 Å, is about 0.10 Å greater than the sum of the covalent radii, corrected for partial ionic character.

[98] J. Landauer and H. McConnell, *J. Amer. Chem. Soc.*, 1952, **74**, 1221.

although parallel, should not have coaxial orientations, but that the strongest binding will occur when they are displaced so that the centres are not directly over each other. This is in accordance with observation. It is also significant that in the p-nitroaniline structure the molecular displacement is such as to bring a nitro-oxygen atom (acceptor) over the benzene ring of an adjacent molecule. In this case a fairly strong intermolecular attraction is clearly evident.

Other types of molecular complex formation are known which

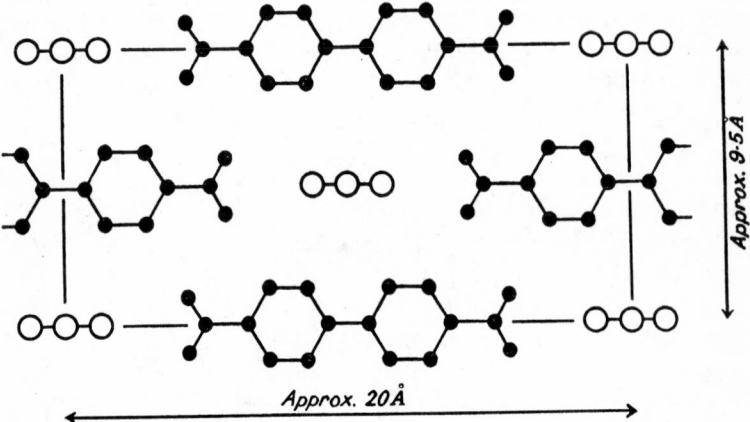

Fig. 103. Idealised structure for the complexes of 4:4'-dinitrodiphenyl (Saunder).

are more directly a function of the crystalline state alone and do not require such interactions as those described above. The complexes which are formed between 4:4-'dinitrodiphenyl and a large number of other diphenyl derivatives appear to be of this kind.[99] In these structures the dinitrodiphenyl molecules themselves may very well be engaged in a type of self-complex formation, but the nature and proportion of the other component seems to be determined by geometrical considerations. The molecular ratio of dinitrodiphenyl to the other substituted diphenyl is 3:1 in the case of the complexes with diphenyl, 4-

[99] W. S. Rapson, D. H. Saunder, and E. T. Stewart, *J. Chem. Soc.*, 1946, p. 1110; D. H. Saunder, *Proc. Roy. Soc.* (London), 1946, **A188**, 31; 1947, **A190**, 508; J. N. van Niekerk and D. H. Saunder, *Acta Cryst.*, 1948, **1, 44.**

hydroxydiphenyl and 4:4'-dihydroxydiphenyl; 4:1 for the complexes with benzidine and $NN:N'N'$-tetramethylbenzidine; and 7:2 or other more complicated ratios for the complexes with various halogen-substituted diphenyls.

X-ray analysis shows that these structures are of the general type illustrated in Fig. 103, where the dinitrodiphenyl molecules, shown by black circles, form a face-centred array with three or four approximately planar molecules overlying each other in each position. The other component molecules are accommodated in the tubular cavities of this structure, their end-on view being represented by the large open circles. The minimum intermolecular distances observed are about the same as in other organic crystals and there is no evidence of localized bonding. The molecular ratios of the two components appear to be determined solely by geometrical considerations, the number of dinitrodiphenyl molecules in the stack depending primarily on the length of the molecules of the other component. The molecular ratios for a number of the complexes can in fact be calculated from this consideration. In some of the structures there is also evidence of a certain randomness in the positioning of the second component in the cavities of the structure.[100]

The clathrates are another important class of molecular compound in which no attractive forces between the components are required. As the name implies, one component is imprisoned by a cagelike structure of the other component, and the forces operating between them are actually repulsive rather than attractive. The researches of Powell and his collaborators[101] in this field have opened out wide possibilities and have led to the prediction and discovery of many new and interesting members of the class.

The basis of these discoveries lay in the X-ray determination of the true structure of the substance formerly known as β-quinol.[102] This proved to be a molecular complex of quinol and methanol, of composition $3C_6H_4(OH)_2 \cdot CH_3OH$, and was shown

[100] R. W. James and D. H. Saunder, *Proc. Roy. Soc.* (London), 1947, **A190**, 518; *Acta Cryst.*, 1948, **1**, 81.

[101] D. E. Palin and H. M. Powell, *J. Chem. Soc.*, 1947, p. 208; H. M. Powell, *ibid.*, 1948, p. 61; D. E. Palin and H. M. Powell, *ibid.*, 1948, pp. 571, 815.

[102] W. A. Caspari, *J. Chem. Soc.*, 1926, p. 2944; 1927, p. 1093.

to be crystallographically similar to other known molecular compounds of quinol, e.g., $3C_6H_4(OH)_2 \cdot M$ where $M = SO_2$, H_2S, HCOOH, HCN, HBr, and HCl.

The true *beta* polymorphic form of quinol was then prepared[103] and shown to have a very complex crystal structure in which the quinol molecules are hydrogen bonded to form cagelike structures so open that two of these structures can interpenetrate and enclose each other. One such cage structure is shown in Fig. 104, where the small open circles represent the oxygen atoms of quinol

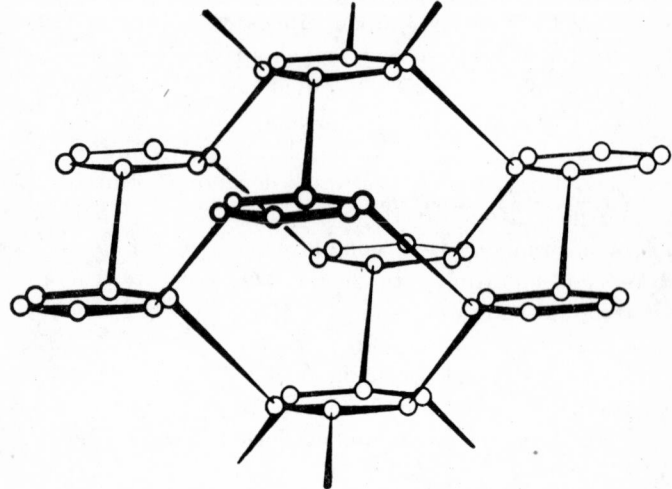

Fig. 104. The quinol cage structure. Circles represent oxygen atoms (Powell).

molecules, these oxygen atoms being hydrogen bonded together in groups of six, to form plane hexagons of side 2.75 Å. The quinol molecules themselves are not shown on the diagram, but the HO $\cdots$ OH axes of the molecules are represented by the longer sloping lines connecting the hexagons at different levels. A second identical cage structure interpenetrates the one shown, being displaced vertically half-way between the top and bottom hexagon. These two frameworks of quinol molecules are thus mutually imprisoned, and the atoms of each are separated from the other by the usual van der Waals non-bonded distances.

Quinol thus forms a type of clathrate compound with itself,

[103] H. M. Powell and P. Riesz, *Nature*, 1948, **161,** 52.

but the important feature of this complex structure is that within the two frameworks there still exist cavities large enough to contain a small molecule not bonded to the surrounding atoms. These cavities lie between the hexagons of oxygen atoms, the sides being formed by benzene rings. Allowing for the contact radii of the enclosing atoms, there remains a roughly spherical unoccupied space about 4 Å in diameter; and there is one cavity of this kind to every three quinol molecules.

The existence of compounds of the general formula $3C_6H_4(OH)_2 \cdot M$ is thus explained. The only conditions for the formation of such compounds are that the molecules M should be small enough to fit the cavity, yet not so small as to escape through possible gaps, and that they should not react chemically with quinol. In addition to the compounds already known ($M = SO_2$, H_2S, HCOOH, HCN, HBr, and HCl), the following new compounds have been prepared and studied:

$$M = MeOH, MeCN, CO_2, C_2H_2, A, Kr, Xe.$$

A detailed study has been made of some of these structures, and it has been shown that the molecules of SO_2, MeOH and MeCN are trapped with their long axes parallel to the c axis of the rhombohedral crystals, and that they are free to rotate about this long axis, but not perpendicular to it. The compounds can be formed by allowing the quinol to crystallise from a solution consisting of the second molecular species, or at least containing a considerable concentration of it. Crystals can sometimes be prepared with a proportion of unoccupied cavities, or with more than one kind of component molecule present.

The rare gas clathrates[104] are of particular interest in establishing the general proposition that it is possible to form a molecular compound without invoking any valency bonding forces. The helium compound has not been made, as the atom is too small and probably escapes through the oxygen hexagons. But compounds have been isolated with the other inert gases present in almost theoretical amount, the formula of the xenon compound being $3C_6H_4(OH)_2 \cdot 0.88$ Xe. These remarkable structures, once formed, are quite stable, the large inert gas atoms being effectively imprisoned in the cages. On solution, the gas is liberated with vigorous effervescence.

[104] H. M. Powell, *J. Chem. Soc.*, 1950, pp. 298, 300, 468.

A second type of clathrate compound has been described[105] where the framework is inorganic, consisting of the ammonia nickel cyanide complex, $Ni(CN)_2NH_3$. Here the nickel atoms are linked by cyanide groups, in extended two-dimensional array, to form a flat network, and from this network the ammonia groups project above and below, from alternate nickel atoms:

$$
\begin{array}{ccccc}
& & & & NH_3 \\
& | & & | & \diagup \\
----Ni---- & CN---- & Ni---- \\
& | & & \diagup \\
& & & NH_3 \\
& CN & & CN \\
& NH_3 & & \\
& \diagup & & | \\
----Ni---- & CN---- & Ni---- \\
\diagup & & & | \\
NH_3 &
\end{array}
$$

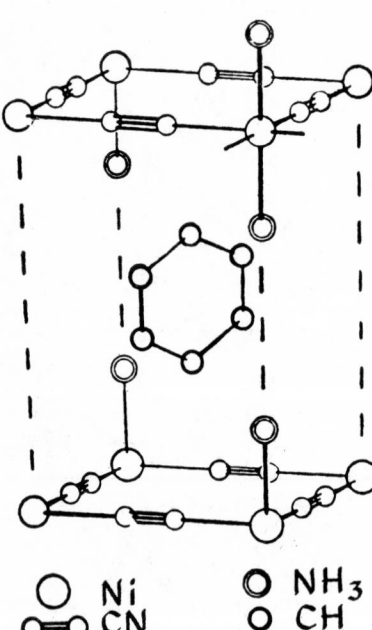

O Ni   O NH₃
O=O CN   O CH

Fig. 105. The benzene ammonia-nickel cyanide complex (Powell and Rayner).

The projecting ammonia groups restrict the approach of successive layers of this kind, and there is too much vacant space for crystallisation to proceed unless these spaces can be filled. This can be achieved if benzene or another molecule of similar size is present. Such molecules be-

[105] H. M. Powell and J. H. Rayner, *Nature*, 1949, **163**, 566; J. H. Rayner and H. M. Powell, *J. Chem. Soc.*, 1952, p. 319. A further important type of molecular compound, the urea-hydrocarbon complex, has been examined by C. Hermann and others. The urea molecules here form a hollow channel structure in which the *n*-hydrocarbon molecules are enclosed. See W. Schlenk, Jr., *Ann.*, 1949, **565**, 204; A. E. Smith, *Acta Cryst.*, 1952, **5**, 224.

come trapped in the cavities between the layers, and with benzene in particular very efficient space filling is obtained, giving rise to the compound $Ni(CN)_2NH_3 \cdot C_6H_6$.   Detailed X-ray analysis of this compound shows that there is no bond between the benzene molecule and the nickel complex, the usual van der Waals approach distances being observed.   The surroundings of the benzene molecule in this structure are shown in Fig. 105. Similar compounds can be formed by a number of other aromatic substances such as pyrrole, furane, thiophene, pyridine, phenol, and aniline, where the molecules have about the same space-filling capacity.

X

Applications to Complex and Partially Known Chemical Structures

1. PROBLEMS INVOLVED AND METHODS AVAILABLE

Most of the crystal structures described in the preceding chapters have been based on formulas well established by the methods of organic chemistry long before the advent of X-ray analysis. In every case the result of the X-ray work has been to confirm the already known chemical structure. In addition, much new knowledge has been gained from the quantitative measurement of bond lengths, valency angles, and intermolecular contacts of various kinds. All this is of fundamental importance in providing a more complete picture of the ultimate structural units with which the chemist is concerned, and in establishing a sound physical basis for the explanation of the properties and behaviour of substances.

Some of the most exciting and attractive problems in X-ray analysis, however, are concerned with the elucidation of chemically unknown, or incompletely known, structures. During the past century progress in organic chemistry has been at least as rapid as in any other science, and the number of known structures is now enormous. But in the continuously advancing field of chemistry there are always fresh problems awaiting solution, and many old ones that have so far defied the standard methods of the science. By the nature of things such unknown structures are usually complex, with molecules containing a fairly large number of atoms arranged in unfamiliar ways. If the compound has received chemical study at all, as must nearly always have been the case in the course of its isolation from some natural source, then something will be known regarding the particular patterns of atoms and groups likely to be present, although their mutual disposition may be quite uncertain.

Even with this partial knowledge the problems which face the X-ray crystallographer in such a case are generally rather baffling in their complexity. Building possible models and testing them against the observed X-ray data might appear to be a feasible method of approach. But unless a structure of fairly rigid configuration is expected, as in the case of the aromatic compounds, this is not in fact a very promising procedure. Even if a chemically correct model can be guessed, it will in general possess many inherent degrees of freedom, due to possible rotations at single bond junctions, and so no quick test leading to its simple rejection or acceptance can usually be made.

At most a general indication of the feasibility of some proposed structure is all that can be looked for by this method of approach. Such indications, especially if supported by independent physical evidence derived, for example, from optical and magnetic data, may be extremely valuable in directing further chemical research into the right channels and narrowing down the possibilities. Perhaps the best illustration of a far-reaching advance achieved in this manner occurred in 1932, when the elucidation of the correct skeleton formula for the sterols by Rosenheim and King[1] was based in one very important respect on the X-ray measurements of Bernal.[2] This work is described briefly in the following section and demonstrates the great importance of even preliminary X-ray measurements on a series of related compounds, when combined with other physical and chemical evidence.

The more detailed kind of X-ray structural analysis, however, with which we are mainly concerned in this book, and which envisages the determination of the precise position of every atom in the molecule, does not readily or immediately follow from such broad surveys. Many chemically unknown, or incompletely known, structures have now been determined successfully in this

[1] O. Rosenheim and H. King, *Chemistry & Industry*, 1932, **51**, 464, 954; *Nature*, 1932, **130**, 513; H. Wieland and E. Dane, *Z. physiol. Chem.*, 1932, **210**, 268.

[2] J. D. Bernal, *Chemistry & Industry*, 1932, **51**, 259, 466; *Nature*, 1932, **129**, 277, 721.

precise manner, but the great majority of these determinations have been the result of what may be termed direct methods of analysis.

If the complexity of the structure is not too great, success can sometimes be achieved by application of the Patterson method, especially in three-dimensional form, or by means of the relationships between the structure factors and their phases discovered by Harker and Kasper and by Zachariasen. With further development it may become possible to apply these methods to structures containing fairly large numbers of light atoms. Up to the present, however, the outstanding successes in the field of unknown or partially known structures have nearly all been attained by application of the direct phase-determining methods based on the use of a heavy atom, and particularly on the more powerful isomorphous replacement method.

A few of these structures are described briefly in the following sections. It will be noted that even in the case of the sterols themselves, although the preliminary survey led to such notable advances, the determinations of atomic positions which have now been made were achieved in large measure by the use of phase-determining methods involving a heavy atom.

2. THE STEROIDS

At the time of the first X-ray measurements on the sterols the skeleton formula generally assigned to these bodies

was known to have serious chemical defects, and many variants of this formula had been considered without much success. Bernal's first measurements[2] of unit cell size for calciferol, ergosterol, cholesterol, and related compounds, combined with optically determined molecular orientations, showed rather conclusively that the space occupied by a sterol molecule must be approximately $5 \times 7.2 \times 17 - 20$ Å. It was further shown that in the crystal a double layer of these molecules must occur, as in the case of the long chain alcohols, and this indicated the presence of a terminal —OH group.

It was clear from these measurements that the accepted formula must be wrong, because the union of three rings at the carbon atom 9 would lead to a molecule much too thick. In addition the attachment of the side chain as shown would make the molecule too short. Other X-ray studies on oestrin and pregnandiol had indicated similar molecular dimensions and pointed to a phenanthrene type of nucleus.

The revised skeleton formula proposed by Rosenheim and King[1] in 1932 explained the chemical properties, particularly the production of chrysene by dehydrogenation of certain compounds in this group, and was strictly compatible with all the features

observed in the early crystal studies. The X-ray evidence at this time, which was before the development of direct phase-determining techniques, could not by any means predict this formula in all its detail; it did, however, perform an essential function in showing that the formula must be of this kind, and in showing that the earlier chemical conclusions must be wrong. The X-ray evidence also showed that, as the molecule must be fairly flat and thin, a general *trans*-configuration for the skeleton was likely, and this was further confirmed by an X-ray comparison of the *cis*- and *trans*-hexahydrochrysenes.[3]

The immense advances in the organic chemistry of the sterols which followed from these discoveries are well known. A full account of the X-ray and optical measurements on over 80 sterol derivatives, belonging mainly to the cholesterol and ergosterol series, has been given by Bernal, Crowfoot, and Fankuchen.[4]

Such general surveys can be of great value in characteriza-

[3] J. D. Bernal, *Chemistry & Industry*, 1933, **52**, 288.

[4] J. D. Bernal, D. Crowfoot, and I. Fankuchen, *Trans. Roy. Soc.* (London), 1940, **A239**, 135.

tion, identification, molecular weight determination, and confirmation of common structural features. With regard to the precise determination of atomic positions, however, and the full elucidation of the geometry of the molecule, it has proved necessary here, as in many other cases, to have recourse to the heavy atom method. Two such determinations have now been completed in this series: for cholesteryl iodide, by Carlisle and Crowfoot,[5] and for a calciferol derivative, by Crowfoot and Dunitz.[6] These determinations have been made directly from the X-ray data and do not involve any chemical theory except in the last stages when an ambiguity imposed by the crystallographic symmetry makes it necessary to choose between possible alternatives for certain co-ordinates of the atoms (see Chapter VI, Section 6A). Once a selection of such alternatives has been made, however, its correctness or otherwise can immediately be tested in a direct manner against the X-ray data.

Only a brief reference to these very elaborate analyses, which involve the use of three-dimensional Fourier series methods, can be given here. The results have not been refined sufficiently to provide bond length measurements of any accuracy, but they are enough to settle many outstanding questions regarding the geometry of these molecules and are in general agreement with later chemical evidence.

Cholesteryl iodide exists in two closely related polymorphic modifications, A and B, both monoclinic, $P2_1$, with two molecules per unit cell. Both forms have been examined, but detailed analysis is confined to the B form, in which at least approximate values have been obtained for all the 84 co-ordinates which govern the structure. It is found that the iodine atom is attached to position 3, with methyl groups at 10 and 13, and a double bond at 5–6. In general, the staggered *trans*-configuration is followed throughout the ring systems and attached chain. The attachment of the side chain at 17 is *cis* to methyl at 13, and the iodine at 3 is *cis* to methyl at 10.

In the case of calciferol it was difficult to find any suitable heavy atom derivative, but finally the very complex calciferyl-4-

    [5] C. H. Carlisle and D. Crowfoot, *Proc. Roy. Soc.* (London), 1945, **A184**, 64. The analysis of a lumisterol derivative has also been carried out by D. Crowfoot Hodgkin and D. Sayre, *J. Chem. Soc.*, 1952, p. 4561.

    [6] D. Crowfoot and J. D. Dunitz, *Nature*, 1948, **162**, 608.

iodo-5-nitrobenzoate, $C_{35}H_{46}NO_4I$, was utilized. The crystals are orthorhombic, $P2_12_12_1$, with four of these molecules in the unit cell. Having obtained the iodine positions by Patterson methods, the work proceeded by the normal heavy atom techniques, followed by successive Fourier series refinements. Owing to the fortunate occurrence of a fairly short axis (6.98 Å), one electron density projection of the structure is found to provide very good evidence regarding the details of this complex mole-

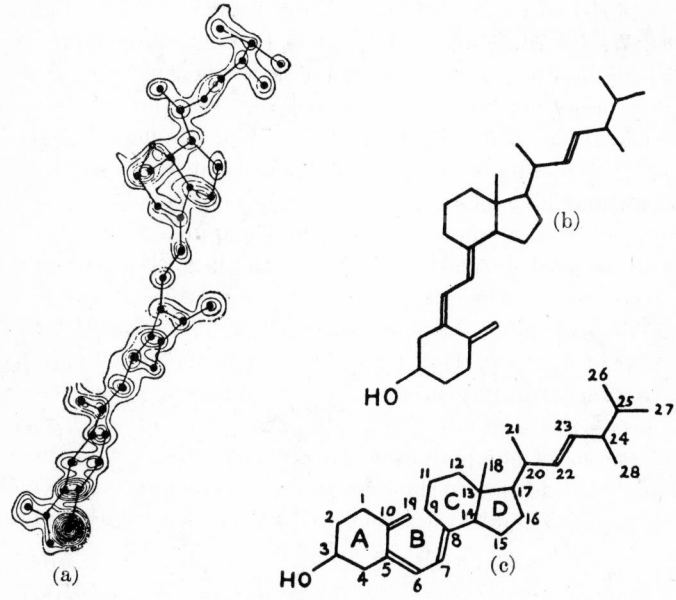

Fig. 106. Structure of calciferyl-4-iodo-5-nitrobenzoate.

cule. This projection is shown in Fig. 106a, with the estimated atomic positions superimposed. The upper part of this map reveals the sterol skeleton in considerable detail, the lower part being merely the nitrobenzoate attachment.

It is at once clear that the structure (106c) deduced from chemical evidence is correct, and that the ring B is broken open. As might be expected, however, the molecule does not maintain the curled-up form suggested by structure (106c), but in this crystal at least it exists in the fully extended form shown in (106b). The stereochemical relations of the atoms can be studied in some detail from this projection, but await final confirmation from three-dimensional work.

3. THE PHTHALOCYANINES

The phthalocyanines are a beautifully crystalline series of macrocyclic organic pigments of extraordinary stability, the copper derivative, for example, subliming without decomposition at 580°. The first member of the series to be discovered was the iron derivative, obtained accidentally in the course of the industrial production of phthalimide at Grangemouth, Scotland, in 1928. When ammonia was passed through molten phthalic anhydride in iron vessels, traces of a dark blue substance of great stability were noticed. The subsequent isolation of this iron phthalocyanine and the preparation and chemical study of many other members of the series have been fully described by Linstead and his collaborators.[7] These compounds include the parent substance, metal-free phthalocyanine, $C_{32}H_{18}N_8$, and numerous metal derivatives of the formula $C_{32}H_{16}N_8M$. With the exception of the platinum compound, these form a very closely isomorphous series, as shown by the crystal constants[8] collected in Table XIII. In addition, many compounds of the closely related porphyrin class have now been synthesised and their properties studied.[7,9]

This series of compounds is of particular interest for several reasons. In addition to their great value as pigments and colouring materials, their close structural relationship to the natural

[7] R. P. Linstead, A. R. Lowe, C. E. Dent, and C. T. Byrne, "Phthalocyanines," pts. I–VI, *J. Chem. Soc.*, 1934, p. 1016; P. A. Barrett, C. E. Dent, and R. P. Linstead (pt. VII), *ibid.*, 1936, p. 1719; E. F. Bradbrook and R. P. Linstead (pt. VIII), *ibid.*, 1936, p. 1744; R. P. Linstead, J. A. Bilton, A. H. Cook, E. G. Noble, and J. M. Wright (pts. IX–XII), *ibid.*, 1937, p. 911; C. E. Dent, *ibid.*, 1938, pp. 1, 546; J. S. Anderson, P. A. Barrett, E. F. Bradbrook, A. H. Cook, D. A. Frye, and R. P. Linstead (pts. XIII–XIV), *ibid.*, 1938, 1151; A. H. Cook, *ibid.*, 1938, pp. 1761, 1845; P. A. Barrett, E. F. Bradbrook, C. E. Dent, R. P. Linstead, J. M. Robertson, and G. A. P. Tuey (pts. XV–XVI), *ibid.*, 1939, p. 1809; R. P. Linstead, P. A. Barrett, J. J. Leavitt, G. A. Rowe, F. G. Rundall, and G. A. P. Tuey (pts. XVII–XIX), *ibid.*, 1940, p. 1070; R. P. Linstead and F. T. Weiss (pts. XX–XXI), *ibid.*, 1950, p. 2975; R. P. Linstead and M. Whalley (pt. XXII, *Conjugated Macrocycles*), *ibid.*, 1952, p. 4839.

[8] R. P. Linstead and J. M. Robertson, *J. Chem. Soc.*, 1936, p. 1736.

[9] I. Woodward, *J. Chem. Soc.*, 1940, p. 601.

TABLE XIII. CRYSTAL DATA FOR THE PHTHALOCYANINES

Compound	Space group	Mol. per cell	a	b	c	β	Density
			(Å)	(Å)	(Å)	(°)	(g./cc.)
$C_{32}H_{18}N_8$	$P2_1/a$	2	19.85	4.72	14.8	122.2	1.445
$C_{32}H_{16}N_8Be$	$P2_1/a$	2	21.2	4.84	14.7	121.0	1.33
$C_{32}H_{16}N_8Mn$	$P2_1/a$	2	20.2	4.75	15.1	121.7	1.52
$C_{32}H_{16}N_8Fe$	$P2_1/a$	2	20.2	4.77	15.0	121.6	1.52
$C_{32}H_{16}N_8Co$	$P2_1/a$	2	20.2	4.77	15.0	121.3	1.53
$C_{32}H_{16}N_8Ni$	$P2_1/a$	2	19.9	4.71	14.9	121.9	1.59
$C_{32}H_{16}N_8Cu$	$P2_1/a$	2	19.6	4.79	14.6	120.6	1.61
$C_{32}H_{16}N_8Pt$	$P2_1/a$	2	23.9	3.81	16.9	129.6	1.97

porphyrins makes it desirable to conduct a careful study of the geometry of these molecules as well as the stereochemistry of the metals with which they exist in combination. They are also of historical interest in connection with the work described in the present chapter, because the closely related series of compounds listed in Table XIII is ideal for the application of the isomorphous replacement and heavy atom methods of direct phase determination. The first complete development and application of these methods to the elucidation of organic structures were in fact conducted on these compounds.

Metal-free and nickel phthalocyanine were selected for the first detailed X-ray studies in this series.[10] The analysis is a perfectly direct one and does not involve any chemical theory. The space group is $P2_1/a$, and viewed along the short monoclinic b axis, the symmetry centres lie on an effectively primitive lattice. As there are only two metal atoms per unit cell, these atoms must coincide with the symmetry centres and make a positive contribution to each X-ray reflection in the $(h0l)$ zone. Assuming complete isomorphism and comparing the absolute values of the structure factors of nickel phthalocyanine (F_{PhNi}) and metal-free phthalocyanine (F_{Ph}), equation 6.18 becomes

$$F_{PhNi} - F_{Ph} = F_{Ni}.$$

In this equation the first two structure factors, which are meas-

[10] J. M. Robertson, *J. Chem. Soc.*, 1935, p. 615; 1936, p. 1195; J. M. Robertson and I. Woodward, *ibid.*, 1937, p. 219.

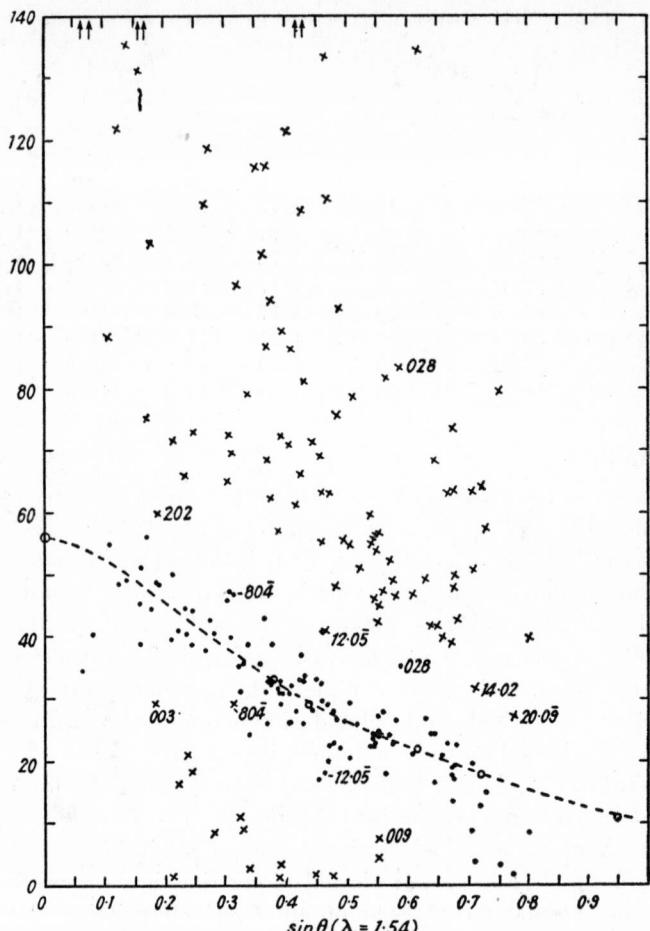

Fig. 107. Values of $F_{\text{PhNi}} - F_{\text{Ph}}$. The resultant curve for F_{Ni} has an upper limit at 56, twice the atomic number of nickel, because there are two molecules in the unit cell.

ured quantities, may be either positive or negative so that four possible combinations of these quantities are involved. But as F_{Ni} must always be positive, only two of these combinations need be considered, and these alternatives are plotted in Fig. 107 for a large number of reflections.

It is found that one choice of signs always gives a value for F_{Ni} which lies near the expected atomic scattering curve, and this

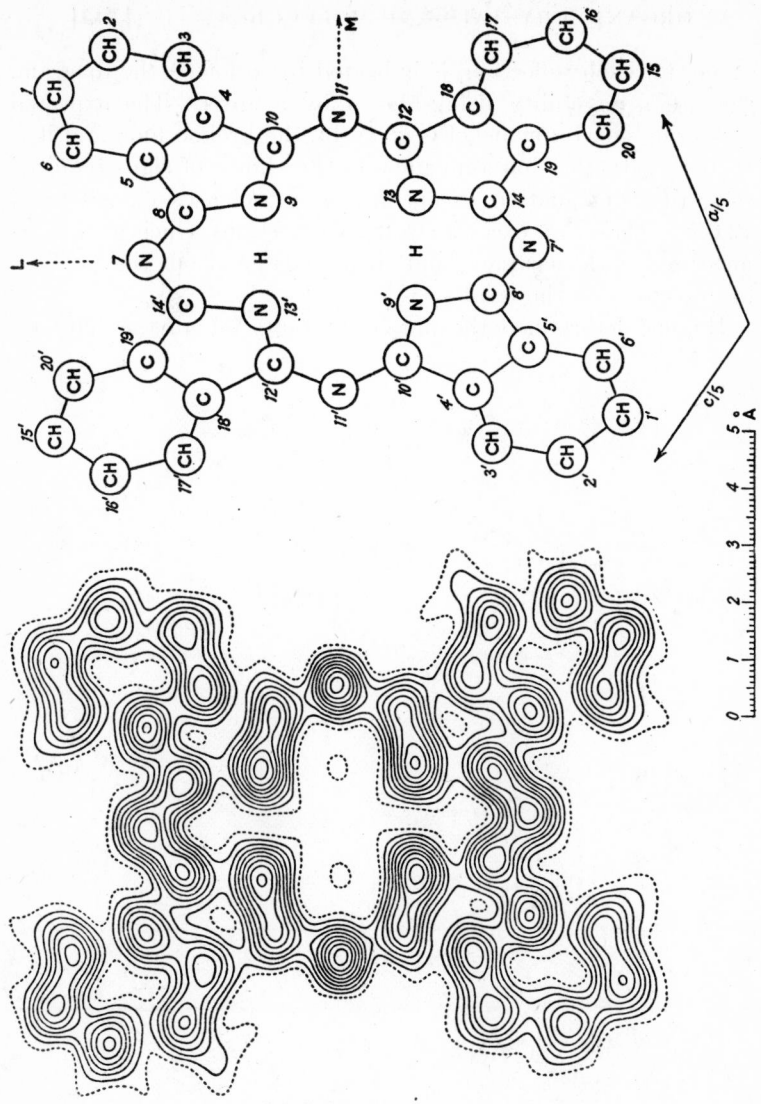

Fig. 108. Electron density projection for metal-free phthalocyanine, and diagram of atomic positions. Contour lines at unit electron intervals, the dotted line being 1 electron per Å².

value, for each reflection, is indicated by a dot on the diagram, the other possibility being shown by a cross. The expected scattering curve for nickel is given by the broken line, and it is clear that there is no ambiguity in the choice of signs required to satisfy the equation, except in one or two cases, e.g., $80\bar{4}$ and $12,0,\bar{5}$. These are excessively faint reflections which cannot be measured with accuracy, and their omission from the analysis has no effect on the results.

Having determined the phase constants for all the $(h0l)$ re-

$\frac{c}{10}$ $\frac{a}{10}$

0 1 2 3 4 5 Å

Fig. 109. Electron density projection for nickel phthalocyanine. Contour scale as before except on the central nickel atom when the increment is 5 electrons per Å² for each line.

flections in metal-free and nickel phthalocyanine in this manner, the electron density maps representing projections of these structures along the short *b* axes can be prepared by direct evaluation of the corresponding double Fourier series. The results are shown in Fig. 108 for free phthalocyanine and in Fig. 109 for the nickel derivative.

Although these are only projections of the structures, they contain enough information to enable the molecular geometry and dimensions to be evaluated almost completely. The four benzene rings at the corners of the large molecule may be used as indicators. By assuming them to be regular planar hexagons, the orientations of the molecules can be deduced accurately, and the values of the third coordinate of each atom (which is not directly observed) can be calculated and then checked against the structure factors in other zones. The molecules appear to be strictly planar, and in each case they are inclined to the projection plane at about 44.2°, the direction M (Fig. 108) making an angle of 45.9° with the *b* axis, and the L direction 2.3°. The perpendicular distance between the planes of successive overlying molecules is 3.38 Å, almost identical with the interplanar distance in graphite (3.40 Å).

In the molecule itself the carbon-nitrogen bond lengths within the great 16-membered central ring are all equal, to within a few hundredths of an Ångstrom. In free phthalocyanine the value obtained is 1.34 Å, and in nickel phthalocyanine, 1.38 Å. The state of complete single bond–double bond resonance required by this result accounts for the great stability of these structures.

The inner ring system is connected to the benzene rings by carbon-carbon bonds which are again all equal, but about 1.47 Å in length. The interpretation of this distance can be reached from a study of the empirical curve connecting interatomic distance and double bond character (Fig. 85, p. 190). This indicates rather less than 20 per cent double bond character for these links. The figure is, of course, very approximate, but it shows clearly that the bonds in question are more nearly equivalent to single bonds than to double bonds. These various facts are consistent with the chemical structure, which can be written as follows, single bond–double bond resonance being implied throughout the entire system:

The four central nitrogen atoms are arranged almost at the corners of a square, the measured lengths of the sides being 2.65 and 2.76 Å in the metal-free compound, and 2.56 and 2.60 Å in the nickel derivative. The formation of bonds to the central nickel atom is thus accompanied by a slight inward shift of the nitrogens. This movement is clearly shown on the enlargement of the central portion of the nickel phthalocyanine structure given in Fig. 110, where the positions of the nitrogens in the metal-free structure are plotted by small crosses. In this portion of the map there is also evidence of well-defined bridges along the directions of the covalent bonds between the nickel and the nitrogen atoms, which stand out above the diffraction trough surrounding the nickel atom.

In all the derivatives that

Fig. 110. Enlargement of central portion of nickel phthalocyanine. The nickel contour levels are now drawn at intervals of 2 electrons per Å².

have been examined the central metal atom is coplanar with the four surrounding nitrogen atoms. The arrangement indicates covalent linkages of the dsp^2 type, the distance Ni—N being 1.83 Å and Pt—N about 2.0 Å. This square planar configuration involving d orbitals is to be expected where the transitional elements are involved. That beryllium should adopt a similar symmetry is surprising, and it appears that in this case the organic portion of the molecule imposes its steric requirements on the metal. The beryllium compound, however, is unstable, and quickly reacts even with moist air to form a dihydrate.[11] No other member of the series behaves in this way except the magnesium compound (which has not been examined by X-rays, but is probably similar to the others in crystal structure).

In the metal-free compound two hydrogen atoms are attached to the central nitrogens, and at first it was suggested that these were probably involved in hydrogen bond formation between neighbouring nitrogen atoms, which are about 2.7 Å apart. It has been pointed out by Donohue,[12] however, that the C—N—H valency angles involved, which are about 80° and 170°, render such an interpretation impossible. These hydrogens must be directed towards the centre of the square, where there is probably just about enough space for them to maintain a planar configuration.

Platinum phthalocyanine is not isomorphous with the other derivatives, but the platinum atom has a sufficiently high atomic number to be completely phase determining for all the reflections, and use of the isomorphous replacement method is not necessary.[13] The result of a perfectly direct Fourier synthesis employing all the 302 structure factors obtained from measurements in the principal zone is shown in Fig. 111. Although serious diffraction effects caused by the intense scattering from the platinum atom are evident, these can be allowed for, and the details of the molecular structure are found to be similar to those of the other derivatives. The larger platinum atom results in a slight outward displacement of its surrounding nitrogen atoms, the Pt—N distance being about 2.0 Å.

[11] P. A. Barrett, C. E. Dent, and R. P. Linstead, *J. Chem. Soc.*, 1936, p. 1720.

[12] J. Donohue, *J. Phys. Chem.*, 1952, 56, 502.

[13] J. M. Robertson and I. Woodward, *J. Chem. Soc.*, 1940, p. 36.

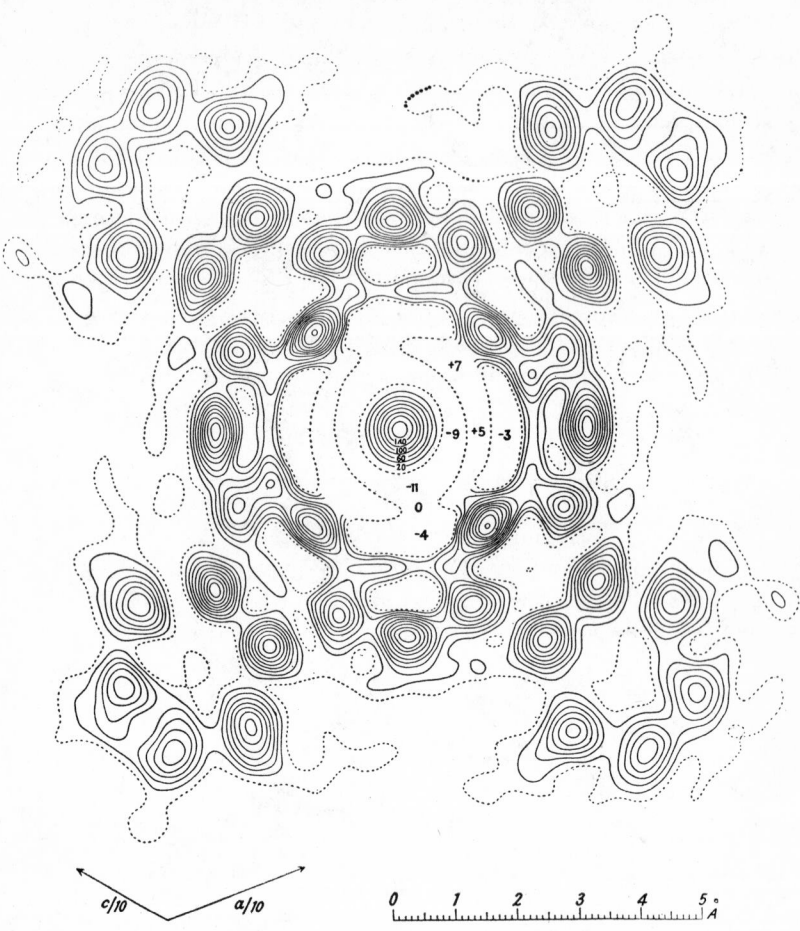

Fig. 111. Electron density map for platinum phthalocyanine. Contour scale as before, but on the central platinum atom it is now 20 electrons per Å² per line.

The orientation of the molecule in the crystal differs considerably from that of the other derivatives, as might be expected from the much shorter b axis (3.81 Å). The molecular plane is now inclined to the projection plane in Fig. 111 at only 26.5° instead of 44°, but the vertical distance between the planes of successive overlying molecules, 3.41 Å, closely maintains the standard graphite interplanar value.

270

4. THE TROPOLONES AND COLCHICINE

A large number of important natural products are now known to contain an unsaturated seven-membered carbon ring system as an essential structural unit.[14] These compounds include the mould products, stipitatic acid, puberulic acid, and puberulonic acid; the thujaplicins, a group of substances extracted from the heartwood of western red cedar which are antibiotic to wood-destroying fungi; purpurogallin, which occurs in the form of diglucosides in various galls; and the tricyclic growth inhibiting alkaloid colchicine, $C_{22}H_{25}O_6N$.

For stipitatic acid Dewar[15] proposed the structure

and concluded that 2-hydroxy*cyclo*heptatrienone, which he named tropolone, should have aromatic properties as indicated by the conjugated structures

At that time the compound tropolone itself was unknown, but it has now been synthesised[16] and its properties and those of many of its derivatives have received very extensive and careful study.[14] The results fully justify Dewar's early predictions regarding the properties of these compounds.

In view of the importance of this series a detailed X-ray examination of the crystal structures is of considerable interest. Tropolone itself is volatile and difficult to handle; its crystal structure is rather complex. It has not yet been fully analysed.

---

[14] For a review with full references see J. W. Cook and J. D. Loudon, *Quarterly Revs., Chem. Soc.*, 1951, **5**, 99.

[15] M. J. S. Dewar, *Nature*, 1945, **155**, 50.

[16] J. W. Cook, A. R. Gibb, R. A. Raphael, and A. R. Sommerville, *J. Chem. Soc.*, 1951, p. 503; *Chemistry & Industry*, 1950, p. 427.

However, the copper salt, cupric tropolone is a very suitable derivative, and its structure has been worked out in considerable

detail.[17]   As an example of the technique of X-ray analysis this structure also provides an interesting sequel to the work on the phthalocyanines described in the last section.

Cupric tropolone is obtained in the form of very fine needlelike crystals of great stability, which melt with decomposition at about 300°. They are monoclinic, space group $P2_1/a$, and the unit cell contains two copper atoms which must lie on the centres of symmetry. An isomorphous series is not available, but owing to the fairly small number of atoms present in the unit cell it can be shown that the copper atom should be almost completely phase determining. The heavy atom method may therefore be applied directly.

There is one very important practical difference as compared with the phthalocyanine structures. In cupric tropolone there is again one very short axis (3.80 Å) along which a good projection of the molecular structure should be obtained. But this axis is now the monoclinic c axis, and viewed in this direction the copper atoms lie on a face-centred lattice. Hence they only make contributions to the $(hk0)$ structure factors when $h+k$ is even. For the planes with $h+k$ odd, the contributions from the two copper atoms are in opposite phase and cancel each other out (see Chapter VI, Section 6A). The projection along the symmetry axis b, corresponding to the phthalocyanine projections, in this case yields no useful structural information because in cupric tropolone this axis is 13.8 Å in length.

It is therefore necessary to proceed with an analysis based on the phase-determined structure factors alone, those for which $h+k$ is even. When the corresponding Fourier series is summed, the result should provide a picture of the structure superimposed on its mirror image. The diagram obtained in this way for cupric tropolone is shown in Fig. 112. There is a centre of symmetry at the copper atom in the centre of the diagram (where the

[17] J. M. Robertson, *J. Chem. Soc.*, 1951, p. 1222.

$\frac{19}{30}b$

$a \sin \beta$

Fig. 112. Symmetrised projection of the cupric tropolone structure.

contour scale is reduced by a factor of five), but in addition to
this symmetry each atom is accompanied by a spurious mirror
image of itself in the adjoining quadrant.   The problem is to
select that group of atoms which constitutes the true structure
and reject the other.   The oxygen atoms can be distinguished by
their heavier peaks, but the seven carbon atoms known to be
present can be selected in 2^7 or 128 different ways.   The correct
solution, once found, will explain the intensities of the $h+k$ odd
reflections and give their phase constants.

In such a case, even if the molecular structure is largely unknown, one can sometimes pick out recognisable fragments,
based on known bond lengths and on the kind of groups likely to
be present.   In Fig. 112 the solution is particularly simple because a seven-membered ring happens to lie rather neatly within
each quadrant.   This selection of peaks is found to give a satisfactory explanation of all the intensities and provides the phase
constants for the odd reflections.

When all the structure factors are now included in the summa-

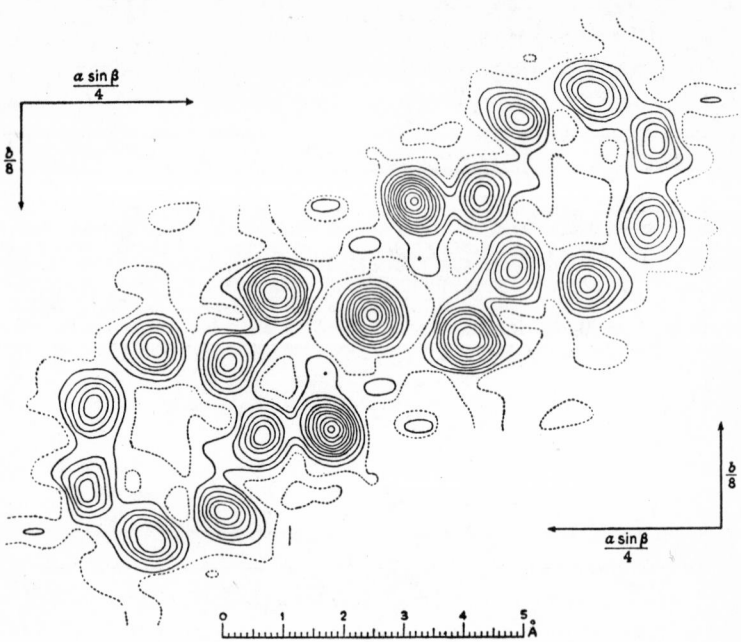

Fig. 113. Electron density projection of cupric tropolone. Contour scale on the central copper atom is reduced by a factor of five.

tion, the electron density map shown in Fig. 113 is obtained, and this reveals the structure of the molecule in considerable detail. The centres of the seven carbon atoms are found to lie quite accurately on an ellipse, of major axis 1.61 Å and minor axis 1.44 Å. This ellipse can be derived from a flat circular ring of carbon-atoms, radius 1.61 Å, inclined at 26.5° to the projection plane, and on this assumption we also obtain the striking result that the perpendicular distance between these circular rings is exactly 3.4 Å, the usual contact distance for large aromatic molecules.

Although only a projection of the structure is available, the evidence for a planar molecule is therefore very strong. Diffraction effects are present and the individual carbon-carbon bond lengths cannot be obtained with very high accuracy. Allowing for the orientation, they are found to vary between 1.37 and 1.44 Å, the average being 1.40 Å. Whether more refined work will show them to be more accurately equal is an interesting question. In any case, these values, together with the apparently planar nature of the rings and their contact distance of about 3.4 Å, provide strong evidence of aromatic character. The distance between the oxygen-bearing carbon atoms does not appear to be significantly different from the distances separating other pairs of adjacent carbon atoms in the ring. This may suggest that conjugation between the hydroxyl and carbonyl groups is relatively unimportant in determining the aromatic behaviour displayed: it is consistent with the view that such aromaticity may reside in the *cyclo*heptatrienone (tropone) system to which tropolone conforms as a hydroxy-derivative.[18]

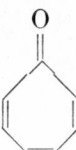

In the centre of the molecule the four oxygen atoms and the copper atom are strictly coplanar, in accordance with the expected dsp^2 type of bond formation. But the two oxygens attached to the seven-membered ring are not exactly equivalent, one, presumably hydroxyl, being at a distance of about 1.34 Å, and the other, presumably carbonyl, at the shorter distance of about 1.25 Å from the carbon atoms. The "hydroxyl" oxygen is also bound more closely to the copper atom, the distance being 1.83 Å, whereas the other copper-oxygen distance is about 1.98 Å. From these results it appears that the two oxygen atoms in tropolone are different in function and are rather like the oxygen atoms in a carboxyl group, which is in keeping with the pronounced acidity and lack of ketonic character in this substance.

Several other tropolone derivatives have been examined by X-rays, but the results have not yet been fully interpreted. These include the natural product nootkatin,[19] $C_{15}H_{20}O_2$, a mono-bromotropolone methyl ether,[20] and α-bromotropolone.[21] An electron density projection of this latter derivative is shown in

[18] H. J. Dauben and H. J. Ringold, *J. Amer. Chem. Soc.*, 1951, **73**, 876; W. E. Doering and F. L. Detert, *ibid.*, 1951, **73**, 876.

[19] H. Erdtman, *Chemistry & Industry*, 1951, p. 12; R. B. Campbell and J. M. Robertson, *ibid.*, 1952, p. 1266; H. Erdtman and W. E. Harvey, *ibid.*, 1952, p. 1267.

[20] T. Watanabe and R. Pepinsky, preliminary communication.

[21] R. B. Campbell and J. M. Robertson, *J. Chem. Soc.* (to be published).

Fig. 114. These analyses, although as yet of a preliminary nature, appear to confirm the main conclusions concerning the tropolone structure mentioned above.

The structure of the alkaloid colchicine is more complex. Extensive chemical studies[22] have shown it to be a tropolone

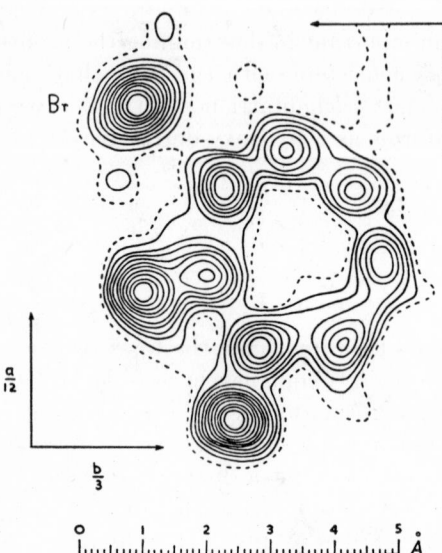

Fig. 114. Electron density projection of α-bromotropolone. Contour scale on the bromine atom is reduced by a factor of five.

derivative and to possess one or other of the following structures, although there is no satisfactory means of distinguishing between them:

I   II

[22] For a review, see J. D. Loudon, *Ann. Repts. on Progress Chem.* (Chem. Soc. London), 1948, **45**, 187.

From the point of view of X-ray analysis there is little prospect of successfully solving a complex structure of this type, which cannot be planar, without the aid of some direct phase-determining mechanism, and suitable heavy atom derivatives are hard to obtain.

A copper salt of the hydrolysed derivative, colchiceine, has been examined by Morrison.[23] In this compound there is evidence of co-ordination of two molecules to give a structure of the type:

but details are obscured by the presence of a number of molecules of water of crystallisation.

Pepinsky and his co-workers[24] have succeeded in obtaining crystalline methylene bromide and methylene iodide addition complexes of colchicine, which are suitable for a fairly straightforward application of the isomorphous replacement method of phase determination. After many refinements on the XRAC computer they have obtained projections of these structures which appear to settle the outstanding points regarding chemical structure. In particular, the presence of the seven-membered ring B is confirmed, and the position of the methoxy-group on ring C is that given by the structure II above, remote from the N-acetyl-group.

5. STRYCHNINE

Another even more complex alkaloid whose structure has been completely solved by the X-ray method is strychnine, $C_{21}H_{22}N_2O_2$, the molecule of which contains no fewer than seven closely interwoven rings of atoms. Two independent X-ray investigations have been made, one making use of the heavy atom method, with

[23] J. D. Morrison, *Acta Cryst.*, 1951, **4**, 69.
[24] M. V. King, J. L. de Vries, and R. Pepinsky, preliminary communication (paper read before Second International Congress of Crystallography, Stockholm, 1951); *Acta Cryst.*, 1952, **5**, 437.

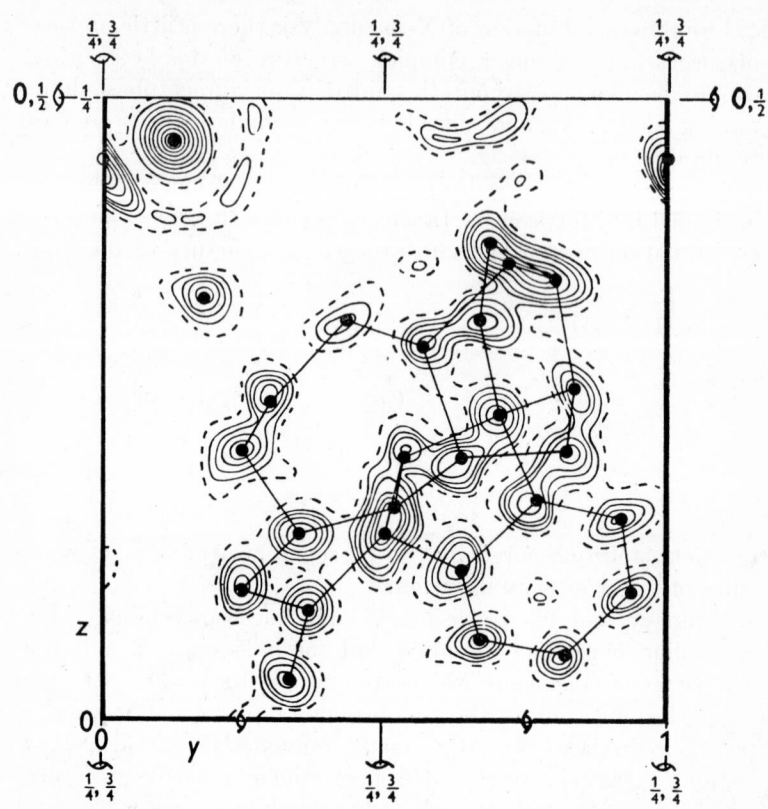

Fig. 115. Electron density map of strychnine hydrogen bromide dihydrate in projection along the *a* axis. Contour scale on the bromine atom (top left) is reduced by a factor of five (Robertson and Beevers).

strychnine hydrogen bromide dihydrate,[25] and the other based on the isomorphous replacement method, with strychnine sulphate and selenate pentahydrates.[26] Both these investigations commenced before the chemical structure was fully known, although many features were, of course, beyond doubt. The final X-ray results succeed in locating all the atoms in the strychnine molecule with an accuracy of about 0.1 Å. They are in complete

[25] J. H. Robertson and C. A. Beevers, *Nature*, 1950, **165**, 690; *Acta Cryst.*, 1951, **4**, 270.

[26] C. Bokhoven, J. C. Schoone, and J. M. Bijvoet, *Proc. Koninkl. Nederland. Akad. Wetenschap.*, 1947, **50**, 825; 1948, **51**, 990; 1949, **52**, 120; *Acta Cryst.*, 1951, **4**, 275.

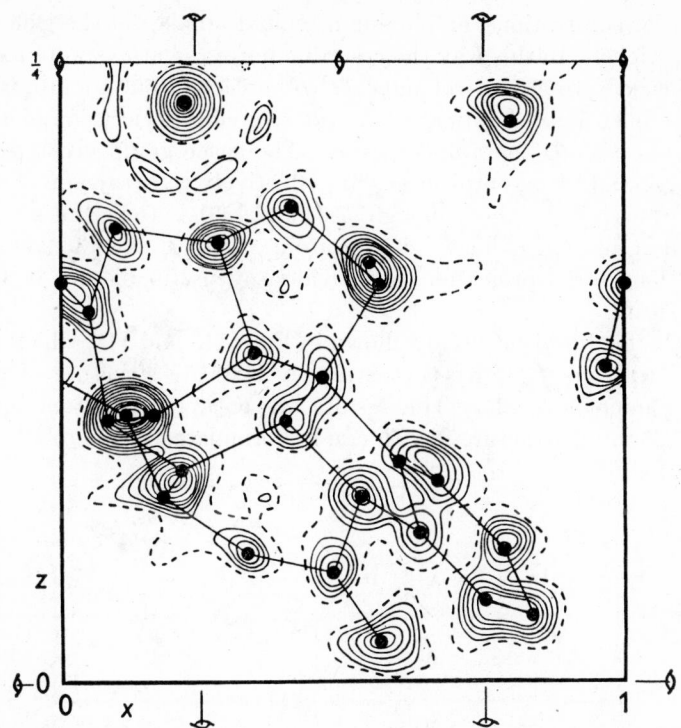

Fig. 116. The structure as in Fig. 115, but projected along the
b axis.

agreement with each other, and with the latest chemical structure, which has now been fully elucidated by the work of Robinson[27] and of Woodward.[28]

In the hydrobromide crystal structure the bromine atoms occupy general positions that can be determined by the Patterson
method.   Their contributions, however, are not phase determining for a sufficiently large number of the structure factors to permit a perfectly straightforward analysis, and it was found necessary to proceed by means of three-dimensional Patterson and

[27] R. Robinson, *Experientia*, 1946, **2**, 28; H. T. Openshaw and R. Robinson, *Nature*, 1946, **157**, 438; R. Robinson, *ibid.*, 1947, **159**, 263; **160**, 18;
R. Robinson and A. M. Stephen, *ibid.*, 1948, **162**, 177.

[28] R. B. Woodward, W. J. Brehm, and A. L. Nelson, *J. Amer. Chem.
Soc.*, 1947, **69**, 2250; R. B. Woodward and W. J. Brehm, *ibid.*, 1948,
70, 2107.

Fourier summations, employing a limited number of the general reflections.   In this way the co-ordinates of the atoms were fixed sufficiently to enable refinements to be carried out by ordinary two-dimensional Fourier summations, giving projections of the structure along the a and b axes.   The space group involved is $P2_12_12_1$, and these projections are effectively centrosymmetrical. The axes are also short enough (7.64 and 7.70 Å) to permit a very good resolution of many of the atoms.   The c axis, 33.2 Å, is too long for a projection to provide any useful picture of the structure.

The final projections are shown in Figs. 115 and 116, which reveal with great clarity the extremely intricate structure of the strychnine molecule.   The results are entirely consistent with the chemical structure, which can be formulated in the following way:

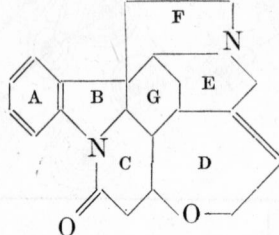

In the crystal it is found that the rings B to G form a compact three-dimensional structure, from which the benzene ring A projects outwards.

In the strychnine sulphate and selenate structures analysed by Bijvoet the space group is $C2$, and the two replaceable heavy atoms in the monoclinic unit cell (S and Se) must be located on the 2-fold axes.   For the b axis projection these are equivalent to symmetry centres, the positions being $(0y0)$ and $(\frac{1}{2}, \frac{1}{2}+y, 0)$, and the analysis proceeds by equation 6.18 as in the case of the phthalocyanine structures.   In the other projection, however, there is no centre of symmetry, and the phase constants must be evaluated by equations 6.19 or 6.20.

The axial lengths are $a = 35.85$, $b = 7.56$, $c = 7.84$ Å, $\beta = 107° 20'$ for the sulphate and $a = 35.9$, $b = 7.58$, $c = 7.90$ Å, $\beta = 107° 40'$ for the selenate.   The isomorphism is thus extremely close, and it will be noted that the dimensions are also very similar to those of the hydrogen bromide derivative.   A very direct projection, showing considerable resolution is therefore obtained in the direction of the b axis. Along the c axis, after calculation of the phase constants, a projection of the structure can be obtained superimposed on its mirror image.

After much refinement this extremely intricate analysis yielded atomic positions for the strychnine molecule which are in complete agreement with the chemical structure, and also with those found by Beevers in his analysis of the hydrogen bromide derivative. The structure of the sulphate pentahydrate is shown in Fig. 117, where columns of sulphate groups and water molecules are arranged between the strychnine molecules. There is prob-

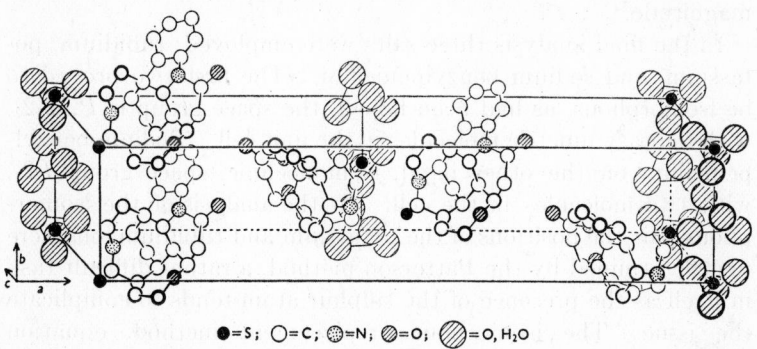

●=S; ○=C; ⊛=N; ⊘=O; ⌀=O, H₂O

Fig. 117. The crystal structure of strychnine sulphate pentahydrate (Bokhoven, Schoone, and Bijvoet).

ably some disorder in the vertical arrangement of the water molecules, which has been disregarded in making this diagram.

6. PENICILLIN

The accurate determination of the structure of penicillin represents one of the outstanding triumphs in the field of X-ray crystallography. The structural formula for benzylpenicillin, which can now be written

$$C_6H_5 \cdot CH_2 \cdot CO \cdot NH \cdot CH—CH \quad\overset{\displaystyle S}{\diagup\diagdown}\quad C(CH_3)_2,$$
$$O{=}C—N———CH \cdot COOH$$

may not at first sight appear unduly complex; but the actual spatial arrangement of the atoms is extremely intricate, and when the X-ray work began the chemical structure was entirely unknown.

This work commenced at Oxford as soon as the first crystalline degradation products became available, and it proceeded step by step with the chemical work until a final solution of the structure was achieved.   A very full account of this collaboration has been given by Crowfoot and others,[29] which well illustrates the application of various X-ray techniques to the solution of a chemically unknown structure, and shows how the form of the molecule gradually becomes clear from a combination of chemical and X-ray evidence, both of which are essential in an undertaking of this magnitude.

In the final analysis three salts were employed: rubidium, potassium, and sodium benzylpenicillin.   The first two proved to be isomorphous, as had been hoped; the space group is $P2_12_12_1$, with four asymmetric molecules in the unit cell.   Sodium benzylpenicillin, on the other hand, is monoclinic, space group $P2_1$, with two molecules in the cell.   In the analysis of the isomorphous pair the positions of the potassium and rubidium ions were first determined by the Patterson method, a rather difficult task in itself as the presence of the sulphur atom tends to complicate the issue.   The isomorphous replacement method, equation 6.18, was then applied, but unfortunately the positions of the heavy atoms are such that only a limited number of reliable phase determinations can be made.   As a result, the projections obtained could not be at all fully interpreted at the time, although evidence regarding the position of the sulphur atom was obtained.

Meanwhile, the sodium salt, which contains no atom heavy enough for direct phase determination, was investigated by testing the X-ray data against various postulated chemical structures.   This process was very greatly assisted by the development of optical diffraction methods for the rapid comparison of trial structures.   Although a measure of agreement was obtained for some structures, it was again not possible to make anything like a full interpretation of the molecular structure of this derivative.

[29] D. Crowfoot, C. W. Bunn, B. W. Rogers-Low, and A. Turner-Jones *The Chemistry of Penicillin*, Princeton: Princeton University Press, 1949, p. 310; G. L. Clark, W. I. Kay, K. J. Pipenberg, and N. C. Schieltz, *ibid.*, p. 367; D. Crowfoot. *Ann. Rev. Biochem.*, 1948, **17**, 115; G. J. Pitt, *Acta Cryst.*, 1952, **5**, 770.

The solution to the problem was finally reached by a brilliant correlation of all the partial resolutions that had been achieved. Certain features of these different projections were invariant, and the molecule was recognised as a very compact, roughly semi-circular arrangement of atoms, which was quite unlike the various structures that had been postulated. The chemical groupings of atoms known to be present could, however, be fitted to this picture. On this basis it was then possible to begin refinements of the structure, a process which was greatly facilitated by an early application of the method of $(F_o—F_c)$ synthesis to sodium benzylpenicillin (see Chapter V, Section 9). From this point onwards the work proceeded rapidly by the usual process of successive approximation until finally it was possible to compute a full three-dimensional Fourier synthesis of the structure.

The final electron density projection along the b axis (6.33 Å) for sodium benzylpenicillin is shown in Fig. 118, while Fig. 119

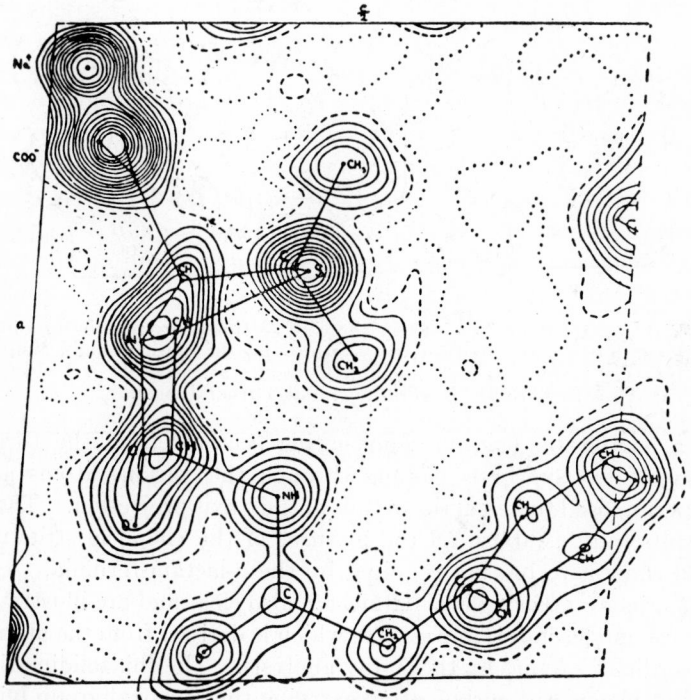

Fig. 118. Electron density projection of sodium benzylpenicillin, along the b axis. Contours at intervals of 1 electron per Å² (Crowfoot, Bunn, Rogers-Low, and Turner-Jones).

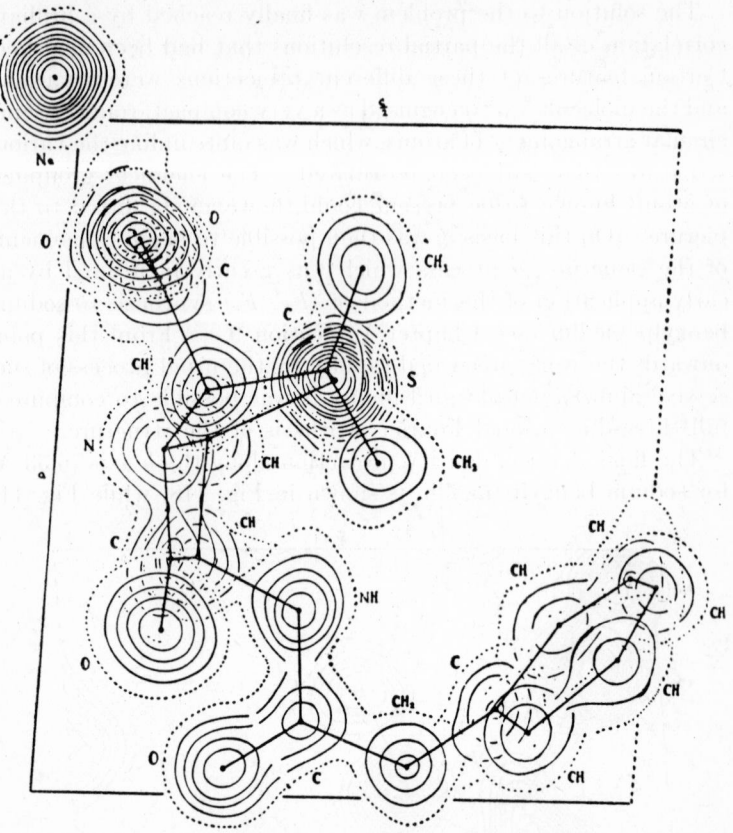

Fig. 119. A composite diagram showing sections through the atoms at different levels, projected on (010), from the three-dimensional Fourier synthesis of sodium benzylpenicillin. Contours at intervals of 1 electron per $Å^3$ (Crowfoot, Bunn, Rogers-Low, and Turner-Jones).

is a composite diagram representing sections from the three-dimensional synthesis passing through the various atoms and superimposed to reveal the details of the *b* axis projection. These results fully establish all the features of the chemical structure and show it to be that required by the β-lactam formula.

Various stereochemical details are also clear and are illustrated by the model shown in Fig. 120, which is derived from the atomic co-ordinates found by the X-ray analysis. The thiazolidine ring is not planar, the carbon atom carrying the carboxyl group lying out of the plane of the other atoms. The β-lactam ring is more nearly planar, but the attached oxygen atom also appears to be bent a little out of the plane of the other atoms.   The carboxyl group and the β-lactam ring lie on opposite sides of the thiazolidine ring; while the amide side chain and the thiazolidine ring lie on the same side of the β-lactam ring.

Chemical confirmation of some of these relations has been obtained.   In particular, the configuration of the two amino acids, penicillamine,

$$(CH_3)_2C(SH)\cdot CH(NH_2)\cdot COOH$$

and the acid

$$CHO\cdot CH(NH_2)\cdot COOH,$$

which may be regarded as the immediate progenitors of the thiazolidine ring structure, have been shown to be D and L re-

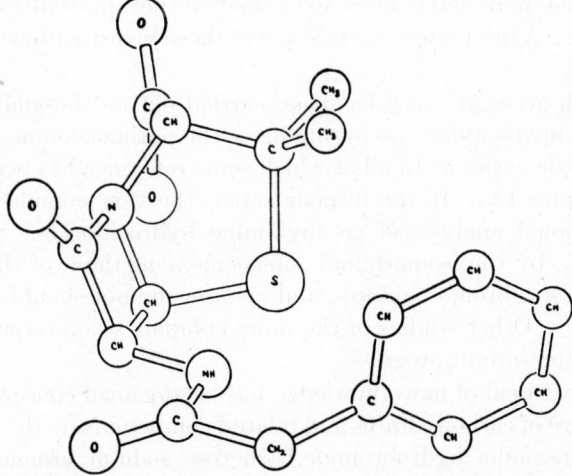

Fig. 120. The arrangement of the atoms in the benzyl-penicillin ion.

spectively.   Such an opposite configuration is demanded by the crystal structure.

The bond lengths and valency angles are not always known with sufficient accuracy to determine bond type.   In the amide side chain, however, a normal amide structure as in acetamide has been established.   The configuration in this part of the molecule, and the position of the benzene ring, appear to be determined by packing considerations.   These are governed by the grouping of the oxygen atoms of the carboxyl group, the lactam ring, and the amide side chain of different molecules around the metal ion, there being six close Na $\cdots$ O approaches in the sodium salt, and seven in the case of the larger ions of the potassium and rubidium salts.

7. OTHER APPLICATIONS

The selection of the complex structures discussed in this chapter has been based on their usefulness in illustrating the analytical methods employed as well as on the chemical interest of the substances themselves.   The potentialities of the X-ray diffraction method in elucidating the structures of partially known chemical substances should be clear from these examples.   In addition, many other complex and important structures which cannot be described here have been examined, either partially or completely.   A brief reference to some of these investigations is given below.

Much accurate work has been carried out on the amino acids and simple peptides, the pyrimidines, the purine adenine, and the nucleoside cytidine, to all of which some reference has been made in Chapter IX.   In the terpene series, the very complete three-dimensional analysis of geranylamine hydrochloride,[30] and the solution, by the isomorphous replacement method, of the structures of d-α-bromo-, chloro-, and cyano-camphor should be mentioned.[31]   Other studies of the more complex sesquiterpenes and triterpenes are in progress.

A great deal of new knowledge has been gained concerning the structure of carbohydrates and related compounds by the analysis of glucosamine hydrobromide,[32] sucrose sodium bromide dihydrate,[33] α-D-glucose,[34] certain fructose derivatives,[35] and kojic

[30] G. A. Jeffrey, *Proc. Roy. Soc.* (London), 1945, **A183**, 388.

[31] E. H. Wiebenga and C. J. Krom, *Rec. trav. chim.*, 1946, **65**, 663.

[32] E. G. Cox and G. A. Jeffrey, *Nature*, 1939, **143**, 894.

[33] W. Cochran, *Nature*, 1946, **157**, 231; C. A. Beevers and W. Cochran, *ibid.*, 1946, **157**, 872; C. A. Beevers and W. Cochran, *Proc. Roy. Soc.* (London), 1947, **A190**, 257.

[34] T. R. R. McDonald and C. A. Beevers, *Acta Cryst.*, 1950, **3**, 394.

[35] P. F. Eiland and R. Pepinsky, *Acta Cryst.*, 1950, **3**, 160.

acid.[36]   In addition to colchicine and strychnine, preliminary studies of the alkaloid ergine[37] and certain thebaine derivatives[38] have been reported.   A number of antibiotic substances have also been examined, in addition to penicillin.   These include derivatives of chloromycetin, aureomycin,[39] terramycin,[40] and potassium isomycomycin.[41]

Finally, the very complex structure of vitamin B_{12} has received extensive study in several places.[42]   In this crystal the space group is $P2_12_12_1$, and unit cell dimensions are roughly $24 \times 21 \times 16$ Å (they vary somewhat with the state of hydration of the crystal).   This cell contains four molecules of approximate composition $C_{63}H_{97}O_{20}N_{14}PCo$.   The cobalt atom is here unfortunately too light to be effective in phase determination, and consequently the analytical problems involved are most formidable, although some progress has in fact been made.   In complexity this structure begins to approach some of the smaller macromolecules referred to in the next chapter.

[36] H. A. McKinstry, P. F. Eiland, and R. Pepinsky, *Acta Cryst.*, 1952, **4**, 285.

[37] J. L. de Vries and R. Pepinsky, *Nature*, 1951, **168**, 431.

[38] J A. Goedkoop, C. H. MacGillavry, and R. Pepinsky, preliminary communication.

[39] J. D. Dunitz and J. E. Leonard, *J. Amer. Chem. Soc.*, 1950, **72**, 4276; J. D. Dunitz, *ibid.*, 1952, **74**, 995.

[40] J. Robertson, I. Robertson, P. F. Eiland, and R. Pepinsky, *J. Amer. Chem. Soc.*, 1952, **74**, 841.

[41] T. Whitehouse, P. F. Eiland, and R. Pepinsky, preliminary communication.

[42] D. Hodgkin, M. W. Porter, and R. C. Spiller, *Proc. Roy. Soc.* (London), 1950, **B136**, 609; D. Crowfoot Hodgkin, J. M. Broomhead, and J. G. White, papers to the Washington meeting of the American Crystallographic Association, 1951, and to the Second International Congress of Crystallography, Stockholm, 1951.

XI

Macromolecules and Biological Applications: Evidence from the Electron Microscope

1. GENERAL SURVEY

It should be apparent from the work described in the last chapter that X-ray analysis can be applied to the problems presented by complex and partially known chemical structures in two quite distinct ways. The first aims at a detailed analysis, leading to the determination of the positions of every atom in the molecule. In this direction outstanding success has been achieved in the great majority of cases only by the use of direct phase-determining methods (the heavy atom or isomorphous replacement method).

The other way in which X-ray analysis can be applied to complex structures does not aim at the absolute determination of atomic positions at all, but is concerned rather with discussing the feasibility of postulated chemical structures. In this connection X-ray evidence alone may be quite misleading. But, if supported by other physical evidence, the results can be of great value in differentiating chemical possibilities, as has been demonstrated in the early X-ray studies of the sterols. Generally speaking, the work on macromolecules and complex biological materials is of this second kind, as neither the methods of analysis nor the extent of the actual X-ray data is sufficient to lead to any completely unambiguous determination of atomic positions.

The extent to which some of the methods now available are likely to be applied successfully in the actual determination of atomic positions in a complex structure has already been discussed in Chapter VI.   The simple heavy atom method is limited by the fact that as the number of light atoms increases, their average contribution to the structure amplitude also increases, as the square root of this number.   The upper limit is likely to be about 100 light atoms (not counting hydrogen) per asymmetric unit, and whether structures of this degree of complexity can be completely solved has still to be demonstrated.   The most complex so far analysed by this type of method fall short of this.   The largest is probably calciferyl-5-iodo-4-nitrobenzoate, described in the previous chapter, which contains 41 atoms in addition to hydrogen.   In this structure the position of every atom has been ascertained with some certainty, at least in projection.

The isomorphous replacement method, and other methods depending on a difference effect, are theoretically not so limited as the simple heavy atom method.   The difficulties here reside rather in questions relating to the accuracy of measurement and also to the attainment in practice of the rather stringent conditions of isomorphism which are necessary.   The method is also dependent to a very large extent on favourable locations for the replaceable atoms.   The number of cases in which it can be applied at full power is therefore probably rather limited.

A molecule even larger than the calciferol derivative mentioned above has been successfully analysed by Vand and Bell.[1]   This is the β-form of trilaurin, $C_{39}H_{74}O_6$, a trilaurate of glycerol, in which the position of every atom has been obtained in a very direct manner.   This analysis introduces another principle, which is important to the present discussion.   Certain molecules may contain inherent structural periodicities that can be used to simplify the analysis.   Examples are provided by the repeated zigzags of a long hydrocarbon chain, a regularly substituted chain, or even the more complicated repeat units in the polypeptide chains of proteins.   In the case of trilaurin a survey of the structure factors revealed the presence of a three-dimensional reciprocal sub-lattice, corresponding to a certain repeat unit, or subcell, within the actual structural unit.   The structure of the subcell, being dependent on a fairly small number of intensities, is easily solved.   Provided that only a comparatively small number of atoms lie in general positions outside the subcell regions, as is the case in trilaurin, the phases of a sufficient number of

[1] V. Vand and I. P. Bell, *Acta Cryst.*, 1951, 4, 465; and later work.

structure factors may be determined to yield the structure of the main cell.   The results can then be refined to any required degree of accuracy in the usual way.   The theory of this method has been discussed by Vand,[2] and it is similar in certain respects to the heavy atom method.

It is conceivable that this type of analysis might be employed to solve the structures of very large molecules indeed, provided that they contain a sufficient degree of internal regularity.   The fibre structures, described in the next section, are in a sense extreme examples of this type of structure, where the whole molecule is built essentially from a comparatively small unit repeated indefinitely in a regular manner.   In the typical or ideal fibre structure, the sub-unit becomes the unit cell, and it is the only part of the structure that can be observed with X-rays.

Apart from these rather special structures, however, there are many examples of macromolecular compounds in the protein series which give rise to quite well formed single crystals, some of which are briefly described later in this chapter.   Instead of a few hundred atoms, several thousand atoms may occur in the unit cells of these crystals, and their analysis by X-rays raises problems of another order of magnitude.   Apart from the difficulties and limitations applying to the analytical methods which have been discussed above, a new and fundamental difficulty now arises from the intensities and range of the observable reflections.

If well-defined X-ray reflections of varying intensity could be recorded out to the limit of, say, copper radiation, then in principle at least these reflections should be capable of determining the positions of all the atoms in a large molecule with as great an accuracy as in a small molecule.   But in practice it appears that the picture is likely to blur and fade out as the number of atoms in the molecule increases.   There is evidence of this state of affairs even in the case of moderately complex structures.   This may be due to the position of the molecule in the crystal, or of the atoms in the molecule, varying slightly from one unit cell to the next, so that high accuracy in the location of atoms cannot be expected.   In cholesteryl iodide, for example, the number of reflections (about 300 for 28 atoms) is amply sufficient to determine the complete chemical structure of the molecule, but is not enough to permit very high accuracy in bond length measurement.   In

[2] V. Vand, *Acta Cryst.*, 1951, 4, 104.

vitamin B_{12}, reflections from planes with spacings smaller than about 1.1 Å cannot be observed.   At this limit it should still theoretically be possible to resolve individual atoms, but not with much accuracy.   For more complex molecules like the proteins the limit of observable spacing is usually in the region of 2.5 Å or more.   While this limitation exists, it rules out the possibility of any actual determination of individual atomic positions, of the kind achieved for the more simple molecules.

Our conclusions, therefore, do not encourage optimism with regard to the ultimate possibility of really detailed analyses of these complex substances notwithstanding the beautiful crystals that can sometimes be obtained.   It is nevertheless of extreme importance to apply the X-ray method in every possible way to these crystals in order to test the validity of the different structures that can be proposed on chemical evidence.   In such work it is essential that other types of physical measurement should also be employed wherever possible to supplement and confirm the X-ray conclusions.   A variety of methods can be used for the estimation of molecular weights and the general shapes of the molecules.   Optical measurements, especially in the infra-red, may give valuable evidence concerning certain parts of the structures.   Finally, an extremely powerful tool is available in the electron microscope.   In the case of fibre structures, viruses, and many proteins it is now possible to photograph individual molecules or unit cells, and the results give striking and direct confirmation of the X-ray measurements, as well as revealing other features of the structures.   Some examples of this kind are given later.

Finally, it may be remarked that although the detailed atomic view of protein structures is at present beyond the reach of the data and the interpretative methods, yet this detailed atomic view is perhaps hardly required of the X-ray approach.   It is already known in outline from the established facts of organic chemistry and from the X-ray analysis of simple molecules and crystals.   The X-ray method, combined with such other approaches as those mentioned above, may be capable of defining the directions and arrangements of the main frameworks into which these known atomic groupings must be fitted.   A very promising start has in fact already been made in the case of several crystalline proteins, and rapid progress may now be expected.

2. FIBRE STRUCTURES

In fibre structures the unit distinguishable by X-rays is only part of the molecule, and a fairly complete analysis of this unit is sometimes possible. But the alignment of the molecules is frequently imperfect and the truly crystalline areas are small (see Chapter IV, Section 8). Only a relatively small number of rather diffuse reflections may be available, and these are frequently not sufficient for an entirely unambiguous structure determination. In spite of these limitations, a very great effort has been devoted to the study and interpretation of fibre structures, because their almost universal occurrence as constituents of living matter gives them a quite outstanding importance. Most of these natural fibre structures are chemically complex and their crystallinity is low. Recently, however, a large number of synthetic polymers have been prepared from simple chemical units, and some of these structures can be studied in considerable detail by the fibre photograph method.

2A. Synthetic fibres.—The simplest example in this class is polyethylene,

$$[-CH_2-CH_2-]_n,$$

which consists of normal hydrocarbon chains several thousand atoms in length. Oriented specimens can be prepared by rolling out sheets of the material, but the very detailed X-ray analysis made by Bunn[3] was carried out mainly from a study of powder photographs. The structure is strictly analogous to the orthorhombic modification of other long chain hydrocarbons, but is more simple because with the indefinitely extended chain the end effects disappear. The unit cell (Table V, p. 169) contains four —CH_2— groups, and the electron density distribution has been studied in some detail by the triple Fourier series method. The zigzag angle is 112°, and the carbon-carbon distance 1.53 Å. The electron distribution around the carbon atoms appears to be far from spherical, an effect probably due to the presence of the hydrogen atoms, together with certain anisotropic thermal motions.

Next in order of complexity is polyvinyl alcohol,

$$[-CHOH \cdot CH_2-]_n,$$

[3] C. W. Bunn, *Trans. Faraday Soc.*, 1939, **35**, 482.

from which oriented specimens can be obtained by drawing fibres.[4] The repeat distance is again that of the zigzag chain, about 2.5 Å, but there is now a double layer of molecules in the unit cell, held together by hydrogen bonds. The system is probably monoclinic, and there may be some randomness in the left- and right-hand positioning of the hydroxyl groups on the chain. There is evidence, from the existence of certain crystalline co- polymers,[5] that a considerable degree of irregularity can be tolerated in such structures.

A number of other more complex polymers, such as polyiso- butene[6] (where the molecule has a helical configuration), poly- butadiene,[7] and the alkyl polyacrylates,[8] have been examined, but without really detailed results. The well-known and very successful synthetic fibres known as nylon, however, give rise to beautiful X-ray diffraction patterns, and a rather detailed analysis of some of these has been accomplished.[9]

The fibre known as nylon 6.6 is polyhexamethylene adipamide,

$$[-NH \cdot (CH_2)_6 NH \cdot CO \cdot (CH_2)_4 CO-]_n,$$

where the value of n is about 50. The identity period of 17.2 Å in this polymer shows that the chain must be a fully extended zigzag. On this basis a model can be set up and adjusted to give a good account of the principal intensities. The unit cell is triclinic and the approximately planar molecules are found to be linked by NH $\cdots$ O hydrogen bonds about 2.8 Å in length to form sheets. This hydrogen bonding involves a lengthwise dis- placement of adjacent molecules which determines the β angle. The hydrogen-bonded sheets of molecules can pack together in two different ways, giving rise to different crystalline modifica- tions, which often occur together.

[4] R. C. L. Mooney, J. Amer. Chem. Soc., 1941, 63, 2828; C. W. Bunn, Nature, 1948, 161, 929.

[5] C. W. Bunn and H. S. Peiser, Nature, 1947, 159, 161; C. S. Fuller, J. Amer. Chem. Soc., 1948, 70, 421.

[6] C. W. Bunn, J. Chem. Soc., 1947, 297.

[7] K. E. Beu, W. B. Reynolds, C. F. Fryling, and H. L. McMurry, J. Polymer Sci., 1948, 3, 465.

[8] H. S. Kaufmann, A. Sacher, T. Alfrey, and I. Fankuchen, J. Amer. Chem. Soc., 1948, 70, 3147.

[9] C. W. Bunn and E. V. Garner, Proc. Roy. Soc. (London), 1947, A189, 39; C. S. Fuller, Chem. Revs., 1940, 26, 143.

Polyhexamethylene sebacamide, or nylon 6.10, has the structure

$$[-NH(CH_2)_6NH\cdot CO(CH_2)_8CO-]_n$$

and is very similar to the 6.6 form in all respects except that the repeating chain unit, or fibre axis, is now found to be 22.4 Å instead of 17.2 Å. This extension is just enough to accommodate the additional —CH_2 groups, and provides further confirmation of the correctness of the structure.

2B. Cellulose.—Cellulose is a typical fibre structure, and its widespread occurrence in the plant kingdom, where it has been estimated to comprise about one-third of all vegetable matter, gives it a special importance. As the name implies, it is the main constituent of the cell walls of most plants. All the native cellulose fibres give essentially the same characteristic X-ray pattern, showing a repeat distance in the fibre direction of 10.3 Å, but varying greatly in the degree of orientation of the crystallites. Typical fibre diagrams of this kind are given by wood, cotton, flax, hemp, jute, ramie, and various grasses and straws.

Cellulose was also the earliest fibre structure studied by the X-ray method, approximate unit cell dimensions being given by Polanyi[10] in 1921. He proposed an orthorhombic cell measuring $7.9\times8.45\times10.2$ Å and containing four glucose residues ($4\times C_6H_{10}O_5$). Later work[11] on highly oriented samples of native cellulose indicates a monoclinic cell of dimensions,

$$a = 8.35, \quad b = 10.3, \quad c = 7.95 \text{ Å}, \quad \beta = 84°,$$

but there is some evidence[12] that the precise values of the parameters may depend to a certain extent on the origin of the sample.

The intensities of about 25 reflections can be estimated from the best photographs. As the unit cell is known to contain 44 atoms in addition to hydrogen, it is clear that no completely unambiguous structure determination, defining the precise position

[10] M. Polanyi, *Naturwissenschaften*, 1921, **9**, 228.

[11] K. H. Meyer and H. Mark, *Z. physik. Chem.*, 1929, **B2**, 115; E. Sauter, *ibid.*, 1937, **B35**, 83; K. H. Meyer and L. Misch, *Helv. Chim. Acta*, 1937, **20**, 232; S. T. Gross and G. L. Clark, *Z. Krist.*, 1938, **99**, 357. For a comprehensive account of cellulose and its derivatives, including X-ray work, see *High Polymers*, New York: Interscience Publishers, 1942, 1943, vols. IV and V.

[12] C. Legrand, *Compt. rend.*, 1948, **226**, 1983.

of every atom, can be expected. In such a MOLECULES
bination of chemical and X-ray evidence may ever, a com-
lishing the structure with a high degree of certaint, in estab-

This combination of evidence, and in particular th
work of Haworth, Irvine, and Hirst, shows that cellulose ical
of a chain of β-glucose residues linked as in cellobiose:

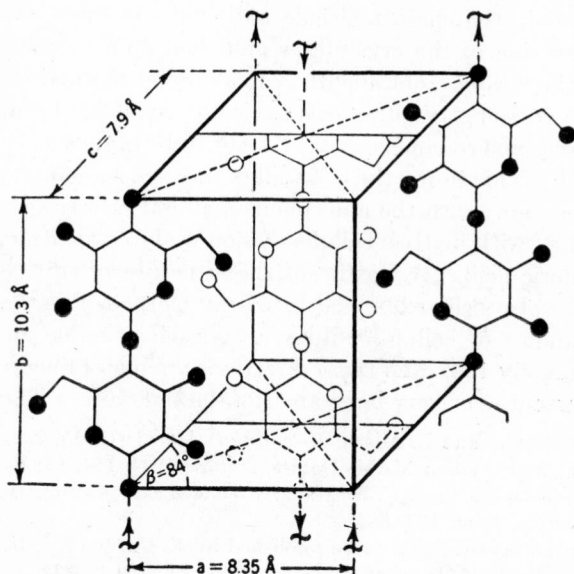

When a three-dimensional model is constructed on this basis with
normal bond lengths and valency angles, it is found that two residues as shown account exactly for the fibre period of 10.3 Å. In
the crystal the chain therefore appears to be fully extended,
although in solution twisted configurations may, of course, occur.

The particular chain arrangement which is required to account

Fig. 121. Structural model for cellulose (Meyer and
Misch).

ORGAN intensities has been carefully studied. The
for the ...*k*0) reflections when *k* is odd shows the presence of
abser screw axis in the *b* direction (C_2^2 is a possible space
a ... The structure proposed by Meyer and Misch[13] and il-
...rated in Fig. 121 places the chain molecules on the screw
axis, alternate chains running in opposite directions and the
mean planes of the glucose rings lying parallel to (002) as required
by the high intensity of the reflection from this plane. A reasonably good agreement with all the observed intensities is obtained
from this model. It also provides a very striking explanation of
the physical properties, e.g., low thermal expansion in the direction of the fibre axis along which there are strong covalent
linkages, as contrasted with much greater expansion in the lateral
directions where only hydrogen bonds and van der Waals forces
operate.

Later modifications of this model have been discussed,[14] and,
although there is some evidence in their favour, the available
data do not appear to be sufficient to permit any really definite
conclusions on points of finer detail.

The X-ray pattern gives no indication regarding the number
of residues in the chain molecule, although evidence concerning
the orientation of the crystallites and their spiral formations in
certain fibres can be obtained. A more direct method of exploring these grosser structural features is now available by means of
the electron microscope. A very beautiful example is shown in
Fig. 122.[15] This shows the cell wall of the alga *Valonia ventricosa*
in surface view, with the amorphous material (always present in
association with native cellulose) removed by treatment with
hydrochloric acid. A feature of the structure of this cell wall,
which has also been examined by X-ray methods,[16] is that overlying bundles of cellulose fibres cross each other at angles of
approximately 120°, and these directions are maintained remarkably constant over very large areas of the structure. The result-

[13] K. H. Meyer and L. Misch, *Helv. Chim. Acta*, 1937, **20**, 232.

[14] W. T. Astbury and M. M. Davies, *Nature*, 1944, **154**, 84; P. H. Her-
mans, J. de Booys, and C. J. Maan, *Kolloid-Z.*, 1943, **102**, 169; F. T.
Peirce, *Nature*, 1944, **153**, 586.

[15] From an investigation to be published by F. C. Steward, K. Mühlathaler, and R. W. G. Wyckoff.

[16] W. T. Astbury, T. C. Marwick, and J. D. Bernal, *Proc. Roy. Soc.*
(London), 1932, **B109**, 443.

MOLECULES

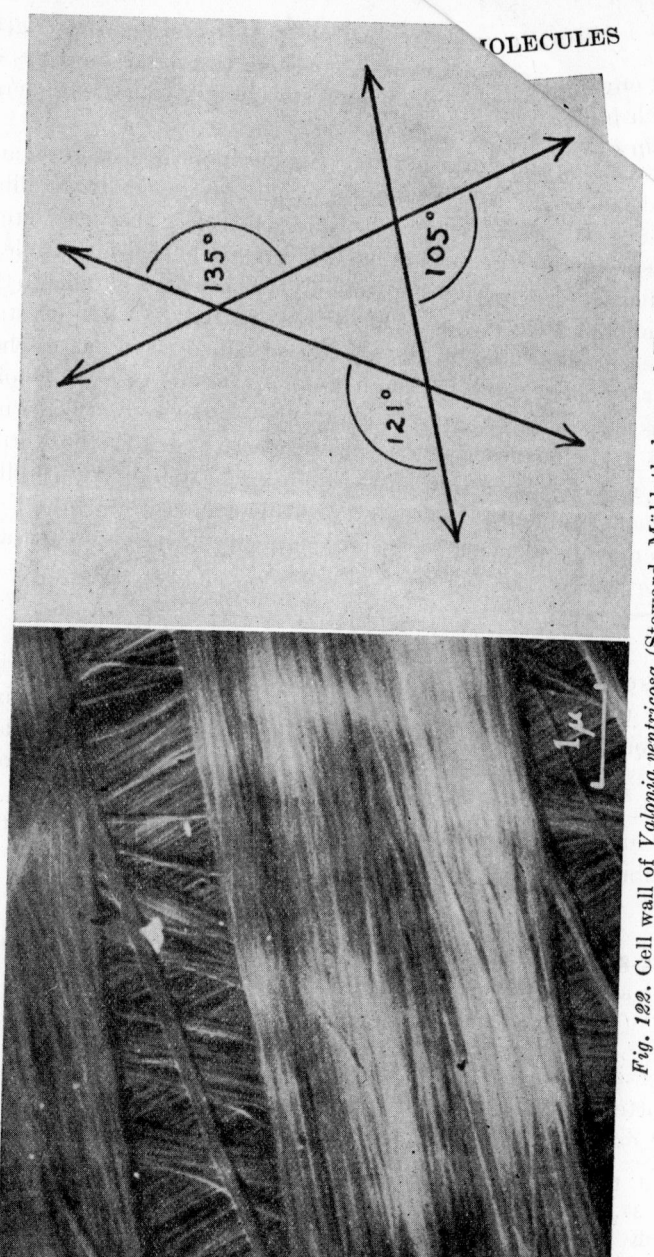

Fig. 122. Cell wall of *Valonia ventricosa* (Steward, Mühlathaler, and Wyckoff).

ORGAN...e of closely woven texture and great strength.
ing env... ual fibre, of course, contains very large numbers of
Each ...cules, which cannot be resolved by the electron micro-
cha...

..C. **Rubber and gutta-percha.**—In mechanical properties and
general behaviour rubber is quite unlike any normal crystalline
material.   It was discovered by Katz,[17] however, that on stretching a specimen to several times its original length an excellent
fibre diagram is obtained, showing crystals oriented parallel to the
direction of stretching.   Diffraction patterns, with crystals
showing random orientation, are also obtained when unstretched
rubber is cooled below 0°.   Gutta-percha, another isoprene polymer, is crystalline at room temperatures and exists in two forms,
α and β.   The naturally occurring α-form is readily converted
to the metastable β-form, and vice-versa.   From drawn or rolled
specimens oriented (fibre) diagrams can again be obtained.

Chemical evidence shows that these substances are essentially
1:4, or head to tail, polymers of isoprene,

$$[-CH_2 \cdot C(CH_3):CH \cdot CH_2-]_n.$$

It was early recognised[18] that the possibility of *cis-trans*-isomerism
about the double bonds in the extended chain molecules might
explain the difference between rubber and gutta-percha.   Later
work by Bunn[19] has succeeded in elucidating the structures in some
detail.

β-Gutta-percha has the most simple structure, with an orthorhombic cell,

$$a = 7.8, \quad b = 11.8, \quad c = 4.72 \text{ Å},$$

space group $P2_12_12_1$, containing four chain molecules parallel to
the c axis.   Rubber is probably monoclinic,

$$a = 12.5, \quad b = 8.9, \quad c = 8.10 \text{ Å}, \quad \beta = 92°,$$

space group, $P2_1/a$, again with four molecules parallel to c.   In
β-gutta-percha, the chain repeat distance of 4.72 Å is close to
that expected for a single isoprene unit, showing in a convincing

[17] J. R. Katz, *Naturwissenschaften*, 1925, **13**, 411; *Kolloid-Z.*, 1925, **36**,
300; **37**, 19.
[18] K. H. Meyer and H. Mark, *Ber.*, 1928, **61**, 1939; C. S. Fuller, *Ind.
Eng. Chem.*, 1936, **28**, 907.
[19] C. W. Bunn, *Proc. Roy. Soc.* (London), 1942, **A180**, 40, 67, 82.

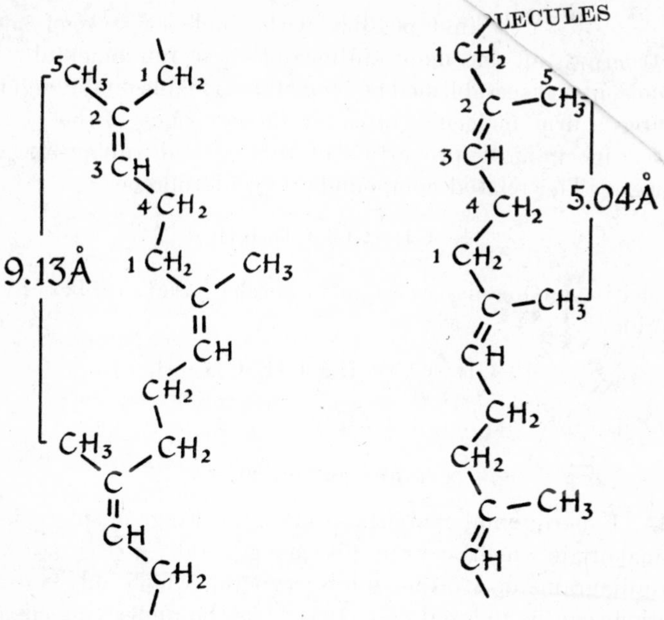

$^5CH_3$ $_1CH_2$
$_2C$
$\|$
$_3CH$
$_4CH_2$
$_1CH_2$ CH_3
   C
   $\|$
   CH
   CH_2
CH_3 CH_2
   C
   $\|$
   CH
   CH_2

9.13Å

$_1CH_2$
$_2C$ —$^5CH_3$
$/\!/$
$_3CH$
$_4CH_2$ 5.04Å
$_1CH_2$
C —CH_3
$/\!/$
CH
CH_2
CH_2
C —CH_3
$/\!/$
CH

Fig. 123. Configuration of the polyisoprene chains in rubber (*cis*) and gutta-percha (*trans*).

manner that the chain structure must be *trans-*. Only a *trans-* model could give a one-unit repeat distance. In rubber, on the other hand, the repeat distance (*c*-axis) approximates to that of two isoprene units.

When planar models with normal bond distances and angles as in Fig. 123 are drawn out, it is found that the repeat distances are somewhat greater than the observed periodicities, in both cases. To explain the observed values and the X-ray intensities of β-gutta-percha, it is necessary to assume that adjacent isoprene units are not coplanar, the 1–4 bond lying out of the plane of the others. The methyl group also appears to be considerably displaced from its expected position in the plane of the adjoining atoms. In rubber the structure appears to be more complicated, and there is evidence that the two isoprene units which make up the repeating unit in the crystal are probably not identical in configuration.

As in the case of cellulose, the available data (about 40 reflections from rubber and 24 from β-gutta-percha, not all separately resolved) are insufficient to permit refinements of these

ORGAN the kind possible with single crystals of simple structu But the main outlines of these fundamental struc-mole ve been established beyond all reasonable doubt, and they tv1de a firm molecular basis for further study of the extraor-.inary mechanical properties of rubber and rubber-like substances. The related compounds polychloroprene,[19]

$$[-CH_2 \cdot CCl : CH \cdot CH_2-]_n,$$

which is analogous to β-gutta-percha, and rubber hydrochloride,[20]

$$[-CH_2 \cdot CCl(CH_3) \cdot CH_2 \cdot CH_2-]_n,$$

have also received detailed study.

3. FIBROUS PROTEINS

3A. Experimental evidence.—The great range of substances of animal origin which occur in this class generally give rise to X-ray fibre diagrams of a rather diffuse type on which only a few reflections can be indexed (Fig. 124). As the underlying chemical structures are known to be vastly more complex than those considered in the previous section, it might be thought that the task of trying to unravel them is almost hopeless. Yet extensive X-ray investigations have been carried out on these structures, and although there is no immediate prospect of determining precise atomic positions, great advances have been made in our knowledge of the molecular configurations involved. The most striking result has been the demonstration that in spite of the many types and varied origins of the fibre structures examined, they appear to belong to no more than two fundamental groups.[21]

[20] C. W. Bunn and E. V. Garner, *J. Chem. Soc.*, 1942, p. 654.

[21] W. T. Astbury, *Fundamentals of Fibre Structure*, Oxford, 1933; W. T. Astbury and A. Street, *Trans. Roy. Soc.* (London), 1931, **A230**, 75; W. T. Astbury, *Trans. Faraday Soc.*, 1933, **29**, 193; W. T. Astbury and H. J. Woods, *Trans. Roy. Soc.* (London), 1933, **A232**, 333; W. T. Astbury and W. A. Sisson, *Proc. Roy. Soc.* (London), 1935, **A150**, 533; W. T. Astbury, S. Dickinson, and K. Bailey, *Biochem. J.*, 1935, **29**, 2351; W. T. Astbury, *Compt. rend. trav. lab. Carlsberg*, 1938, **22**, 45 (Sørensen jubilee volume); *Trans. Faraday Soc.*, 1938, **34**, 377; H. J. Woods, *Proc. Roy. Soc.* (London), 1938, **A166**, 76; W. T. Astbury, *J. Intern. Soc. Leather Trades Chemists*, 1940, **24**, 69 (First Procter Memorial Lecture); W. T. Astbury

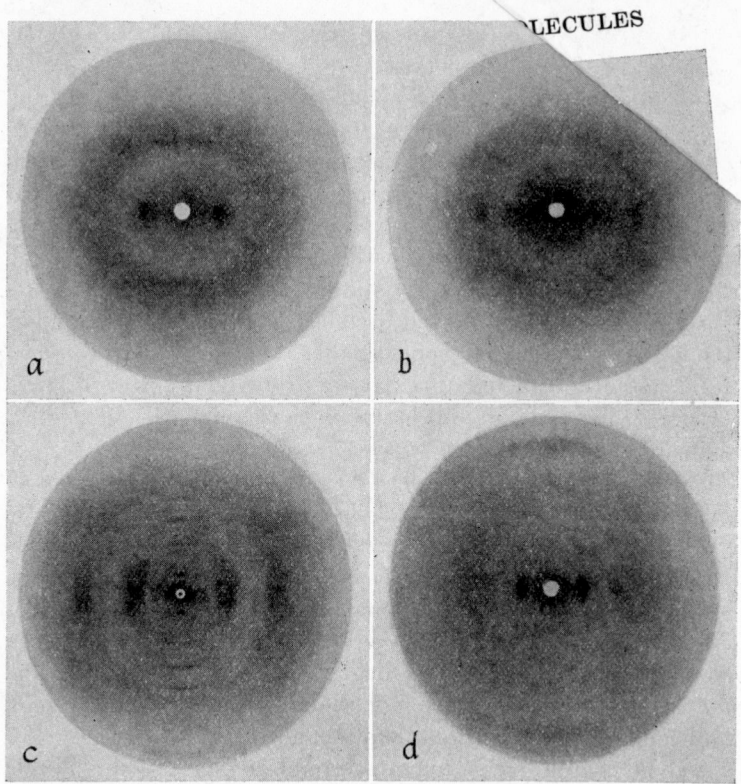

Fig. 124. X-ray diffraction patterns from fibrous proteins (Astbury).
(a) Unstretched Cotswold wool (α-keratin). (b) Stretched (90%) Cotswold
wool (β-keratin). (c) Seagull quill. (d) Rat tail tendon (collagen).

The first is known as the keratin-myosin group, and this includes fibrous substances such as hair, wool, horn, nails, and generally the fibrous proteins of the epidermis of mammals, amphib-
ians, and certain fishes. A typical X-ray diagram is that given
by unstretched hair or wool (Fig. 124a) in which the principal
features are a broad equatorial spot of about 9.7 Å spacing and a
composite meridian arc (possibly representing a fibre period) of

---

and S. Dickinson, *Proc. Roy. Soc.* (London), 1940, **B129**, 307; W. T. Astbury, *Chem. and Ind. Rev.*, 1941, **60**, 491; *J. Chem. Soc.*, 1942, p. **337**;
Advances in Enzymology, 1943, **3**, 63; K. Bailey, W. T. Astbury, and
K. Rudall, *Nature*, 1943, **151**, 716; W. T. Astbury, *Proc. Roy. Soc.* (London), 1947, **B134**, 303 (Croonian Lecture); and other papers.

ّ pacing. In this unstretched condition the fibrous about 5s known as α-keratin.

substretching the fibre to about twice its original length, an ~ly different X-ray photograph is obtained (Fig. 124b). In ~ ~is condition it is known as β-keratin, and two strong equatorial spots appear, which now have spacings of about 10 and 4.5 Å, while the fibre period is about 6.66 Å. The transformation from α- to β-keratin is reversible in the presence of water, although when exposed to steam or hot water the β-keratin fibre loses its power of elastic recovery and becomes "set." A further important observation is that keratin fibres, in the α- or unstretched state, can be contracted, on exposure to steam or hot water, by a further 30 per cent or so below their normal length. This phenomenon is referred to as *supercontraction*.

The fibrous protein myosin, which is the chief component of muscle and presumably responsible for its elastic properties, also gives an X-ray photograph very closely similar to that of α-keratin. Myosin also undergoes a spontaneous contraction in hot water, similar to the supercontraction of α-keratin, and it can be stretched to give the β-keratin diagram. The analogy in elastic properties and X-ray diagrams is therefore striking and indicates that an explanation of the behaviour and structure of keratin may throw light on the mechanism of muscular action.

Also belonging to the β-keratin group are the proteins fibroin and fibrin and more complex structures such as feathers and other avian and reptilian keratin. Extremely complex diagrams containing a great deal of fine structure are obtained from some of these materials (Fig. 124c), but they are essentially of the β-keratin type.

The second great division of the fibrous proteins is the collagen-gelatin group. This family includes the fibres of connective tissue, tendons, skin, cartilage, and the like, also the artificial product gelatin, which retains much of the collagen structure. Collagen fibres are of great strength, and they do not show the long range reversible extensibility of keratin. In water, however, they undergo spontaneous contraction and recovery, which is somewhat analogous to the supercontraction of α-keratin.

The collagen diffraction patterns (Fig. 124d) are of quite a different type from those described above. On the equator there is a very diffuse reflection at a spacing of about 4.4 Å (the "backbone" reflection of Astbury) and a strong ("side-c. ECULES of spacing about 11.5 Å at ordinary humidity, a reflection 10.4 Å on thorough drying. A strong meridian arc aing to remains very constant even in material derived from wide6 Å ferent sources and subjected to various treatments, and i- appears to represent a fundamental fibre period.

The collagen photographs, especially those from kangaroo tail tendon,[22] also display a large amount of fine structure, from which various very long spacing measurements in the direction of the fibre axis have been obtained. These indicate that the complete molecular pattern along the collagen chains must measure several hundred Ångstroms. Values of 103 Å,[22] 432 Å,[23] and 640 Å,[24] with many intermediate orders, have been reported. The precise value is found to vary with the relative humidity of the specimen over the range 628 to 672 Å.[25]

In this region the X-ray work overlaps with electron microscope studies, and a typical electron micrograph of collagen is shown in Fig. 125. The collagen fibrils generally vary in width between 200 and 2500 Å. They also display large-scale striations of an average period about 640 Å, which correspond to the X-ray long spacing measurements. In addition, a considerable amount of finer detail can be detected on certain photographs.[26]

In addition to the naturally occurring fibrous proteins described above, a number of synthetic polypeptides have now received detailed X-ray and infra-red study. These products include certain homogeneous polypeptides made from a single amino acid species as well as copolymers made from two or more amino acids.[27] Oriented films and fibres of poly-γ-methyl-L-glutamate

[22] R. W. G. Wyckoff, R. B. Corey, and J. Biscoe, *Science*, 1935, **82**, 175; R. B. Corey and R. W. G. Wyckoff, *J. Biol. Chem.*, 1936, **114**, 407; R. W. G. Wyckoff and R. B. Corey, *Proc. Soc. Exptl. Biol. Med.*, 1936, **34**, 285.

[23] G. L. Clark, E. A. Parker, J. A. Schaad, and W. J. Warren, *J. Amer. Chem. Soc.*, 1935, **57**, 1509.

[24] R. S. Bear, *J. Amer. Chem. Soc.*, 1942, **64**, 727; 1944, **66**, 1297.

[25] B. Wright, *Nature*, 1948, **162**, 23; *Amer. Mineral.*, 1948, **33**, 780; P. Kaesberg, H. N. Ritland, and W. W. Beeman, *Phys. Rev.*, 1948, **74**, 71.

[26] C. Wolpers, *Macromol. Chem.*, 1948, **2**, 37; C. E. Hall, M. A. Jakus, and F. O. Schmitt, *J. Amer. Chem. Soc.*, 1942, **64**, 1234.

[27] W. T. Astbury, C. E. Dalgliesh, S. E. Darmon, and G. B. B. M. Sutherland, *Nature*, 1948, **162**, 596.

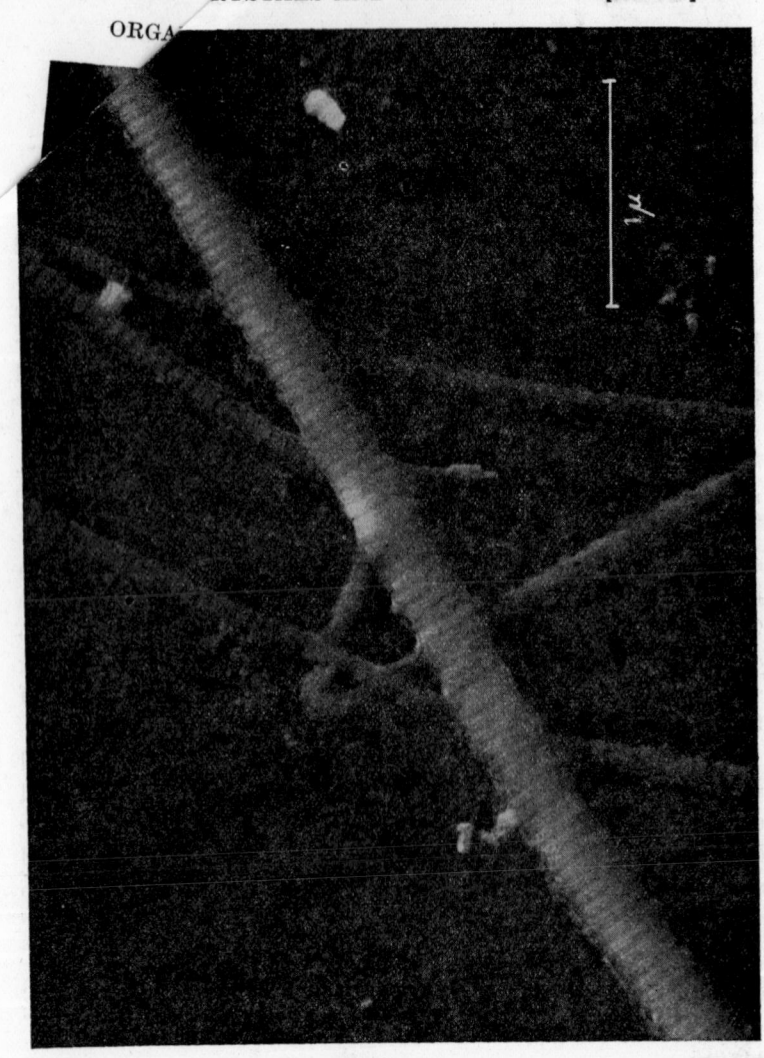

Fig. 125. Electron micrograph of collagen fibrils (I. M. Dawson).

and poly-γ-benzyl-L-glutamate[28] give some excellent X-ray photographs which show a repeat distance along the fibre axis of about 5.50 Å.   They also undergo a reversible $\alpha \rightarrow \beta$ transformation analogous to the keratin type.   The infra-red studies provide a valuable structural clue in giving evidence of the occurrence of hydrogen bonds in the direction of the fibre axis.   This work is only in a preliminary stage but will no doubt prove of great value in the elucidation of protein structure.

3B. Interpretation.—The outstanding result of these various X-ray studies on the fibrous proteins has been the demonstration that the underlying molecular configurations must conform to only a few fundamental patterns.   Of these the most important are the α- and β-keratin and collagen types.   It is not implied, of course, that the various members of these groups have anything like identical structures; but the X-ray evidence indicates that the different molecules which occur within any one of the groups must have somewhat similar shapes.

Since the pioneer chemical researches of Emil Fischer it has been known that the proteins are built from α-amino acids to form polypeptide chains of the type

$$\begin{array}{ccccccc} & R' & & & & R''' & \\ & | & & & & | & \\ CO & CH & NH & CO & CH & \\ \diagdown NH \diagup & & \diagdown CO \diagup & \diagdown CH \diagup & NH & \diagdown CO \diagup \\ & & & & | & \\ & & & R'' & & \end{array}$$

where the groups R', R'', etc., denote the side chains of the various amino acids.   In the β-keratin type of structure the polypeptide chain must be extended, but not necessarily fully extended with all the chain atoms lying in one plane as in Fig. 126. A structure of this type would have a chain period of about 7.23 Å, but other configurations of somewhat reduced period may be derived by varying the orientations of the bonds at the α-carbon atom.   In β-keratin the fibre period is 6.66 Å, and in silk fibroin about 7.0 Å.

In the transition from β- to α-keratin the fibre contracts to about half its original length, and there is strong evidence that

[28] C. H. Bamford, W. E. Hanby, and F. Happey, *Proc. Roy. Soc.* (London), 1951, **A205**, 30; E. J. Ambrose and A. Elliott, *ibid.*, 1951, **A205**, 47.

this reversible change is due to some manner of folding of the polypeptide chain. The precise value of the fibre period in this folded condition is not very clearly indicated in the various X-ray diagrams. The strong meridian arc at 5.1 Å in α-keratin may indicate the reduced fibre axis period, but this interpretation has been questioned, and there is evidence of a fibre period of about 5.65 Å in certain myosin preparations,[29] and 5.50 Å in the synthetic polypeptides. In any case, the folding must be such as to reduce the chain length to about half.

The nature of this intramolecular folding and the precise configurations adopted by the polypeptide chains in protein structures are the central problems in this field. Many solutions have been proposed, by Astbury and Bell,[30] Huggins,[31] Bragg, Kendrew, and Perutz,[32] Bamford, Hanby, and Happey,[28] and Pauling and Corey.[33] The main difficulty is to obtain sufficient X-ray evidence of a clear-cut nature to afford a conclusive test of any model structure proposed. The complexity of the problem is enormous, especially when we consider the great diversity of side chains that are known from chemical evidence to be attached to the main chain in largely unknown sequence. One most hopeful line of approach, which is being actively pursued, lies in the study of synthetic polypeptides (such as those mentioned earlier), which have a simpler constitution than the natural products.

The stereochemical requirements and general metrical conditions that must be fulfilled in the production of acceptable models have been studied most critically by Pauling and Corey.[33] From their careful and detailed studies of the crystal structures of various amino acids and simple peptides they find the most probable values for the bond lengths and angles in the polypeptide chain to be those shown in Fig. 126.

[29] R. O. Herzog and W. Jancke, *Naturwissenschaften*, 1926, **14**, 1223; W. Lotmar and L. E. R. Picken, *Helv. Chim. Acta*, 1942, **25**, 538.

[30] W. T. Astbury and F. O. Bell, *Nature*, 1941, **147**, 696.

[31] M. L. Huggins, *Chem. Revs.*, 1943, **32**, 195.

[32] W. L. Bragg, J. C. Kendrew, and M. F. Perutz, *Proc. Roy. Soc.* (London), 1950, **A203**, 321.

[33] L. Pauling and R. B. Corey, *J. Amer. Chem. Soc.*, 1950, **72**, 5349; L. Pauling, R. B. Corey, and H. R. Branson, *Proc. Nat. Acad. Sci. U. S.*, 1951, **37**, 205; L. Pauling and R. B. Corey, *ibid.*, 1951, **37**, 235, 241, 251, 256, 261, 272, 282, 729; *ibid.*, 1952, **38**, 86.

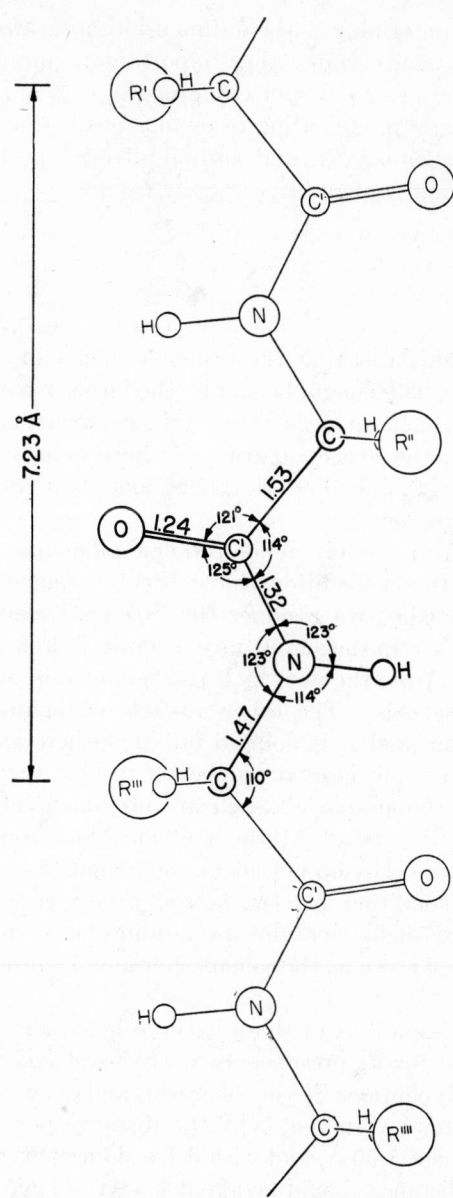

Fig. 126. Dimensions of the polypeptide
chain (Pauling, Corey, and Branson).

As well as conforming to these dimensions a satisfactory model must comply with certain other important requirements. One arises from the observed shortening of the carbon-nitrogen link in the amide group, indicating resonance of the double bond between the carbon-oxygen and carbon-nitrogen positions. This requires a planar configuration for the residue

$$\begin{array}{ccc} H & & C_\alpha \\ \diagdown & & \diagup \\ & N-C & \\ \diagup & & \diagdown \\ C_\alpha & & O \end{array}$$

A second condition is that the principle of maximum hydrogen bonding (see p. 239) should be met by the formation of a hydrogen bond between each nitrogen atom and the oxygen atom of some other residue, the expected distances here being about 2.8 Å, with the line of the bond not deviating more than about 30° from the N—H direction.

In an exhaustive study of the problem all configurations compatible with these conditions have been examined for a single polypeptide chain with residues that are equivalent except for the side chains. In the simple model shown in Fig. 126, one residue is derived from the next by a rotation of 180° and a translation along the axis. The other possible configurations are all helical in form, and it is pointed out that there is no need for them to contain an integral number of residues per turn of the helix. (In previous models such an integral number was generally assumed.) Most of the configurations found must be rejected because they do not adequately comply with the hydrogen-bonding condition, but two helical structures survive. One, known as the α-helix, contains 3.7 residues per turn of the helix, and the other, known as the γ-helix, contains 5.1 residues per turn of the helix.

The most important of these is the α-helix, and evidence has been adduced for its presence as a structural feature in certain synthetic polypeptides,[28,33] in α-keratin, and even in some of the globular proteins. In this helix the distance per residue along the axis is about 1.50 Å, and with 3.7 residues per turn, the fibre axis repeat distance should be about 5.5 Å. (The precise values for the distance per residue and the number of residues per turn depend upon the distance assumed for the hydrogen bond.) Each amide group is hydrogen bonded to the third amide group lying

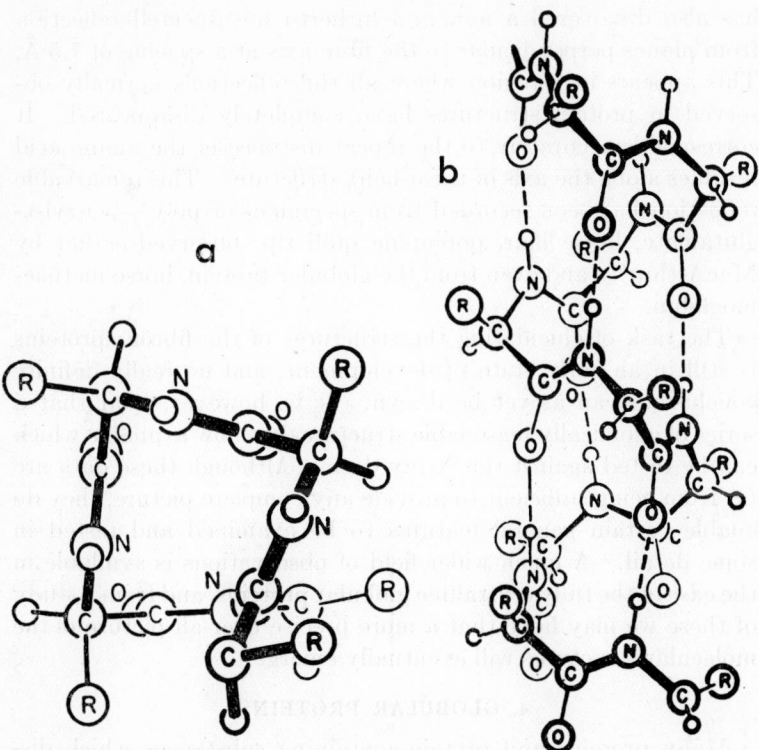

Fig. 127. The α-helix of 3.7 residues per turn, (a) seen in plan, (b) in elevation, including about two and a half turns with nine residues (Pauling and Corey).

beyond it in the helix. A plan and an elevation of the structure are shown in Fig. 127. All the configurational requirements are fully satisfied by this model, and the hydrogen bond vectors lie within about 10° of the N—H direction.

In α-keratin and the synthetic polypeptides the fibre period and other observed spacings are compatible with the α-helix structure. A more detailed examination of the intensity distributions to be expected from the helical structures in general has been made by Cochran, Crick, and Vand,[34] and this appears to be in agreement with the main features observed on some of the α-keratin and synthetic polyglutamate photographs. Perutz[35]

[34] W. Cochran, F. H. C. Crick, and V. Vand, *Acta Cryst.*, 1952, **5**, 581.
[35] M. Perutz, *Nature*, 1951, **167**, 1053; see also L. Pauling and R. B. Corey, *ibid.*, 1951, **168**, 550.

has also discovered a new and hitherto unsuspected reflection from planes perpendicular to the fibre axis at a spacing of 1.5 Å. This appears in a region where all the reflections normally observed in protein structures have completely disappeared. It corresponds accurately to the repeat distance of the amino acid residues along the axis of the α-helix structure. This remarkable reflection has been recorded from specimens of poly-γ-benzyl-L-glutamate, horse hair, porcupine quill tip (observed earlier by MacArthur[36]), and even from the globular protein, horse methaemoglobin.

The task of elucidating the structures of the fibrous proteins is still in an early state of development, and no really definite conclusions can as yet be drawn. It is, however, clear that a series of chemically reasonable structures are now available which can be tested against the X-ray data. Although these data are far from being sufficient to provide any complete picture, they do enable certain general features to be examined and tested in some detail. A much wider field of observations is available in the case of the fully crystalline globular proteins, and from a study of these we may hope that a more precise over-all picture of the molecular structures will eventually emerge.

4. GLOBULAR PROTEINS

Many proteins and protein-containing substances which dissolve in aqueous solution can be obtained in the form of true single crystals. In the wet condition these crystals are remarkably perfect and generally consist of well-formed prisms and simple polyhedra, like the crystals built from ordinary small molecules and ions. This class now covers a great range of biologically important substances of extremely high molecular weight, including some enzymes, viruses, and antibodies, in which the basic component consists of protein. In most of these substances the ultimate particle or molecule appears to be of globular form, and presumably the polypeptide chains must now be folded together in some compact way, in contrast to the fibrous proteins in which they are extended to form chainlike molecules.

The first X-ray diffraction pattern from a protein single crystal was obtained from pepsin by Bernal and Crowfoot in 1934.[37]

[36] I. MacArthur, *Nature*, 1943, **152**, 38.
[37] J. D. Bernal and D. Crowfoot, *Nature*, 1934, **133**, 794.

A number of spots lying close to the centre of the photograph indicated, by their positions and relative intensities, an enormous unit cell containing relatively dense, globular molecules, apparently separated by large spaces containing water. In addition, a definite pattern of reflections of small spacing was obtained, pointing to the existence of a well-defined atomic arrangement within the large molecules themselves. These striking results were obtained from the wet crystal, in contact with its mother liquor. On drying, most of the pattern faded and was replaced by a vague blackening. This must be due to some disorientation of the molecules, possibly accompanied by disorganisation within the molecules themselves.

The work of Astbury and others[38] has later shown that on denaturation, by heating or treatment with various chemical reagents, the polypeptide chains can be liberated from the compact globular proteins. By suitable subsequent treatment materials have been obtained which resemble hair and wool, exhibiting elastic properties and giving X-ray diagrams like those of the fibrous proteins. For example, from the vegetable globulin edestin, fibres were obtained which, on stretching, gave a photograph that was unmistakably of the β-keratin type.[39] Such processes are now the basis of industrial production of the fibre "ardil," obtained from the denatured globulin of ground nuts.

It appears, then, that in the single crystals of globular proteins the basic protein structure is present in a fully organised, three-dimensional form. The great wealth of X-ray data which are available from such crystals provide the possibility of a much more complete elucidation of the structure than is possible with fibrous proteins. Since the early X-ray experiments described above, many of these fully crystalline proteins have been examined, but the interpretation of the data provides problems of extraordinary complexity and magnitude. These are now being actively studied in many centres. At present the subject is in a state of rapid development, and with frequently changing hypotheses it is not suitable for a detailed review. Some of the main features are briefly mentioned in the following pages.

[38] W. T. Astbury and R. Lomax, *Nature*, 1934, **133**, 795; 1936, **137**, 803; *J. Chem. Soc.*, 1935, p. 846; W. T. Astbury, *Nature*, 1937, **140**, 968.

[39] W. T. Astbury, S. Dickinson, and K. Bailey, *Biochem. J.*, 1935, **29**, 2351.

TABLE XIV. CRYSTAL DATA ON PROTEINS

Substance	Ref.	Space group	State	a (Å)	b (Å)	c (Å)	α (°)	β (°)	γ (°)	Mols. per cell	Mol. wt.×10⁻³
Ribonuclease	[40]	$P2_1$	wet dry	30.9 29.1	38.8 30.1	54.1 51.0		106.0 114.0		2	13.4
Insulin	[41] [42]	$R3$	wet dry	49.4 44.4			114.3 114.5			3	3×12.5
Acid insulin sulphate	[43]	$P2_12_12_1$	dry	44	51.4	30.4				4	~12
Lysozyme I		$P2_12_12$	wet	56.3	65.2	30.6				4	
Lysozyme II	[44]	$P4_12_1$	wet dry	79.1 71.2		37.9 31.4				8	14
Metmyoglobin of horse	[45]	$P2_1$	wet dry	57.3 51.5	30.8 28	57 37		112 98		2	17
Metmyoglobin of whale	[45]	$P2_12_12_1$	wet	97.4	39.8	42.5				4?	17?
Chymotrypsin	[46]	$P2_1$	wet dry	49.6 45	67.8 62.5	66.5 57.5		102 112		4	32
β-Lactoglobulin	[47] [48,49]	$P2_12_12_1$	wet dry	69.3 60.7	70.4 61	154.5 112.4				8	35.4
Pepsin	[37] [50]	$C6_22$	wet dry	67.9 60.5		292 268				12	~40
Met-, oxy-, and carboxy-haemoglobin of horse I	[51]	$C2$	wet dry	109 102	63.2 56	54.4 49		111 134		2	66.7
Met-, oxy-, and carboxy-haemoglobin of horse II	[51]	$P2_12_12_1$	wet dry	122 95	82.4 72	63.7 61				4	66.7
Met-, oxy-, and carboxy-haemoglobin of horse III	[52]	$P2$	wet	64.5	101	104		102		4	66.7

Substance	Ref.	Space group	State	a	b	c	α	β	γ	n	M
Reduced haemoglobin of horse	52	$C3,2$	wet / dry	56.1 / 47.4		354 / 308				6	66.7
Methaemoglobin of adult sheep	53	$C2$	wet	164	70	66		94.5		4	68
Methaemoglobin of foetal sheep I	53	$B22,2$	wet / dry	112 / 78	108 / 99	56 / 54				4	68
Methaemoglobin of foetal sheep II	53	$P2_1$	wet / dry	55 / 48	83 / 71	64 / 58		102 / 101		2	68
Met-, oxy-, and carboxy-haemoglobin of man I	54	$P2_12_12_1$	wet / dry	85.2 / 70.4	77 / 73.3	86 / 77				4	68
Met-, oxy-, and carboxy-haemoglobin of man II	52	$P2_12_12_1$ (?)	dry	50	55	123				4	68
Met-, oxy-, and carboxy-haemoglobin of man III	55	$P4_12_1$	wet / dry	53.7 / 47.4		193.5 / 174				4	68
Reduced haemoglobin of man I	52	$P2_1$	wet	66	98	110		98		4	68
Reduced haemoglobin of man II	52	$P2_1$	wet / dry	62.5 / 59	83.2 / 70	52.8 / 47.5		98 / 97.5		2	68
Excelsin	56	$R3$	dry	85.5			60.5			1	294
Tobacco seed globulin	57	Cubic F	dry	123						4	300
Tobacco necrosis virus	58	$P1$	wet	179	219	243	87.5	97.5	97.5	1	1850
Tomato bushy-stunt virus	59	Cubic I	wet / dry	386 / 314						2	~10000
Turnip yellow mosaic virus	60	Cubic F	wet / dry	706 / 528						8	

Reference notes appear on the following page.

313

Table XIV gives a summary of the principal dimensions obtained from the crystalline proteins and related substances. By far the most extensive studies have been carried out on the haemoglobins, but neither within this group nor in the various other substances listed do the unit cell sizes give any immediate evidence of a common structural unit. The dimensions vary from one substance to another in what appears to be an erratic way, and there are also many examples of polymorphism.

Several of these crystal structures, especially ribonuclease, insulin, and the haemoglobins, have now been studied in considerable detail, mainly by the use of vector maps constructed by the

[40] C. H. Carlisle and H. Scouloudi, *Proc. Roy. Soc.* (London), 1951, **A207**, 496; J. D. Bernal, *Nature*, 1952, **169**, 1007.

[41] D. Crowfoot, *Nature*, 1935, **135**, 591; *Proc. Roy. Soc.* (London), 1938, **A164**, 580; D. Crowfoot and D. Riley, *Nature*, 1939, **144**, 1011.

[42] D. Wrinch, *J. Chem. Phys.*, 1948, **16**, 1007.

[43] B. Low, *Nature*, 1952, **169**, 955.

[44] K. J. Palmer, M. Ballantyne and J. A. Galvin, *J. Amer. Chem. Soc.*, 1948, **70**, 906.

[45] J. C. Kendrew, *Acta Cryst.*, 1948, **1**, 336.

[46] J. D. Bernal, I. Fankuchen, and M. Perutz, *Nature*, 1938, **141**, 522.

[47] D. Crowfoot, *Chem. Revs.*, 1941, **28** 215.

[48] F. R. Senti and R. C. Warner, *J. Amer. Chem. Soc.*, 1948, **70**, 3318.

[49] I. M. Dawson, *Nature*, 1951, **168**, 241; D. P. Riley, *ibid.*, 1951, **168**, 241.

[50] M. F. Perutz, *Research*, 1949, **2**, 52.

[51] J. D. Bernal, I. Fankuchen, and M. F. Perutz, *Nature*, 1938, **141**, 523; M. F. Perutz, *ibid.*, 1939, **143**, 731; 1942, **149**, 491; J. Boyes-Watson and M. F. Perutz, *ibid.*, 1943, **151**, 714; M. F. Perutz, *Trans. Faraday Soc.*, 1946, **42B**, 187; J. Boyes-Watson, E. Davidson, and M. F. Perutz, *Proc. Roy. Soc.* (London), 1947, **A191**, 83; M. F. Perutz, *ibid.*, 1949, **A195**, 474. See also Ref. 32.

[52] W. L. Bragg, E. R. Howells, and M. F. Perutz, *Acta Cryst.*, 1952, **5**, 136; W. L. Bragg and M. F. Perutz, *ibid.*, 1952, **5**, 277, 323.

[53] J. C. Kendrew and M. F. Perutz, *Proc. Roy. Soc.* (London), 1948, **A194**, 375.

[54] M. F. Perutz and O. Weisz, *Nature*, 1947, **160**, 786.

[55] M. F. Perutz, A. M. Liquori, and F. Eirich, *Nature*, 1951, **167**, 929.

[56] W. T. Astbury and F. O. Bell, *Tabulae Biol.* (Haag), 1939, **17**, 90.

[57] D. Crowfoot and I. Fankuchen, *Nature*, 1938, **141**, 522.

[58] D. Crowfoot and G. M. J. Schmidt, *Nature*, 1945, **155**, 504; F. C. Bawden and N. W. Pirie, *Brit. J. Exptl. Path.*, 1942, **23**, 314; P. Cowan and D. Crowfoot Hodgkin, *Acta Cryst.*, 1951, **4**, 160.

[59] C. H. Carlisle and K. Dornberger, *Acta Cryst.*, 1948, **1**, 194.

[60] J. D. Bernal and C. H. Carlisle, *Nature*, 1948, **162**, 139.

Patterson method. With thousands of atoms present in the asymmetric crystal unit it would, of course, be quite hopeless to attempt anything like a detailed analysis of these vector distributions. Nevertheless, some extremely important information can be obtained from a general survey of the vector maps. For example, it was shown in early studies, especially of insulin,[41] that the vector maps obtained from wet and air-dried crystals contained certain invariant features, differing only in orientation. This clearly pointed to the existence of some definite, organised unit of structure, or molecule, which survived in all essentials even in the dry state.

Again, there are certain resemblances between the vector structure of a crystal and the real structure which can be utilized. The presence of chainlike molecules will give rise to concentrations of interatomic vectors parallel to the chain directions, and any repeat of molecular pattern along the chain should be indicated by recurring maxima in these directions. Such evidence can provide some information regarding the main features of the structure.

Ribonuclease is probably the most simple protein structure that has received detailed study.[40] From careful density measurements and the dry cell dimensions the molecular weight is calculated as 13,400, assuming two molecules in the unit cell (space group $P2_1$). A study of the vector maps shows a hexagonal distribution of peaks separated by 9.5 Å, which are thought to correspond to five polypeptide chains running parallel to the c axis and about 46 Å in length. A vector repeat of 4.9 Å in the chain directions is also observed. The over-all dimensions of the molecule are estimated to be about $18 \times 30 \times 48$ Å.

The trigonal crystal of insulin was one of the earliest to be studied.[41] Vector maps have been prepared, from the wet and dry crystals, and there is some evidence for polypeptide chains running parallel to the trigonal axis, although other interpretations have been suggested.[42] Acid insulin sulphate[43] crystallises in the orthorhombic system, and in this structure there is evidence of concentrations of density which may indicate chains of atoms lying parallel to the a axis.

The lactoglobulin structure[47] is crystallographically more complicated, and no interpretation of the Patterson maps has been attempted. Very accurate studies of the cell dimensions and molecular weight have been made.[48]   A high-resolution electron micrograph of this crystal is shown in Fig. 128, on which the individual unit cells can be seen and actually measured with some degree of accuracy.[49]   It is clear that this crystal has a very well-defined structure which extends over quite large areas.

Very detailed X-ray studies on the various forms of haemo-globin have been made by Perutz and his co-workers,[51-55] and our knowledge of structure is now more extensive for this protein than for any of the others.   Several independent methods have been utilized in approaching this structure, and most of the work has been carried out on the various forms of horse haemoglobin.

The wet crystals contain over 50 per cent of liquid of crystallisa-tion, and an extensive study has been made of their swelling and shrinkage properties (especially those of Table XIV, form I).   As long as complete drying is avoided, a number of well-defined re-versible transitions between different lattice states are observed, each stable at a particular pH or vapour pressure.   Patterson projections calculated for these different states show that the crystal possesses a structure with alternate layers of protein and liquid of crystallisation parallel to (001), there being one layer of each kind per unit cell.   The molecules themselves do not shrink or swell, except on complete drying, and they are not penetrated by the liquid of crystallisation.

The salt content of this replaceable liquid can be varied be-tween wide limits.   When heavy ions are introduced in the liquid, some of the low order diffraction spots are found to change in intensity.   This observation can be utilized as a means of applying the isomorphous replacement method to these low order reflections.   This technique has been employed to estimate the approximate outer form of the hydrated (horse) haemoglobin molecule.   The dimensions appear to be about 55 Å in the b and c directions (Table XIV, form I) with a length of about 65 Å in the a direction.   It has also been shown that the molecules of the other haemoglobins from different animal species are roughly similar in external form.[52]

The signs of some of the $(00l)$ structure factors can also be determined by comparing intensities at different degrees of swelling.   This gives a means of evaluating the molecular Fourier transform, by plotting the intensities against $\sin\theta/\lambda$.   The points where the curve passes through zero may then be taken

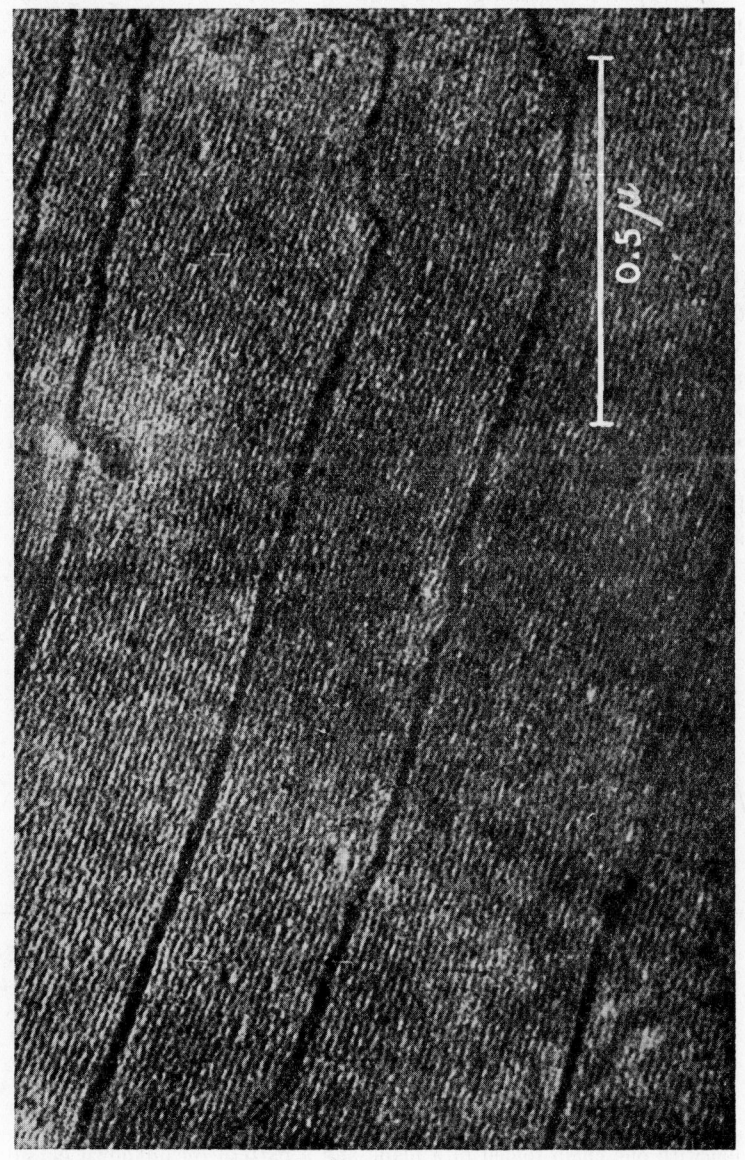

Fig. 128 Electron micrograph of a β-lactoglobulin crystal (I. M. Dawson).

to correspond to changes in sign of the F function. By this
interesting method it has been possible to evaluate a one-
dimensional Fourier synthesis along a line normal to (001), which
shows four prominent concentrations of scattering matter just
under 9 Å apart. (Later and more detailed work[52] has shown,
however, that there must be an uneven number of layers, probably consisting of three heavy central and two lighter outer layers
of chains, as indicated in Fig. 129 b and c.)

In addition to these studies, a very complete three-dimensional
Patterson synthesis has been evaluated.[51] This shows a hexagonal
arrangement of rods of high vector density, about 10.5 Å apart,
lying parallel to the a axis. Within these rods there are also
maxima which recur lengthwise at intervals of 5 Å. This vector
picture provides strong evidence for molecular chains running
parallel to the a axis, and it may be correlated with the chemical
evidence for the existence of polypeptide chains in the molecule.
These may be folded backwards and forwards in some form of
large-scale zigzag, the neighbouring folds lying about 10.5 Å

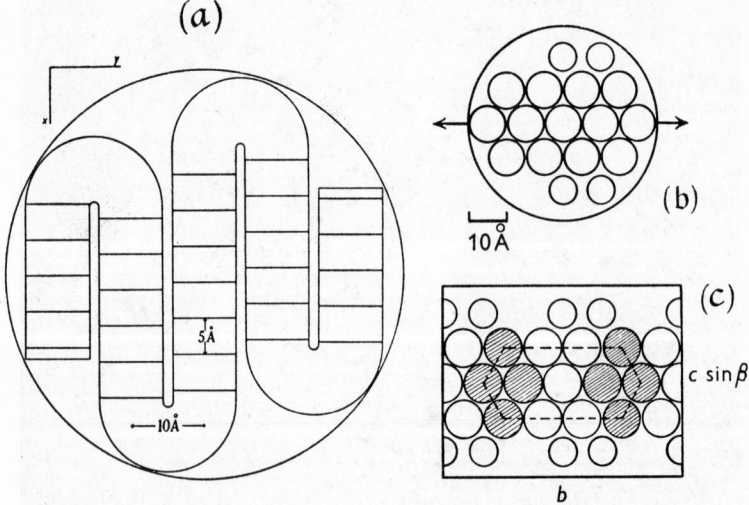

Fig. 129. (a) Idealised picture of type of haemoglobin structure compatible with the Patterson synthesis, showing a section through the molecule with one chain folded in a plane and a pattern (of unknown detail)
repeating at 5 Å intervals along its length (Perutz).
(b) View of the haemoglobin molecule seen along the chain direction.
(c) View of the c-face-centred unit cell of haemoglobin. Overlapping
molecules are shaded.

apart and containing a molecular pattern which repeats at about
5 Å intervals along the chain directions. The various parts of
Fig. 129 and Fig. 130, which are due to Perutz and his co-workers,
illustrate these features of the haemoglobin structure in a very
diagrammatic form.

The features of the polypeptide chain which are suggested by
this work bear a striking resemblance to the α-keratin chains
studied by Astbury. These show a repeat in the chain direction
of about 5.1 Å and another spacing of 9.7 Å at right angles to
this direction. This relationship points to a fundamental resemblance between the polypeptide chains in the two cases.

The haemoglobin molecule, of molecular weight about 67,000
is an excessively complex structure, although the fact that it lies
on a 2-fold axis of symmetry in one of the crystals examined
(Table XIV, horse haemoglobin I), and apparently consists of

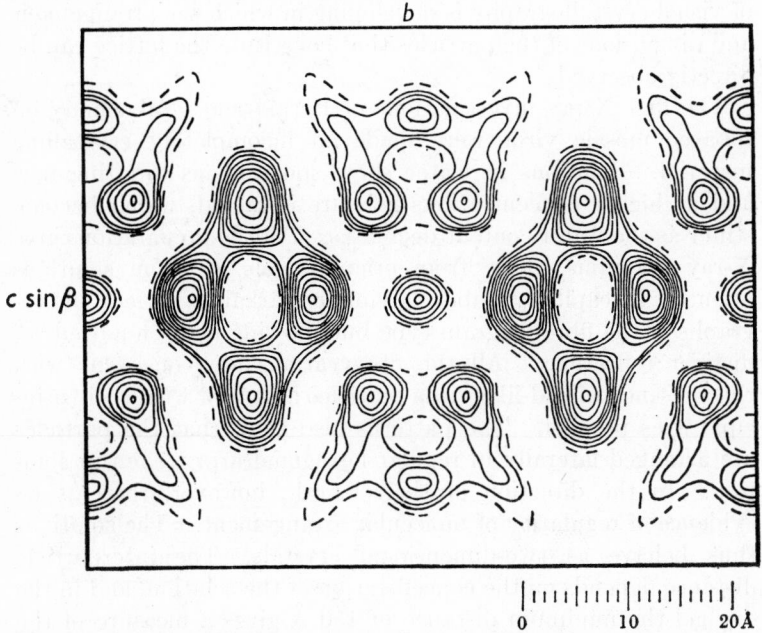

Fig. 130. Fourier map (projection) of the unit cell of haemoglobin, showing a possible electron density distribution. As the Fourier peaks are only
one-third of the height expected, it is not likely that the chains run
straight through the entire molecule. A considerable fraction of the space
must be taken up by the chains turning corners or running in directions
other than the a axis (Bragg, Howells, and Perutz).

two structurally identical halves, makes for some simplification. The myoglobin (muscle haemoglobin) protein has a smaller molecule of weight about 17,000, and Patterson projections of these crystals[45,50] show rodlike features of the same general type as in haemoglobin, indicating a basic structural similarity. The haemoglobin molecule is known to contain four iron-bearing pigment groups or haems, while myoglobin contains one haem. But the positions of these groups in the molecules have not been determined.

5. CRYSTALLINE VIRUSES

Molecules of a still higher order of complexity are encountered in the elementary virus particles, and many truly crystalline preparations of these substances have now been obtained. In this field the electron microscope is rapidly becoming an even more important tool than X-rays. As a consequence a new type of visual crystallography is developing in which the arrangement and dimensions of the particles that constitute the lattice can be directly observed.

The first X-ray investigations of Bernal and Fankuchen[61] on tobacco mosaic virus were made on incompletely crystalline material. Solutions in water show spontaneous birefringence, and at higher concentrations gels are obtained, which become stiffer as the water content decreases. All the preparations give X-ray diagrams, those from oriented gels, or from solutions oriented in capillary tubes, giving remarkably perfect photographs of the fibre diagram type but showing many hundreds of distinct reflections. All the observations are consistent with the presence of rod-like virus particles arranged with their principal axes parallel. The patterns also show that the particles are arranged laterally in regular hexagonal array, even in solution. In the direction of their length, however, there is no evidence of regularity of molecular arrangement. The solutions thus behave as two-dimensional crystals. The interparticle distance depends on the concentration of the solution, and in the dry gel the minimum distance of 150 Å gives a measure of the diameter of one particle. The X-ray measurements do not give any direct evidence concerning the particle length.

[61] F. C. Bawden, N. W. Pirie, J. D. Bernal, and I. Fankuchen, *Nature*, 1936, **138**, 1051; J. D. Bernal and I. Fankuchen, *ibid.*, 1937, **139**, 923; J. D. Bernal and I. Fankuchen, *J. Gen. Physiol.*, 1941, **25**, 111.

The strictly parallel and equidistant arrangement of the particles shown in the solution is certainly extraordinary. It appears that this arrangement is probably a result of the ionic atmosphere surrounding the particles.[62] For very long particles, the energy arising from this source must be greater than that of the thermal motion. At some particular mutual distance depending on the ionic concentration, attractive and repulsive forces will balance and stable equilibrium will occur. As a consequence, a regular two-dimensional lattice arrangement results, with the particles widely separated.

The X-ray evidence also shows that each individual particle must have an inner regularity of a crystalline type. There is some evidence for a sub-unit within the virus particles of about 11 Å cube, fitted into a hexagonal or pseudo-hexagonal lattice with $a = 87$ Å and $c = 68$ Å. The detailed pattern also shows certain dominant reflections from spacings of 11 Å and 4.5 Å, indicating some basic internal structure resembling the proteins.

This virus preparation has since been frequently examined by means of the electron microscope, and the conclusions reached fully confirm the X-ray analysis of Bernal and Fankuchen. A typical electron micrograph is shown in Fig. 131. The rodlike particles have a diameter of about 150 Å. Even when examined by high-resolution electron microscopy,[63] however, they do not reveal any periodic surface structure, either transverse or longitudinal. The normal rods appear to exist in the form of elongated hexagonal prisms. With regard to length, a careful study of many samples[64] has shown that the majority of the particles exist as monomers, with a very uniform length of 2980 ± 10 Å, or as dimers of twice this length. Shorter particles are occasionally found.

A number of virus preparations are now known which give rise to true three-dimensional single crystals. The first X-ray study of such a crystal was made by Crowfoot and Schmidt[58] on tobacco necrosis virus. Quite large crystals were obtained, and from these the enormous unit cell listed in Table XIV was measured. The reflections are naturally so crowded together that they can

[62] S. Levine, *Proc. Roy. Soc.* (London), 1939, **A170**, 145, 165; and later papers.

[63] R. C. Williams, *Biochim. et Biophys. Acta*, 1952, **8**, 227.

[64] R. C. Williams and R. L. Steere, *J. Amer. Chem. Soc.*, 1951, **73**, 2057.

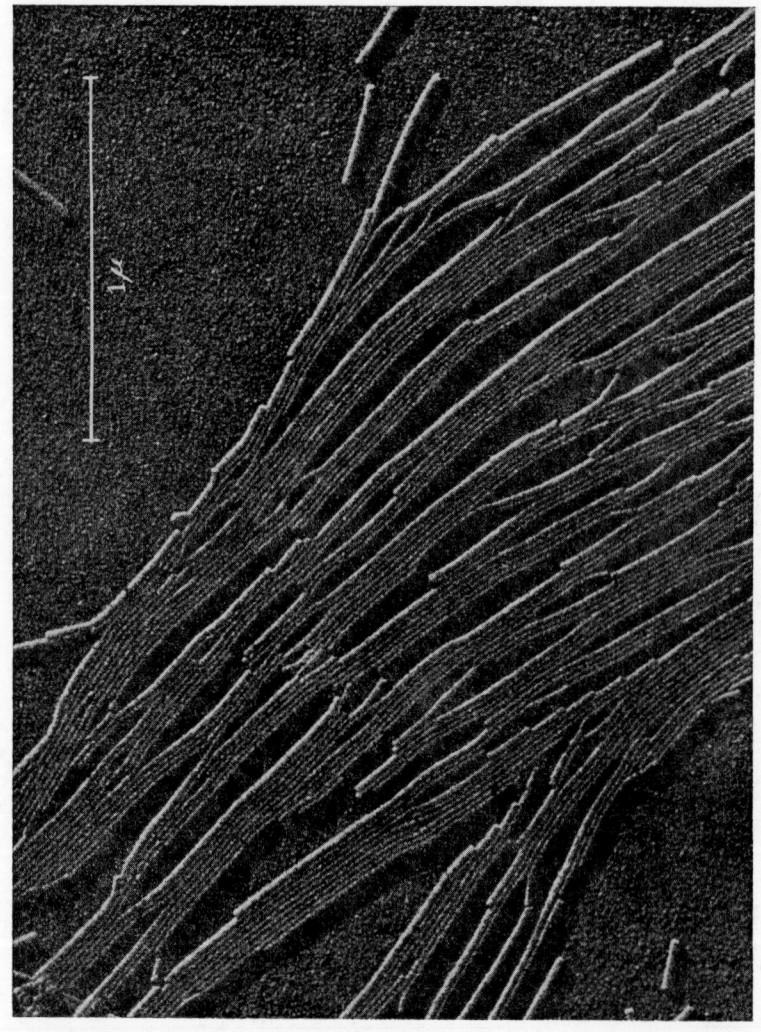

Fig. 131. Electron micrograph of tobacco mosaic virus (I. M. Dawson).

scarcely be resolved, and even when a stationary crystal is exposed to a nearly parallel monochromatic X-ray beam, large numbers of planes are in a position to reflect. "Still" photographs taken in this way show concentric rings of reflections corresponding to the points where the sphere of reflection intersects the reciprocal lattice points. From the positions of these rings the cell dimensions can be deduced.

In spite of the overwhelming complexity of the structure, reflections have been observed down to spacings of 2.8 Å, indicating a high degree of internal regularity in the molecules. These molecules are probably of a somewhat spherical shape, with a radius of between 80 and 100 Å, but little is known about the dimensions in detail.

Tobacco necrosis virus preparations have also been examined with the electron microscope,[65] and a number of very striking photographs of the crystals have been obtained. One of these is shown in Fig. 132. The large molecules are well within the resolution limit of the microscope, and they are seen to form a regular array in what is apparently a cubic, close-packed lattice. The diameter of the particle is here estimated to be about 240 Å. It is not certain, however, that the tobacco necrosis virus used in these electron micrographical studies belonged to exactly the same strain as the one used by Crowfoot and Schmidt in their X-ray work.

Crystals of tomato bushy-stunt virus are cubic, body centred, with a still larger unit cell, the length of the cube edge being 386 Å in the wet crystals and 314 Å in the dry state. An early investigation by the powder method[66] has now been supplemented by single crystal studies[59] employing the method of "still" photographs referred to above. From the wet crystals reflections have been obtained extending to spacings of about 7.5 Å, indicating quite a high degree of internal regularity. In the dry state, however, the reflections fade and there is apparently a considerable measure of disorientation.

Very remarkable results have been obtained both by X-ray diffraction[60] and by electron microscopy[67] from crystalline turnip

[65] R. Markham, K. M. Smith, and R. W. G. Wyckoff, *Nature*, 1947, **159**, 574; 1948, **161**, 760; R. W. G. Wyckoff, *Acta Cryst.*, 1948, **1**, 292.

[66] J. D. Bernal, I. Fankuchen, and D. P. Riley, *Nature*, 1938, **142**, 1075.

[67] V. E. Cosslett and R. Markham, *Nature*, 1948, **161**, 250; R. Markham, R. E. F. Matthews, and K. M. Smith, *ibid.*, 1948, **162**, 88.

Fig. 132. Electron micrograph of tobacco necrosis virus crystal. (*Electron Microscopy*, by R. W. G. Wyckoff. Copyright 1949, Interscience Publishers, New York-London.)

yellow mosaic virus. In the X-ray work powder photographic methods were employed, and the reflections recorded correspond to a cubic, face-centred lattice, of edge 706 Å for the wet crystal and 528 Å for the dry. They also show that the particles are arranged on a diamond type lattice, with an inter-particle distance of 306 Å. This arrangement has been confirmed from the electron microscope studies. Such an open tetrahedral type of co-ordination is most unexpected where virus particles are involved. It generally only results from the action of strongly directed valency bonds, between separate atoms or small molecules.

On drying, the inter-particle distance appears to diminish from 306 to 228 Å. If the virus particles are assumed to be rigid, this would mean that in the wet state they are separated by no less than 78 Å of liquid. A rather similar situation exists in the case of the tobacco mosaic virus, but the capacity for maintaining equilibrium at a distance is more easily understood when the particle has the form of an elongated rod. In other proteins a layer of water between the molecules is a common feature, but it is generally very much thinner.

Nucleic acid-free (non-infective) crystals of the turnip yellow mosaic virus type have also been examined, and these give spacings which indicate a slightly *larger* unit cell (725±25 Å, wet; 550 Å, dry). This paradoxical result is hard to explain, but appears to be confirmed by the electron micrographs, which show a slightly larger particle in the nucleic acid-free material. The presence of the nucleic acid does not affect the inter-particle forces but seems to draw the protein material of the particles themselves into smaller volume. As yet these matters are imperfectly understood.

Index of Names

INDEX OF NAMES

Index of Subjects